SOILS IN THE BRITISH ISLES

SOILS
IN THE BRITISH ISLES

L. F. Curtis, F. M. Courtney and S. T. Trudgill

LONGMAN
London and New York

Longman Group Limited London

*Associated companies, branches and representatives
throughout the world*

*Published in the United States of America
by Longman Inc., New York*

© Longman Group Limited 1976

First published 1976

Library of Congress Cataloging in Publication Data

Curtis, Leonard F.
    Soils in the British Isles.

    Bibliography: p.
    Includes index.
    1. Soils — Great Britain. 2. Soils — Irish Republic.
    I. Courtney, F. M., joint author. II. Trudgill, S. T.,
    joint author. III. Title.
S599.4.G7C87 631.4'941 75–46527
ISBN 0 582 48556 8 paper
ISBN 0 582 48555 X cased

Set in IBM Journal 10 on 12pt
and printed in Great Britain by
William Clowes & Sons Ltd,
London, Colchester and Beccles

CONTENTS

PREFACE

This book is directed towards the growing numbers of students who are concerned with environmental planning, ecological problems and the earth sciences. These include students at University or Polytechnic institutions reading Geography, Botany, Biology, Geology, Environmental Science, Planning and Landscape Architecture as well as Advanced Level school students in Geography, Biology and Environmental Studies.

The aim of the text is to interest the reader in the evolution and uses of British soils by adopting a series of viewpoints. The authors hope, also, that the content as a whole provides a useful overall view of the regional variations in British soils. In particular those readers with working interests in the fields of agriculture, land-use, ecology and conservation may find this synoptic approach helpful. The book is not intended for research workers in soil science who can find more appropriate information in various journals and regional soil monographs.

The text begins with an introduction to those concepts and terms in soil science which are necessary for reading and understanding the wide literature on British soils that is now becoming available. It then takes up a series of themes and examines these with the aid of local examples. It is, of course, impossible to give an even treatment to all areas of the country. Indeed, the amount of information available is not evenly spread and does not permit exhaustive treatment. The themes have, therefore, been chosen to cover a variety of local and national issues of different kinds and levels of importance. They all seek to demonstrate the part played by soil conditions in both man-made and natural ecosystems.

The experienced reader will quickly recognise that many of the examples are derived from the work of the Soil Surveys of Great Britain and Ireland. The authors of this work are heavily indebted to the writers of various individual Soil Survey publications in which detailed accounts of local field conditions can be found. These publications are referenced in the text and we feel that, together with the maps published by the Surveys they have much to offer a wider readership. If the present volume leads to wider appreciation and use of the publications of the Soil Surveys of Great Britain and Ireland we shall feel that we have succeeded in some measure.

A large number of individuals have assisted directly or indirectly in the preparation of this volume and we are deeply indebted to them all. In particular we would like to thank the following for their help at various stages: Dr T. C. Atkinson, Mr B. W. Avery, Dr D. J. Briggs, Mr K. E. Clare, Mr D. W. Cope, Mr D. C. Findlay, Dr E. A. Fitzpatrick, Dr R. Glentworth, Mr D. W. King, Mr D. Mackney, Mr N. J. Sneesby, Mr S. J. Staines, Mr J. Tew, Dr R. B. G. Williams.

We would also like to thank Ann P. Barham for her work in compiling the index. Catherine Courtney and Helga and Brian Lawfield together with Diana Curtis are thanked for their encouragement and support which enabled us to complete the book on time.

L. F. Curtis, F. M. Courtney and S. T. Trudgill

ACKNOWLEDGEMENTS

We are grateful to the following for permission to reproduce copyright material:

Academic Press Inc. (London) Ltd and the author for Fig. 1.1; Macmillan Publishing Co. Inc. for Figs 1.3 and 3.1; United States Department of Agriculture for Figs 1.4 and 13.8; the author for Fig. 1.6; the author for Fig. 1.7; the Controller of Her Majesty's Stationery Office for Figs 1.8, 5.10, 8.4, 8.5, 8.7 and 8.8; and for Figs 7.19, 8.1 and 11.2 (based on Geological Survey material); Oxford University Press and the editor, *Journal of Soil Science*, for Figs 1.9, 2.1 and 2.4; Oxford University Press for Appendix 2 (from Clarke's *Handbook of the Soil Survey of England and Wales*, 1940, as reprinted in *The Study of Soil in the Field* by G. R. Clarke, 4th edition, 1957); The Twentieth Century Fund for Figs 3.3 and 3.4; the Cambridge University Collection for Figs 4.1, 9.10 and 13.6; Cambridge University Press and the author for Figs 4.2 and 4.3; Cambridge University Press for Fig. 13.1 and Appendix 3; University College of Wales, Aberystwyth, for Fig. 4.4; the Yorkshire Grassland Society for Fig. 4.5; the Horticultural Education Association and the Soil Survey of England and Wales for Fig. 4.6*a*; the editor and the Devonshire Association for Figs 4.7 to 4.10 and 9.1; the Ministry of Defence (Air Force Department) for Figs 5.4 and 8.2; the author for Fig. 5.9; the Forestry Commission for Fig. 5.11; the editors, *East Midland Geographer* for Figs 6.2 and 13.9; the Field Studies Council for Fig. 6.3; the Geologists' Association and the author for Fig. 6.6; Blackwell Scientific Publications Ltd for Fig. 6.2; An Foras Taluntais (The Agricultural Institute), Dublin, for Figs 6.26, 8.9 to 8.13, 8.15 to 8.17, and Tables 8.2 and 8.3; Longman Group Ltd for Figs 7.1 and 12.1; Associated Book Publishers Ltd for Figs 7.3 and 7.18; the author for Figs 8.3 and 9.23; the editor of *Irish Geography* for Fig. 8.14; the authors and the *Biuletyn Peryglacjalny* for Figs 9.4, 9.6 and 9.13; Svenska Sällskapet för Antropologi och Geografi for Fig. 9.5; R. B. G. Williams for Figs 9.11 and 9.12; the Royal Scottish Geographical Society and the authors for Figs 9.21, 9.22 and 9.27; the author for Figs 9.24 to 9.26; the Institute for British Geographers for Figs 11.4, 11.7, 15.6, 16.3, Tables 11.1 and 11.2; Ransomes, Sims and Jefferies Ltd, Ipswich, for Fig. 12.3; Somerset County Council for Figs 13.3 to 13.5; N. J. Sneesby for Fig. 13.10; G. P. Wibberley for Table 14.1; the Ministry of Agriculture, Fisheries and Food for Figs 14.1 and 15.3; the Ordnance Survey and the

Soil Survey of England and Wales for Figs 14.2, 14.4 and 14.5; John Wiley and Sons Inc. for Table 15.1; The Times Newspapers Ltd for Fig. 15.1; Gustav Fischer Verlag Jena for Fig. 15.2 and Table 15.2; the editor, *Journal of Soil Science* for Figs 15.4 and 15.5; the authors and the Third International Peat Congress for Fig. 16.1; Aerofilms Ltd for Fig. 16.2; Michael Joseph Ltd for Fig. 16.4; and the Soil Survey of England and Wales for Figs 1.2, 1.10, 2.2, 2.3, 4.6*b*, 4.11, 5.1 to 5.3, 5.5, 5.6, 5.8, 6.19, 7.2, 7.4 to 7.17, 7.20, 7.21, 9.2, 9.3, 9.7, 9.15 to 9.20, 10.2 to 10.5, 12.2, 12.4 to 12.7, 13.7, 13.11 to 13.13, 14.3, 14.6, Appendix 1, Appendix 4, and Tables 5.1, 5.2, 7.2, 9.2, 9.3, 10.2, 12.1 to 12.3, 13.1 to 13.3 and 14.5.

1 SOIL GENESIS IN BRITAIN

§1.1 SOIL CONSTITUENTS AND SOIL PROCESSES

1.1.1 Introduction

This book attempts to present to the reader a number of viewpoints on the soils of
Britain and their distribution. Before one can do this it is necessary to understand
something of the way in which soils are formed. The initial land surface may consist of
bare rock, sand or clay but this is gradually transformed into soil by the processes of
weathering and by modifications brought about by plants and animals. Providing
climatic conditions are favourable to plant growth, the first colonising plants on bare
rock, sand or clay are followed by a succession of later plant communities. Each stage
in the plant succession is normally a progression towards larger and more demanding
plant communities. Eventually, however, a complex and integrated climax plant community is established which Clements (1916)* considered to be in equilibrium with
the environmental conditions, i.e. with climatic and soil conditions.

Each plant (seral) succession is accompanied and paralleled by gradual transforma-
tion of the original mineral particles of the regolith. These particles become weathered
and break down both physically and chemically. As a result of physical weathering and
the creation of fine particles of silt and clay, the moisture holding properties of the
regolith gradually change and become more effective. Meanwhile chemical weathering
provides nutrients for the development of plants and the plant cover leads to a gradual
accumulation of plant debris at the surface of the soil which encourages the development of soil microflora (e.g. bacteria) and microfauna (e.g. protozoa). These micro-
organisms feed on the plant debris and decompose the organic matter to form humus
layers at the surface of the soil.

1.1.2 The soil profile

Eventually orderly and meaningful layers are formed beneath the soil surface. These
layers or soil horizons reflect the interaction of the environmental factors operating at

* Full details of the references quoted in this and the following chapters are given in the
Bibliography, on p. 327. Square brackets refer to notes at the end of chapters.

the present time and can also show some features related to former environmental conditions. If a vertical section is exposed in a mature soil it normally displays a number of horizons. These can be sub-divided into three major categories each of which may be further sub-divided (Fig. 1.1) At the surface the A horizon is that in

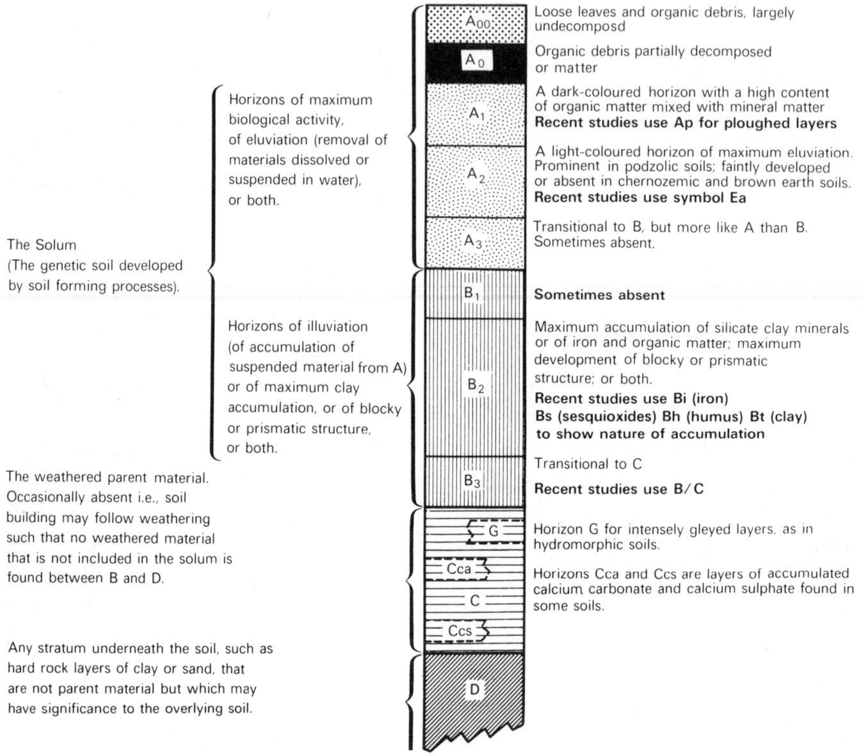

Fig. 1.1 The soil profile and horizon nomenclature. (Source: Curtis, 1973)

which maximum biological activity takes place and from which material may be removed by the leaching effect of soil water. Beneath the A horizon there is the B horizon in which material translocated from above may be deposited. Materials moved downwards from the surface may include iron (Fe_2O_3), aluminium (Al_2O_3), clay and humus. This horizon may also be one in which carbonates (e.g. calcium carbonate, $CaCO_3$), sulphates (e.g. calcium sulphate $CaSO_4$) or chlorides (e.g. sodium chloride, NaCl) may accumulate. At its lower boundary the B horizon gradually merges with the C horizon or parent material horizon. The parent material consists of weathering regolith material. This may be weathering country rock *in situ* or it may be weathering debris consisting of transported material laid down by water, wind or ice. Where the soil is weathered directly from *in situ* country rock it is usually termed a sedentary soil.

1.1.3 The soil constituents

A soil consists of four constituents: mineral matter, organic matter, soil air and soil water. The mineral matter includes particles of clay (less than 2 μm diameter), silt (2–50 μm diameter) and sand (50 μm–2 mm diameter) together with gravel and stones of larger size where present. The percentages of clay, silt and sand present in a soil determine what is termed the soil texture. These percentages can be determined by laboratory analysis and plotted on a triangular diagram on which certain combinations of percentages can be delineated and given special names. Thus the area marked sandy clay in Fig. 1.2 includes soils with more than 35 per cent clay, less than 20 per cent silt and more than 45 per cent sand. Each of the names shown on the triangular texture diagram (Fig. 1.2) are soil textural names and they enable one to indicate the approximate particle size composition of a soil.

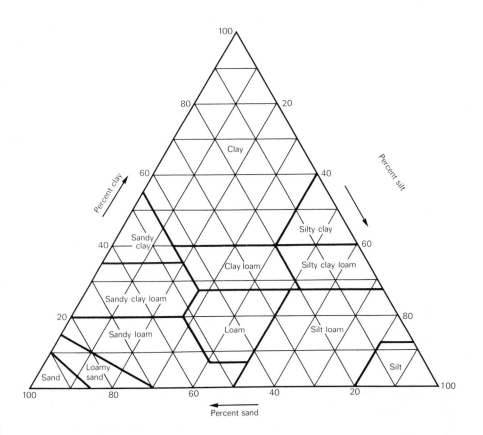

Fig. 1.2 Soil texture diagram. (Source: Crampton, 1972)

1.1.4 Cation exchange processes

Soil texture is important because it is closely correlated with other soil properties such as water holding capacity, porosity, cation[1] exchange. Clay soils normally display high water-holding capacity, high total porosity and high cation exchange capacity. In contrast sandy soils normally display opposite characteristics, that is low cation exchange capacity, low water-holding properties and low total porosity. The clay fraction of a soil has a profound effect on the chemical and physical characteristics of soils. In particular it largely controls the mechanism of cation (base) exchange through which plant nutrients are made available. The negatively charged external surfaces of the clay particles are able to attract and hold positively-charged bases (cations) such as calcium (Ca^{++}), potassium (K^+), magnesium (Mg^{++}) and sodium (Na^+) occurring in the soil solution. The bases may, however, be taken up by plant roots or leached from the clay by water moving through the soil. They are then normally replaced (exchanged) by other cations (bases) released in weathering.

The cation-exchange capacities of soils vary considerably. This is mainly due to variations in the humus content and the amounts and kinds of clay present. The cation-exchange capacity of representative silicate clay is approximately 0.5 milli-equivalents[2] for each per cent clay. Well-humified organic matter may give an average of 2.0 milliequivalents per cent. Table 1.1 gives typical ratings of cation-exchange for British soils.

Table 1.1 Cation exchange capacity ratings

Rating	Cation exchange capacity (m.e./100 g)	Total exchangeable cations (m.e./100 g)	Percentage base saturation
Very high	40	25	80−100
High	25−40	15−25	60−80
Moderate	12−25	7−15	40−60
Low	6−12	3−7	20−40
Very low	6	3	0−20

Whereas hydrogen and aluminium cations move the soil towards acidity the remaining bases (e.g. Ca, Mg, K, Na) act towards alkalinity. The proportion of the cation-exchange capacity occupied by bases other than hydrogen and aluminium is termed percentage base saturation. As the base saturation is reduced so the hydrogen ion concentration is increased. Since acidity of a soil is expressed in terms of the hydrogen ion concentration and increasing hydrogen ion quantity means higher acidity it follows that any fall in base saturation leads to greater acidity.

The hydrogen ion concentration is measured and expressed by the pH[3] value of the soil. Natural soils generally have pH values in the range 3.5 (very acid) − 8.5 (very alkaline), neutrality being indicated by a pH value of 7 (Fig. 1.3).

Although clay plays an important part in determining cation-exchange and water-holding capacity it has been shown that when organic matter is present in sufficient

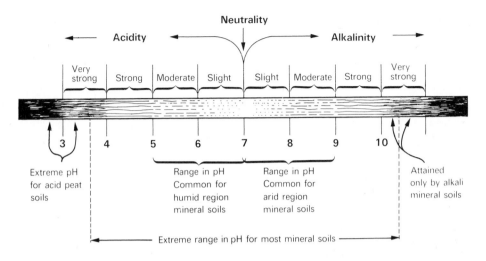

Fig. 1.3 Diagram showing the extreme range in pH for most mineral soils and the range commonly found in humid-region and arid-region soils respectively. The maximum alkalinity for alkali soils is also indicated, and the minimum pH likely to be encountered in very acid peat soils. (Source: Buckman and Brady, 1974)

quantity it may have greater effects than clay. The amount of organic matter present in the soil depends on many internal soil factors and external environmental factors. In British soils the proportion of organic matter expressed as a percentage of the total mass of soil typically shows the ranges shown in Table 1.2.

Table 1.2

	% organic matter		% organic matter
Very high	13	Moderate	3—8
High	8—13	Low	3

As a result of the exchange properties of clay and humus greater or lesser amounts of bases can be held within the soil. In British soils the elements of Ca^{++}, Mg^{++}, K^+ and Na^+ have the exchangeable ratings shown in Table 1.3.

Table 1.3 Exchangeable cation ratings (m.e./100 g) Ca^{++}, Mg^{++}, K^+, Na^+

Rating	Calcium	Magnesium	Potassium	Sodium
Very high	20	6	1.2	2
High	10—20	3—6	0.8—1.2	0.7—2
Moderate	5—10	1—3	0.5—0.8	0.3—0.7
Low	2—5	0.3—1	0.3—0.5	0.1—0.3
Very low	2	0.3	0.3	0.1

1.1.5 Soil structure and its formation

It is important to understand that soils are dynamic systems in which there are constantly changing conditions in respect of temperature, moisture, microbial life and chemical status. The changes in moisture status are largely responsible for the shrinking and swelling of soils and consequent development of internal stresses. These stresses often lead to the formation of fractures and pressures which in turn result in the production of soil structures. The various forms of the soil structure are described in Appendix 1, but soil structures can be regarded as the architecture of soil, i.e. its build. Structure categories seek to describe the ways in which individual particles of mineral matter and organic matter become aggregated together (Fig. 1.4). In some

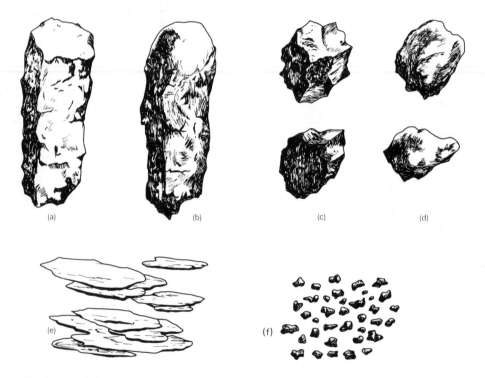

Fig. 1.4 Drawings illustrating some of the types of soil structure: a, prismatic; b, columnar; c, angular blocky; d, sub-angular blocky; e, platy; and f, granular. (Source: USDA, Soil Survey Staff, 1951)

loamy soils the aggregates are small and crumb-like (e.g. *Sherborne* series developed on Oolitic limestone), in others such as the clay soils the aggregates are large and prism-like (e.g. *Charlton Bank* series developed on Lias Clay). The importance of soil structure lies in the fact that it largely controls the spaces within the soil. Thus it is along structure faces that water frequently moves through the soil. Similarly the spaces between aggregates are important in allowing air to penetrate the soil. Good structural development is, therefore, necessary to obtain a well-drained and well-aerated soil.

The textural and structural characteristics of soils are mainly responsible for their handling qualities when they are brought under cultivation. Some are sticky and intractable when wet and very hard when dry. Others are friable and easily tilled over quite a wide range of moisture content. The term soil consistence is used to describe the handling characteristics of a soil as determined by the kind of cohesion and adhesion (Hodgson, 1974).

1.1.6 Soil colour

Although soil features such as texture, structure, pH, cation-exchange capacity and consistence can vary from horizon to horizon within a soil profile they are often partially correlated. For example, texture, structure and cation exchange capacity are often closely correlated and quite often these characteristics are, in turn, related to an easily observed feature such as soil colour. In these circumstances colour can be used as a useful field guide to other qualities of the soil. The colours of soil horizons normally change from the surface downwards (see colour plates between pp. 42 and 43, and 74 and 75), the surface horizons being darkest due to their organic matter content. When soil colours are being described the Munsell system of colour notation is usually employed. This contains a combination of spectral colour (Hue), departure from absolute whiteness (Value) and purity of hue or departure from greyness (Chroma) as the basis of a colour matching system. For example, a Munsell notation of 5YR 5/6 representing Hue/Value/Chroma describes a colour which is given the Munsell name of Yellowish-Red.

§1.2 SOIL-FORMING FACTORS

The early nineteenth-century Russian studies of the Dokuchaiev school (see Chapter 2) established the science of pedology by recognising that soils display consistent and distinctive morphological characteristics which are the result of the integrated effects of climate, parent rock, vegetation and associated organisms, relief of the landscape and time. Some of the quantitative relationships between soil properties and single features of the environment have been established by Jenny (1941). A general expression for soil formation in terms of soil-forming factors can be given as follows:

$S = C \cdot P \cdot B \cdot R \cdot T.$

Where

S = soil formation
C = climatic factor
P = parent material factor
B = biotic factor (the influence of vegetation, animals and man)
R = relief factor
T = time over which the factors have operated.

Fig. 1.5 Banded layers of organic matter (dark) and bleached sand on an eroding surface at Levisham Moor. (Photograph: L. F. Curtis)

Each of the accepted soil-forming factors is, in fact, a complex group and we will consider them briefly in relation to the British environment. In doing so a distinction can be drawn between the uplands and lowlands of Britain. In lowland Britain centuries of land-use have resulted in many soils becoming essentially man-made. In these areas the effects of natural soil-forming factors are frequently masked or over-ridden as a result of cultivation, drainage and fertilising practices. In consequence many lowland soils have been turned into what might be termed 'agricultural brown earths'. Such soils are, of course, affected by the factors of climate, parent material and relief but man often intervenes on such a scale that his influence can be deemed to be dominant. For example, by irrigating or draining the land the effects of climate are substantially amended. Likewise the addition of fertilisers largely over-rides the weathering properties of the parent material. Also the intensity of cultivation together with the adoption or neglect of conservation practices such as contour ploughing considerably affects the influence of relief on soil formation. Some of the problems associated with the massive intervention by man in the soil dynamics of lowland soils

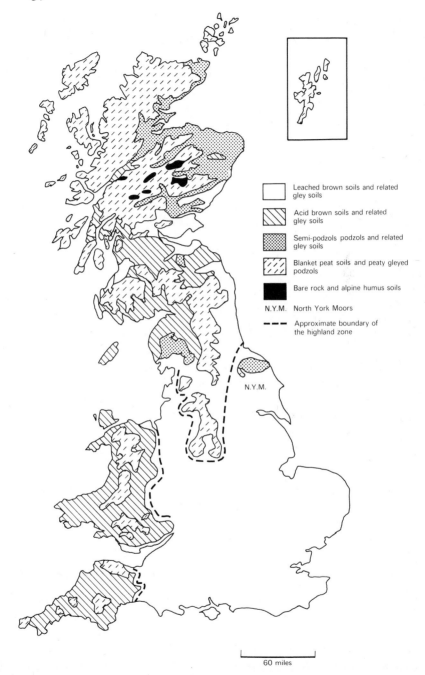

Leached brown soils and related gley soils

Acid brown soils and related gley soils

Semi-podzols podzols and related gley soils

Blanket peat soils and peaty gleyed podzols

Bare rock and alpine humus soils

N.Y.M. North York Moors

Approximate boundary of the highland zone

N.Y.M.

60 miles

Fig. 1.6 Soil regions of Britain. The bold broken line indicates the approximate position of the Highland Zone defined by Fox, 1932. (Adapted from Burnham, 1970)

are discussed in Chapter 16. At this point it is convenient to concentrate upon the factors of soil formation in relation to soil development in upland Britain; where they can still be seen to be operating on relatively unaltered, natural and semi-natural soils. For this purpose one may adopt the divisions into upland and lowland Britain made by Sir Cyril Fox (1932) in his classic work on the influence of the environment on inhabitant and invader in prehistoric and early times (Fig. 1.6). The North York Moors can also be regarded as part of the upland zone.

§1.3 CLIMATIC FACTORS INFLUENCING SOIL FORMATION IN UPLAND BRITAIN

The climatic conditions in upland Britain exhibit a marked weather gradient due to increasing altitude. A major feature of this gradient is increasing precipitation with increasing height above mean sea level. This increase in precipitation is mainly due to the elevation and buckling of frontal structures in the weather systems. Manley (1945) points out, however, that not only quantity of rain but also frequency and duration of rain increases with altitude.

With increasing rainfall there is increasing cloud cover and persistence, relatively lower cloud bases, high humidity and liability to mist and fog. In relation to these conditions evaporation and sunshine amounts are reduced.

Unlike rainfall, temperatures normally become lower with increasing elevation except under conditions of inversion. This decrease is of the order of 1°F per 270 ft (0.55°C per 82 m) and the altitude at which the maximum temperature exceeds 50°F (10°C) for two months is approximately 2 200 ft (670 m) in Britain. This represents the height at which trees are unlikely to survive and so a tree line at approximately 2 200 ft (670 m) may be postulated for much of Britain.

The effect of altitude on temperature is mainly seen in lower maximum temperatures during the summer months. Thus the growing season becomes progressively restricted with increase of altitude. If the commonly accepted figure of 42°F (5.5°C) is taken for the threshold at which growth begins and continues then comparisons between upland and lowland situations can be made (Fig. 1.7).

Under British conditions there is not only a sharp shortening of the length of the growing season with increasing height but also a decrease in the rate of growth and yield of crops. It has been estimated that yields of oats in the Northern Pennines may be approximately half those obtained at sea level nearby when grown at altitudes of 1 000 ft (300 m).

The lower temperatures in winter are accompanied by increasing incidence of frosts, snowfall and snow cover (Manley, 1947). Indeed, the greatest extremes of temperature probably occur in the upland valleys which experience cold air drainage from snow-covered uplands.

Exposure to wind also increases with altitude. Gloyne (1960) gives figures indicating that up to about 1 000 ft (300 m) wind speeds inland in Scotland do not exceed those at low level near to the coast. At about 2 000 ft (600 m), however, wind speeds in northern Britain are of the same order as those over the open sea.

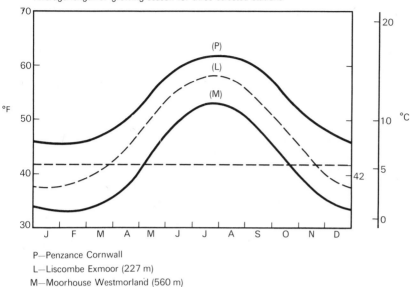

P—Penzance Cornwall
L—Liscombe Exmoor (227 m)
M—Moorhouse Westmorland (560 m)

Fig. 1.7 Average length of growing season for three selected stations. (Source: Curtis, 1974)

The contrasts between upland and lowland weather vary according to the types of air masses prevailing and their accompanying weather systems. In general the effect of seasonal variations is much more marked on higher ground. Under calm anticyclonic weather in summer the uplands may show greater positive benefit than the lowlands.

Thus upland climate in Britain implies a conspicuously late, slow, short growing season with low growth potential, becoming increasingly restrictive with increasing altitude. In distinguishing the Highland Zone and Lowland Zone of Britain, Fox (1932) suggested that the soil factor was the controlling factor in respect of prehistoric settlement of the lowlands, whereas in the highland zone, although soil conditions exerted a powerful influence, they were often over-ridden by climate.

In relation to soil formation the following features of upland climates are significant:

(*a*) Higher rainfall amounts are accompanied by low temperatures; increased cloudiness and low solar radiation lead to an excess of precipitation over evaporation. As a result the leaching factor is high and waterlogging occurs on sites with poor surface and sub-surface drainage.

(*b*) The lower temperatures experienced in upland areas lead to a reduction of biological activity and slower breakdown of plant debris by biological reducers. This in turn leads to the accumulation of plant debris in the form of litter (L and F) horizons and may ultimately lead to the formation of organic mor horizons and peat.

(c) Exposure to wind in the uplands can lead to periodic dessication of the surface horizons of soil. This is particularly the case on sites with good surface drainage exposed to dominant winds. Steep hill slopes are liable to dry out during dry spells, especially if south facing.

(d) Temperature variations in upland regions may be large, and at higher levels frost incidence is high. As a result soils unprotected by vegetative cover will experience ice formation in surface layers with consequential frost disruption of surface structures and some degree of frost heave.

(e) South- and west-facing slopes are warmer and dryer than north- and east-facing slopes. Taylor (1958) showed that growth potential (as indicated by soil temperature) on a south-facing slope may be estimated as 33 per cent higher than north-facing slopes. West-facing slopes were estimated to show a 24 per cent advantage and east-facing a 7 per cent advantage over north-facing slopes.

Some upland soil features are, however, the product of climatic conditions existing at periods before the present. Many of the periglacial features remaining in soils (Fitzpatrick, 1956a) are fossil features relating to the Glacial episodes or to the latter part of the Allerod oscillation (Zone III). In the Boreal period weather was more anticyclonic and in the Sub-Boreal winters were drier and summers warmer so that the tree line probably then stood at 3 000 ft (910 m) in Scotland. In the Sub-Atlantic period deterioration of climate with lower temperatures and higher rainfall is indicated by the vegetational history (Godwin, 1956). Late Glacial and Post Glacial fluctuations have been discussed by Manley (1961) and Lamb (1961) has described climatic change within the historic period.

§1.4 PARENT MATERIAL FACTOR

The principal properties of parent materials which affect soil formation in upland areas are base content and particle size composition. Fine textured materials are slow draining and moisture retentive, thereby predisposing the soil towards waterlogging and processes of gley formation (see pp. 24, 50). Coarse-textured materials are free draining and predispose the soil to podzolisation through effective leaching both vertically and laterally. High base content leads to larger populations of micro-organisms and higher rates of biological activity (Russell, 1973). As a result, breakdown of plant debris and its humification and incorporation in the surface horizons proceeds more rapidly in soils of high base content. Conversely, low base contents generally lead to impoverished populations of plants and animals and a lower rate of biological decay and recycling of nutrients. On low base status soils plant debris may accumulate at the surface forming a deep litter layer. Incorporation of organic material into the soil is often slow and peat formation can take place (Fig. 1.8).

The parent materials of the upland zone of Britain have two principal characteristics. Firstly, they are mainly derived from relatively hard and resistant Palaeozoic

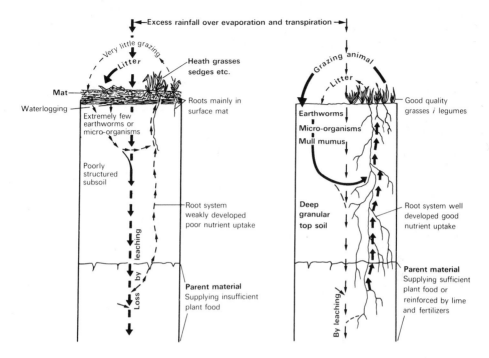

Fig. 1.8 Diagrammatic representation of soil—pasture relationships: (left) peaty mat formation under poor pasture and wet heath; (right) mull humus under good pasture. (Source: Crompton, 1953)

rocks; secondly, they are mainly of low base content. Those upland areas in which granite, slate, schist or arenaceous sandstone predominate generally give rise to parent materials yielding moderately coarse textured soils. Those areas in which fine-grained sandstones or fissile slates and shales occur generally provide parent materials in which silt or clay predominates.

Glacial drift deposits can give rise to widely different textural classes of parent materials within a particular region (Ragg, 1960). Normally, however, the fine material contained in a drift deposit is closely related to the country rock over which it has passed. Thus Glentworth (1954) reports that in north-east Scotland till shows a displacement of about 1 mile from the contributing formations. Similar observations have been made in respect of drifts associated with Carboniferous and Triassic deposits in Yorkshire (Crompton and Matthews, 1970) and Derbyshire (Bridges, 1966).

Within till derived from a particular bed rock formation there may be variation in textural range that can be related to topographic position. Shallow loamy-textured material is usually found on steeper slopes and hill tops. Finer textured materials are generally confined to gently sloping ground and valley bottoms (Ragg, 1960). Resorting of till by surface creep, hill wash and possibly by lateral lessivage[4] processes produce such topographic gradients in textural composition of soil materials.

Major amendments and additions to the parent materials of upland soils have been made by periglacial processes. Peltier (1950) discusses the geographic cycle in periglacial regions and the processes which contribute to the formation of periglacial deposits. Examples of the varied periglacial forms occurring in upland Britain are reviewed by Fitzpatrick (1956a, 1958). Loessial additions to the surface horizons have been reported by Perrin, 1956, and Findlay, 1965. The occurrence of screes, block fields and tors have been described by Palmer and Neilson (1962), Williams (1968) and Ball (1966). Layering in periglacial head deposits has been discussed by Waters (1961, 1964, 1965, 1966); Stephens (1966); Dylik (1966). The effects of periglacial activity on soil parent materials can be summarised as follows:

1. Redistribution of regolith materials, especially removal of fines from the summit areas with consequential formation of stony uplands (in some cases block fields).

2. Formation of head deposits of varying thickness and composition.

3. Scree development.

4. Formation of stone layers within the regolith.

5. Deposition of wind-blown silt (loess) over wide areas.

6. Micro-patterning of regolith materials as a result of freeze—thaw processes (polygons, stripes, ice-wedge phenomena).

§1.5 RELIEF (TOPOGRAPHIC) FACTORS

Relief affects soil formation chiefly through its effects on soil movement and receipt of solar radiation. There is a direct effect on the loss of soil material downslope by variations of the three component factors of slope, i.e. degree, length and shape (see Fig. 1.9). Many studies have shown that both run-off and erosion increase with angle of slope (Baver, 1938), but the effect of length of slope depends on other variables such as rainfall, intensity soil character, vegetation cover and steepness of slope. Under certain conditions, however, doubling the slope length can increase soil losses threefold (Zingg, 1940).

Landscapes formed by sub-aerial agents of denudation often have frequently occurring slope elements. In a fully developed terrain Wood (1942) distinguished four elements: the waxing slope, free face, debris slope and pediment (waning slope). Savigear (1960) states that any fully developed slope is divisible normally into three sections (i) the crestslope, extending from the interfluve to where the slope steepens, (ii) the backslope, consisting of the steepest morphological units of the profile, (iii) the footslope, a gently inclined section extending from the backslope to the valley centre. Savigear's basic divisions are adopted in this book and can be modified in polyphase or polycyclic profiles (Curtis, Doornkamp and Gregory, 1965), as shown in Fig. 1.9.

Soils occur on each of these elements of the landscape and are part of them, i.e. 'Soils are landscapes as well as profiles' (USDA Soil Survey Manual, 1951). Therefore

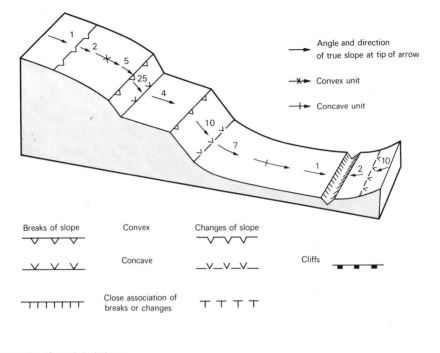

Construction of morphological map

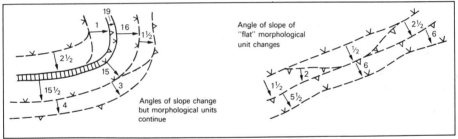

Interrelationships of breaks and changes of slope

Fig. 1.9 Construction of morphological map; inter-relationships of breaks and changes of slope. (Source: Curtis *et al.*, 1965)

soil profiles may be studied in relation to the evolution of the elements of the landscape. Where soil profiles occur in a repeated manner and are geographically associated with relief features they are termed a soil catena. The term soil catena was first proposed by Milne (1936) for topographically linked groups of soils in East Africa which occurred on the sides of major valleys. Different soils were found to occupy (i) the upland (crestslope) and upper backslope, (ii) the lower backslope, (iii) the alluvial footslope. Since they were linked together in much the same fashion as a hanging chain the term catena (Latin = chain) was used to describe the distribution.

Milne stated that profiles changed character along the traverse in accordance with conditions of drainage and past history of the land surface. Two variants of the soil catena can be distinguished in the field. In one the topography is modelled from geological formations essentially similar in lithological character. Soil differences in such a catena are brought about by drainage conditions, differential transport of eroded material and the leaching, translocation and redeposition of mobile chemical constituents (Fig. 1.10).

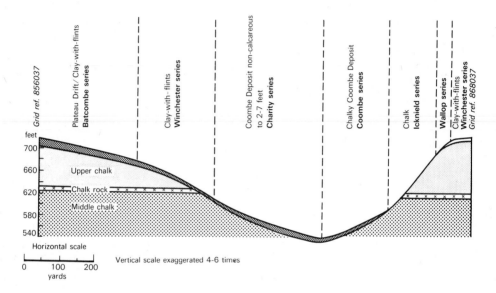

Fig. 1.10 Cross-section of the valley east of Manor Farm, Little Hampden. (Source: Avery, 1964)

In the other variant the topography is carved out of two or more geological formations which differ lithologically. In such a catena such factors as drainage and transport operate upon a more complex range of parent materials, though still forming a linked series of soil profiles. Thus physiographic history and geomorphic evolution of the landscape are involved in the catena concept (Ruhe, 1960).

In Scottish Soil Survey publications the classification of soils is often based on variations in drainage characteristics. The soils are frequently grouped into associations termed hydromorphic associations. In some instances these associations of soils resemble catenary sequences, as when freely drained soils occupy the crestslopes, imperfectly drained soils are on backslopes and poorly drained soils on footslopes.

If sub-aerial erosion proceeded at a steady rate one could assume that soil development is a continuous process. It is clear, however, that soil development and erosion have been periodic rather than continuous in most parts of the world. Therefore Butler (1959) has suggested that there is a case for developing a framework of soil studies on the basis of landscape periodicity. It is suggested that in many soil landscapes it is possible to distinguish a number of soil cycles each of which comprises an alternation of phases of instability and stability.

In the unstable (u) phase erosion of soils from one area and burial of soils elsewhere takes place. A considerable diversity of processes and events may take place within one unstable phase, e.g. mass movement, slope washing, loess deposition. Erosion and deposition is likely to be scattered and patchy in a landscape depending on such variables as slope, aspect, exposure, surface cover and character of soil material. In some cycles erosion and deposition may spread from points of initiation to involve large areas. More often some areas will persist in a stable soil development phase through being unaffected by erosion.

In the stable (s) phase there is a period of soil development and a relative pause in active erosion and deposition. Butler has used the term K cycle to indicate the time unit in which each period of instability occurs. Thus a number of cycles can be identified extending back from the present (Ko), e.g. K1 (s), K1 (u); K2 (s), K2 (u); K3 (s), K3 (u), etc. (Fig. 1.11).

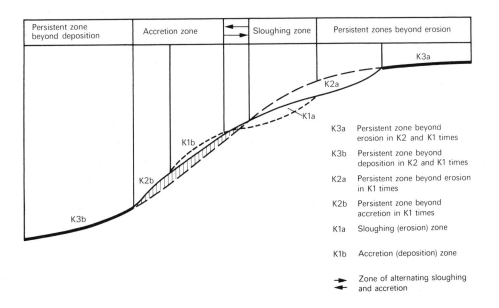

Fig. 1.11 Schematic representation of K1, K2, and K3 ground surfaces on a hill slope illustrating the relationships of the sloughing and accretion zones. (Developed from Butler, 1959)

The evidence for such cycles is found in stone layers and in discontinuities in the weathering material forming the negolith. Contemporary formation of buried horizons on upland heaths in Britain is frequently the result of instability of the soil surface induced by burning of moorland. Observations made on Levisham Moor, North York Moors (Curtis, 1965), shows that recently burned areas frequently carry a network of meandering channels where the peaty surface has been eroded to a depth of a few inches to expose the underlying albic horizon. Following rainstorms these channels carry a load consisting of bleached sand grains and granules of peat (Fig. 1.5). In sites where the flow of water is temporarily checked deposition of material occurs during which heavier sand grains are deposited first, followed by the peat granules. Samples of these channel deposits show laminations made up of bleached sand alternating with thinner laminations of redeposited peat. The sand layers are normally 1–2 cm in thickness, whereas the organic layers may be measured in millimetres.

Soil profiles have been examined in adjacent areas where the surface does not show evidence of contemporary erosion. The profiles consist of 2–3 in (5–7.5 cm) of peaty loam over an Ea (albic) horizon composed of 'wash' layers of bleached sand and organic matter overlying a thin iron pan (Fig. 1.12). The laminations in the Ea horizon closely resemble the depositional layers observed in the channels of presently eroding moorland surfaces. Therefore it is suggested that such profiles provide evidence of an unstable phase (K1 u) in an earlier period (Curtis, 1975).

At the heads of valleys which incise upland moorland surfaces buried soil profiles commonly occur. For example, on the North York Moors peaty gleyed podzols, humus–iron podzols and peaty gleys have been buried beneath approximately 1 m of eroded material. The depositional (unstable) phase was dated by means of artifacts (mainly chinaware) found incorporated in a soil section at Dundale Griff. This showed that the onset of the unstable phase was within the past 100 years. It is considered, therefore, that this K1 (u) phase was related to burning of the moorland within the last 100 years (Fig. 1.13).

The occurrence of the buried soil profiles and evidence of surface wash deposits at the heads of valleys accords well with evidence of overland flow elsewhere on upland heaths such as those of Mendip, Somerset, observed by Stagg (1973). Waterlogging of the surface layers of soil above the iron pan leads to overland flow, and where the surface vegetation cover is broken or reduced by firing or overgrazing the soil surface is eroded and redeposited lower down the valley slopes.

The formation of landscape elements under the conditions outlined above is often accompanied by the formation of stone layers (Ruhe, 1959). These stone layers vary considerably in thickness, particle size and persistence (Parizek and Woodruffe, 1957). Three profiles showing evidence of stone or gravel in upland soils of Wales have been described by Ball (1963). Some of these profiles were examined further by Ball (1967) and it was found in some instances that there was a fairly closely packed line of stones forming a conspicuous concentration in a layer approximately parallel to the ground surface, long axes of individual stones being parallel to the general slope of the stone pavement as a whole.

§1.6 BIOTIC FACTORS

The addition of organic matter to the soil by plants and animals can be regarded as the most important factor in soil formation. As a result of organic matter and its

Fig. 1.12 Thin iron pan soil on Levisham Moor. Note the banded layers at a depth of about 15 cm. The card is scaled at 5 cm intervals. (Photograph: L. F. Curtis)

Fig. 1.13 Buried thin iron pan soil in Pigtrough Griff, Levisham Moor. (Photograph: L. F. Curtis)

decomposition a number of soil properties are modified. These include the type and stability of soil structure, bulk density, porosity, carbon dioxide content, temperature response, consistency, infiltration capacity and available water capacity. In addition the organic acids and humus-clay colloids derived from organic matter play an important part in soil profile development. In particular the solution, precipitation, oxidation and reduction of iron in the soil have been specifically related to microbial action (Bloomfield, 1949).

Soil fauna are principally important in terms of the comminution of organic material and the formation of channels and burrows which promote aeration and the movement of water. The soil fauna usually display great variety but normally the earthworm is an important member. Hopp and Slater (1948) found that earthworms increased the infiltration rate on fine textured soils by a factor of four. Jeanson and Monnier (1965) found that earthworms contribute towards stability of soil aggregates by intermixing organic matter with mineral soil. It is important, therefore, to note that many unreclaimed upland soils have few, if any, earthworms present. Thus mixing of soil horizons is reduced and infiltration rates are little affected by earthworm activity.

The action of organic bonds between mineral particles in the soil may lead to water stable structures (Emerson, 1959), and this is an important by-product of biological processes. However, some aggregates formed rapidly by the action of grass-root pressures (Low, 1955) are considered by Rogowski and Kirkham (1962) not to be water stable.

Apart from its significance as a source of organic matter vegetation is important in respect of the protection it affords the soil surface. Baver (1956) summarised the effects of vegetation on run-off and erosion from soils as:

(a) The interception of rainfall by the vegetative canopy.

(b) The decreasing of velocity of run-off and of the cutting action of water.

(c) Biological activities associated with vegetative growth and their influence on soil porosities.

(d) The effects of roots increasing granulation (aggregation).

(e) Transpiration of water leading to subsequent drying out of the soil.

Interception of rainfall by vegetation is difficult to measure accurately and depends on intensity and duration of rainfall as well as the type of vegetation cover (Penman, 1963). Luchshev (1940) gave mean values of interception of 15–22 per cent for oak canopy. Kittredge (1948) gives values for a number of tree covers as being of the order of 0.007–0.31 in (0.2–8 mm) retainable water. Stahlfelt (1944) gave values of 9–18 per cent interception for moss and litter in areas of Spruce. Information on interception by agricultural crops is scanty but Baver (1956) has a short summary table in which clover is stated to have interception values ranging from 26–56 per cent.

The hydrological importance of plant litter is not only in its ability to store water but in its maintenance of good surface structure, thus acting as a very efficient water filter and protecting the surface against raindrop splash. Lowdermilk (1930) found that forest litter kept soils in a state of greater absorptive capacity. This is borne out by comparisons made by Nordbye and Campbell (1951) who measured infiltration rates on New Zealand hill pastures brought into use after felling of trees, burning, sowing and sheep grazing. One 30-year-old pasture tested had lost most of the forest litter and humus layer (30 cm in depth) and the infiltration rates measured on grass were 0.5 in (1.25 cm) per hr compared with 3.5 in (9 cm) per hr under forest.

Musgrave (1947) analysed a large amount of data concerning erosion losses under different vegetation covers. Losses from continuous row crops such as cotton and corn and from fallow ground were taken as 100 per cent and relative erosion of other types of cover were calculated as percentages. Forest litter and grass sod had relative erosion of 0.001 and 1.0 per cent on the basis of the data presented.

In the upland areas of Britain rainfall amounts greatly exceed the losses by evaporation and transpiration. Thus leaching is effective and soluble constituents are gradually removed and the soils become acid in reaction. In consequence the vegetation is dominated by plants that are tolerant of acidity, while earthworms tend to disappear.

Leaf litter falling from acid-tolerant plants tends to be poor in bases and, in the absence of incorporating fauna like earthworms, remains as a slightly decomposed mat. During its partial decomposition organic acids are produced which further accentuate the leaching process and impoverish the soil beneath (Crompton, 1953). The surface mat absorbs water readily, and under suitable conditions of site and precipitation a wet surface condition is maintained. This further inhibits decomposition and encourages the growth of sedges and rushes. In this way plant debris tends to accumulate, even on sites with normal run off, and wet, acid, blanket peat can develop.

This natural state of impoverishment of upland soils can be modified if the base content of the soil is changed by management (e.g. addition of lime and slag) and the grazing animal is introduced in appropriate numbers. Under such management the former tendency toward peat development can be halted and a relatively stable dynamic system established. In this system grasses grown on the soil are incorporated and mineralised, either directly or via the grazing animal, to provide circulation of nutrients. Crompton (1953) has shown that such differences occur naturally on acidic and calcareous parent materials at altitudes approaching 2 000 ft (610 m) in the Pennines. Some management effects in the Exmoor area are discussed in Chapter 5.

§1.7 THE TIME FACTOR: PERIOD OF SOIL FORMATION

Few examples of soil formation over known periods of time have been recorded in Britain. Changes in the chemical status of sand dunes at Southport over a period of more than 100 years were described by Salisbury (1925). Soil profile development over shorter periods has been noted by Bridges (1961) in restored open-cast ironstone workings near Corby. Curtis (1965, ibid.) has recorded development of 3—4 in (7.5—10 cm) of peaty loam over acid brown earths on upland soils within a period of

100 years. For the most part, however, evidence of the time factor in soil formation is obtained by soil pollen analysis. The technique of pollen analysis (Dimbleby, 1961*b*) depends upon extraction of pollen by treatment of soil samples with hydrofluoric acid.

In a monograph on British heathlands and their soils Dimbleby (1962) brings forward evidence to show that the present soils on the uplands have largely assumed their present characteristics as a result of processes which began about 2000–3000 B.C. The nature of the evidence and the role of man in initiating changes in soil development are discussed in Chapter 4.

§1.8 DEVELOPMENT PROCESSES: WEATHERING AND TRANSLOCATION

Just as each of the soil-forming factors is in fact a complex group, so it is justifiable to regard the principal soil-forming processes as complex interactions of several mechanisms. The processes of organic cycling, erosion and deposition have been discussed on pp. 12–22. Weathering and translocation of elements merit further consideration, especially in respect of those aspects relevant to upland soil formation.

Polynov (1937) demonstrated the varying rates of mobility of elements leached from rocks in various catchment areas. He established a mobility sequence which has been largely confirmed, although within any one phase orders of mobility may not always be as he indicated (Perelman, 1955). Taking chloride, the most mobile, as 100, Polynov suggested the following relative rates of mobility:

Phase I	Cl^-	...	100
	SO_4^{--}	...	57
Phase II	Ca^{++}	...	3.00
	Na^+	...	2.40
	Mg^{++}	...	1.30
	K^+	...	1.25
Phase III	SiO_2	...	0.20
Phase IV	Fe_2O_3	...	0.04
	Al_2O_3	...	0.02

The elements clearly form groups with very much greater similarity in mobility within a group than between one group and another. Polynov, therefore, concluded that the removal of elements in Phases I and II should occur while the greater part of the silica remained. The eventual loss of non-quartz silica would leave the least mobile sesquioxides (together with inert minerals) as a final residue to suffer only exceedingly slow removal.

The general pattern of mobility of elements in the upland areas of cool temperate regions can be expected to result in the 'soluviation' of the normal Polynov sequence. However, Swindale and Jackson (1956) have indicated that where an acid mor humus develops at the soil surface the movement of iron and aluminium is greatly accelerated. This is the result of the process of 'cheluviation' in which chelates (organic–mineral complexes) are leached through the soil. The term 'cheluviation' therefore implies a mechanism for the phenomena of podzolisation in soils of free drainage.

The term podzol was used first by early Russian workers to describe an ashy-grey soil horizon found beneath a dark surface organic layer. Muir (1961) points out in his comprehensive review of podzol soils that no reference was made originally to specific sub-soil morphology below the ashy-grey horizon. Furthermore the soils were not confined only to those which were freely drained. Later writers, however, placed emphasis on the B horizon and the accumulation of iron to form the so called 'ortstein' layer.

Thus the terms podzol and podzolic have been loosely used in the literature (Hallsworth *et al.*, 1953; Damann, 1962) and there remains considerable confusion in nomenclature. Most writers agree, however, that some form of raw humus or peat is associated with the presence of the albic (Ea) horizons (Robinson, 1949; Bloomfield, 1953; Crompton, 1960; Mackney, 1961). Muir (1961, ibid.) indicates that the Russian workers found widespread evidence of burning. This suggested that a distinct organic layer might well have existed widely over all the podzolic soils. The role of organic matter in producing podzolisation was studied by Bloomfield (1953, 1954, 1955) using aqueous extracts of Scots Pine needles and leaves of other species. He demonstrated that the iron—aluminium compounds in soil were reduced and held in an organic—mineral complex. From the point of view of reduction and mobility of iron, therefore, Bloomfield showed the close similarity between the processes of podzolisation and gleying. Later Bloomfield (1957) described the dispersing effect that organic extracts can have on clay suspensions.

Stobbe and Wright (1959), in a review of modern concepts of the genesis of podzols, emphasise the importance of an initial conditioning process prior to podzolisation. In this preliminary stage raw humus (mor humus) can form and many writers have considered raw humus to be a prime agent in the processes of translocation of iron, aluminium and clay which are characteristic of podzolised soils. How far the raw humus is an initiating factor and how far it is merely a reflection of the processes taking place is difficult to determine. However, the role of complex organic substances such as the polyphenols or fulvic acid are now receiving increasing attention. Recent work shows that the translocation of iron and aluminium may be closely linked with the action of these leaching agents (Schnitzer, 1969).

The problem of explaining the mechanism of re-precipitation of the translocated oxides of iron and aluminium has also been studied by Bloomfield (1953, 1955). He suggested that microbial action may be largely responsible for the re-precipitation and this view was supported by Crawford (1956). It was also suggested that no more than partially aerobic conditions may be sufficient to account for re-oxidation of iron in the B horizon. Sorption of ferrous iron on ferric oxide nucleii was found to produce surfaces capable of sorbing more material. Thus in the B horizon the process may be self-regenerating.

In soils of poor drainage, in which periodic waterlogging occurs, the phenomenon of gleying is encountered. The gleying process is essentially a process of reduction and in the case of iron the ferrous state of iron becomes mobile, usually together with manganese and some other trace elements. The gleying process may lead to translocation of iron (particularly when lateral water movement occurs) and to an increase in the Si/Fe ratio. However, the Si/Al ratio may be little altered unless the pH is very low when aluminium may also become mobile (Bloomfield, 1951, 1954).

The processes of podzolisation and gleying are the dominant mechanisms of soil formation in upland Britain. Thus podzolic and gley soils are important categories in the classification of upland soils.

§1.9 FACTORS OF SOIL FORMATION IN LOWLAND BRITAIN: THE IMPACT OF MAN

Most of lowland Britain was originally clothed in deciduous forest. The main characteristics of the vegetation pattern have been summarised by Tansley (1953) and Willis (1973). Existing grassland, heathland and arable areas are mainly the product of centuries of land-use. There were, however, marshy areas which carried vegetation developed in a hydroseral succession. For example in the Fenlands there were large areas in which marsh plants were dominant, e.g. sedges (*Cladium mariscus*); reeds (*Phragmites communis*); grasses (*Molinia caerulea*) and mosses (*Sphagnum* spp.).

Under the deciduous forest the soils were subject to a nutrient recycling process (Fig. 1.14) and there was little decline in base content over a period of time. The soils beneath the deciduous forest were principally Brown Earths on well-drained sites and Gley soils on ill-drained sites or where slow drainage occurred in the soil profile.

The gradual development of settlement and agriculture in lowland areas led to the progressive removal of the nutrient cycling process and in its place an extractive system was introduced. For example, in the mediaeval period soils were sometimes included in the three-field system and sometimes used for grazing. The soils in the three-field farming system received little return of nutrients, except perhaps those fields owned by the Lord of the Manor on which sheep might be placed occasionally, thereby providing some return of organic matter. Thus the fields under arable cultivation gradually declined in fertility in the Middle Ages. In contrast the fields grazed by sheep were fairly stable in their condition. Indeed, one of the effects of the development of the wool trade and enclosure was to increase the area of sheep grazing and thereby improve the soil structure and organic matter levels of soils formerly in arable cultivation.

The intensity of agricultural land-use has increased through the centuries so that the soils in the lowlands today are largely man-made. Also the soil dynamics came increasingly under man's control following the introduction of farm rotations. Perhaps the greatest advances were made in the nineteenth century with the invention of artificial fertilisers which have allowed the farmer to control the chemistry of the soil to a large extent (see Tables 1.4, 1.5 and 1.6). Alongside these developments there has been a rapid extension in the use of agricultural machinery, allowing soils to be cultivated and drained of excess water. Likewise moisture deficiencies in arable soils have been remedied by irrigation.

Thus the natural factors of soil formation can no longer be deemed to be

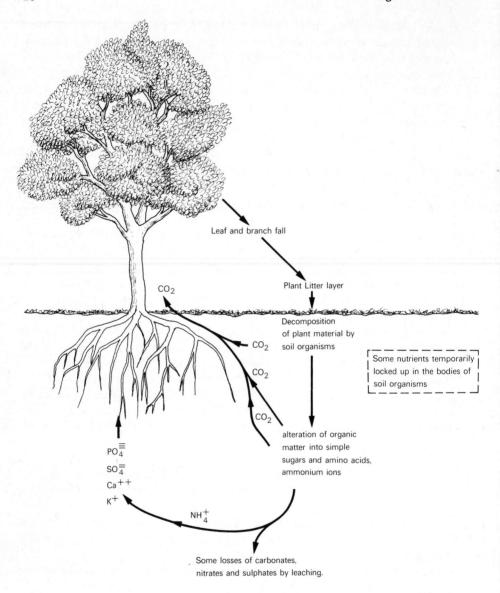

Leaf and branch fall

CO_2

Plant Litter layer

Decomposition
of plant material by
CO_2 soil organisms

CO_2

┌─────────────────────────────┐
│ Some nutrients temporarily │
│ locked up in the bodies of │
│ soil organisms   │
└─────────────────────────────┘

CO_2

alteration of organic
matter into simple
sugars and amino acids,
ammonium ions

$PO_4^{\equiv}$
$SO_4^{\equiv}$
Ca^{++}
K^+

NH_4^+

Some losses of carbonates,
nitrates and sulphates by leaching.

Fig. 1.14 Nutrient recycling process of soils under deciduous forest.

operating freely in many lowland areas. In particular the original vegetation cover has
been replaced by crops and a new biotic balance introduced. The constraints of
climate have been modified by irrigation, drainage and protection of soils by hedgerows. Even the effects of topography have been altered somewhat by the introduction
of field drains and barriers to surface movement of soil in the form of boundary banks

Table 1.4 Inorganic nitrogen carriers

Fertiliser	Chemical form	Source	Approximate per cent nitrogen
1. Sodium nitrate	$NaNO_3$	Chile saltpetre and synthetic	16
2. Ammonium sulphate	$(NH_4)_2SO_4$	By-product from coke and gas, and also synthetic	21
3. Ammonium nitrate	NH_4NO_3	Synthetic	33
4. Urea	$CO(NH_2)_2$	Synthetic	42—45
5. Calcium cyanamid	$CaCN_2$	Synthetic	22
6. Anhydrous ammonia	Liquid NH_3	Synthetic	82
7. Ammonia liquor	Dilute NH_4OH	Synthetic	20—25
8. Nitrogen solutions	NH_4NO_3 in NH_4OH or Urea in NH_4OH	Synthetic	27—53
9. Ammo-phos	$NH_4H_2PO_4$ and other ammonium salts	Synthetic	11 (48% P_2O_5)
10. Diammonium phosphate	$(NH_4)_2HPO_4$	Synthetic	21 (53% P_2O_5)

Table 1.5 Potash fertiliser materials

Fertiliser	Chemical form	Percentage expressed as	
		K_2O	K
Potassium chloride*	KCl	48—60	40—50
Potassium sulphate	K_2SO_4	48—50	40—42
Sulphate of potash-magnesia†	Double salt of K and Mg	25—30	19—25
Manure salts	KCl mostly	20—30	17—25
Kainite	KCl mostly	12—16	10—13
Potassium nitrate	KNO_3	44 (and 13% N)	37

* All of these fertilisers contain other potash salts than those listed.

† Contains 25 per cent of $MgSO_4$ and some chlorine.

around fields. Furthermore, as a result of deep cultivation the original horizons have often been mixed or destroyed.

Although the soils of lowland Britain can be regarded as largely anthropogenic in their present state there are several environmental factors which man cannot modify substantially and permanently. Among these soil texture and internal soil drainage are particularly resistant to permanent modification. Additions of clay to sandy soils (usually termed 'marling' because calcareous clays were often used) have been made in attempts to increase the moisture-holding capacity and nutrient exchange properties of sandy fields. The effects were however often slight and short lived. Similarly the installation of mole[5] drainage has been employed to improve the permeability of

Table 1.6 **Phosphorus carriers**

Fertiliser	Chemical form	Approximate percentage of available P_2O_5	Per cent P
Superphosphates	$Ca(H_2PO_4)$ and $CaHPO_4$	16—50	7—22
Ammoniated superphosphate	$NH_4H_2PO_4$ $CaHPO_4$ $Ca_3(PO_4)_2$ $(NH_4)_2SO_4$	16—18 (3—4% N)	7—8
Ammo-phos	$NH_4H_2PO_4$ mostly	48 (11% N)	21
Ammonium polyphosphates	$(NH_4)_4P_2O_7$ and others	58—60 (12—15% N)	
Diammonium phosphate	$(NH_4)_2HPO_4$	46—53 (21% N)	20—23
Basic slag*	$(CaO)_5 . P_2O_5 . SiO_2$	15—25	7—11
Steamed bone meal†	$(Ca_3PO_4)_2$	23—30	10—13
Rock phosphate	Fluor- and Chlor-apatites	25—30	11—13
Calcium metaphosphate‡	$Ca(PO_3)_2$	62—63	27—28
Phosphoric acid	H_3PO_4	54	24
Superphosphoric acid	H_3PO_4 and $H_4P_2O_7$	76	33

* The formula of this fertiliser is very uncertain. Basic slag is intensely alkaline because of the presence of large amounts of the hydroxide and carbonate of lime.
† The steamed bone meal is cooked under pressure and the fat and oil removed. The bone is left open and porous.
‡ Synthesised from rock phosphate or limestone and $P_2O_5Ca_3(PO_4)_2 + 2P_2O_5$ —— $3Ca(PO_3)_2 +$ impurities or $CaCO_3 + P_2O_5$ —— $Ca(PO_3)_2 + CO_2$ + impurities.

slow-draining soils. This method is effective but after a period of time the mole drain either collapses or becomes blocked with silt. Repeated endeavour is therefore required if environmental conditions adverse to crop growth are to be kept under control. The problems of maintaining our soil resources and the impact of modern farming are discussed in Chapters 14, 15 and 16.

FURTHER READING

Buckman, H. O. and Brady, N. C. (1974). *The Nature and Properties of Soils*, Macmillan.
Curtis, L. F. (1975). 'Landscape periodicity and soil development', in R. Peel, M. Chisholm and P. Haggett (eds), *Processes in Physical and Human Geography*, Heinemann, pp. 247—65.
Foth, H. D. and Turk, L. M. (1972). *Fundamentals of Soil Science*, J. Wiley and Sons.
Willis, A. J. (1973). *Introduction to Plant Ecology*, George Allen & Unwin.

NOTES

1. Cations are positively-charged ions: Some cations carry one positive charge ($^+$) others may carry more than one charge ($^{++}$). Anions are negatively charged.

2. Milliequivalent = one milligram of hydrogen or the amount of any other ion that will combine or displace it. The abbreviation m.e. will be used to express milliequivalent, e.g. 20 m.e./100 g = 20 milliequivalents per 100 grams of soil.

3. pH = log $(1)/(H^+)$. At neutrality the hydrogen ion concentration has been determined as 0.000 000 1 or 1×10^{-7} gram of H^+ per litre of solution. Thus the pH = log $(1)/(0.000\ 000\ 1) = 7$. See Appendix 1.

4. Lessivage = washing of particles through pore spaces in a soil, i.e. mechanical movement of fine particles.

5. Mole drainage is achieved by drawing a metal torpedo-like object through the soil beneath the surface. It results in a circular drainage channel similar to a mole channel. Hence the term.

2 MAPPING AND ASSESSING SOILS

§2.1 SOIL CLASSIFICATION

2.1.1 Soil: three, four or five dimensions?

In the natural landscape soil changes in its characteristics *vertically*, from the land surface downwards to rock, and *laterally* from one place to another. Each time a different characteristic is used to describe soil, like colour or wetness, in one sense another dimension is being added to the concept of soil variation. To further complicate the issue some of the characteristics — for example, wetness — change through time. One day the soil may be relatively dry while after a cloudburst it may be totally waterlogged. We have now added yet another dimension to the soil system — the temporal or time dimension.

Viewed on the scale of the variation in a small area — say one field — classification appears complex. Viewed on a national or international scale the classification problem appears almost intractable. How is it possible to describe soils at all, let alone make maps of them?

Although this book deals with the soils in Britain, British soil classification has always been closely linked with international developments. Therefore, in order to understand British systems it is also necessary to consider developments elsewhere in the world.

2.1.2 Early attempts at soil classification

Soil scientists have been occupied by devising systems of soil classification since the origins of agricultural science and as crop treatment and fertiliser usage became more sophisticated more precise soil classifications were needed. Because of this and also because of other deficiencies in the earlier classifications (which will become clearer later in this chapter) detailed classifications are still being produced today.

In Britain in the late eighteenth and early nineteenth centuries a series of County Agricultural Reports were commissioned by the Board of Agriculture (see for example Marshall, 1809). Many of the original reports contained both written soil information and a rudimentary soil map. In most cases the soil classification was couched in very simple terms. For example, Thomas Rudge (1807) used the phrase 'shallow, brashy loam' to describe the brown calcareous soils of the Cotswold Hills (now termed the *Sherborne* series — see p. 208).

2.1.3 The Dokuchaiev classification

The first major advance in soil classification was the scheme proposed by the Russian pedologist V. V. Dokuchaiev at the end of the nineteenth century. Dokuchaiev was concerned principally with large-scale variation and his classification depended heavily on the relationship which he discovered existed between soil, natural vegetation and climate in Russia. The work emphasised that the processes involved in the development of different soil layers (or *horizons*) were governed by environmental factors, soils developing continually. In other words, soils are dynamic and not static. The hypothesis was that soils could reach a 'mature' state of equilibrium, in the same way that Tansley and others later suggested plant communities reached a climax.

Dokuchaiev proposed three soil classes: normal, transitional and abnormal. These were sub-divided into a total of thirteen soil types. Later a close collaborator of Dokuchaiev — Sibertsiev — renamed the classes zonal, intrazonal and azonal respectively. Because many of the Russian names are still in use today it is necessary for us to consider the classes in detail.

The *normal* or *zonal* class contained soils which had developed in particular climatic and/or vegetational regimes. With this class the influence of climate and vegetation was seen to be the most important factor in the development of soil characteristics. Dokuchaiev proposed seven soil types in this class:

Zone		*Soil type*
I	Boreal	Tundra (dark brown) soils
II	Taiga	Light grey podzolised soils
III	Forest-steppe	Grey and dark grey soils
IV	Steppe	Chernozem
V	Desert-steppe	Chestnut and brown soils
VI	Aerial or desert zone	Aerial soils, yellow soils, white soils
VII	Subtropic and zone of tropical forests	Laterite or red soils

In the *transitional* or *intrazonal* class local physiographic or lithological factors modify or 'override' the zonal factors in influencing soil development. Here there are three soil types:

VIII	Dryland moor soils or moor-meadow soils
IX	Soils containing carbonate (rendzina)
X	Secondary alkaline soils

The *abnormal* or *azonal* class contains those soils where erosion and depositional processes dominate the more usual soil processes. The soil types are:

XI	Moor soils (for example, moorland peats)
XII	Alluvial soils (for example, riverine 'wet-land' soils)
XIII	Aeolian soils (for example, sand-dune soils)

2.1.4 Developments of the Dokuchaiev model

The work of the Dokuchaiev school in Russia stimulated the development of soil classification in various parts of the world. Baldwin *et al.* (1938) produced a zonal— intrazonal—azonal classification of North American soils based on similar distinguishing criteria and making the same underlying assumptions about the influence of the present environment on soil development. Other classifications were also produced based on other distinguishing criteria, a leading example of which was Kubiena's work (Kubiena, 1953) which emphasised the importance of soil moisture regimes.

These other classifications were, of course, making implicit assumptions about different relationships which were supposed to exist in the soil system. The main problem was that the assumptions about the various relationships were usually implicit and rarely either stated or proven. A different but equally important practical problem that had arisen by the late 1950s was the plethora of soil types described by a single name. (Tavernier and Smith, 1957, showed, for example, that the name *Braunerde* (brown forest soil or brown earth) applied to at least eleven different kinds of soil in different countries.)

Clearly this confusing situation could not be allowed to continue and some fundamental questions were being asked about the assumptions underlying the various classification schemes (for example see Leeper, 1956, and Butler, 1958). In particular, as Avery (1969) pointed out, the influence of climate and vegetation had been very much overstressed and it was seen that soil parent material plays an important, continuing role in soil development. Also, many of the world's agricultural soils have been influenced for centuries by man's activity and are only in a very limited sense 'natural'. Duchaufour (1959) and Manil (1956, 1963) showed that soils under a semi-natural vegetation have a sub-surface horizon which is strongly influenced by the prevailing land-use, while Henin *et al.* (1960) demonstrated how, under intensive agriculture, the soil progressively acquires new morphological and chemical characteristics.

Because of our very limited state of knowledge about soil development processes it has become clear that *typological* classifications, based on inferred genetic factors, are suspect. For that reason recently devised classification systems have been *definitional* and based on recognisable soil properties. However, before we proceed to definitional classifications we will examine the working classification that was in use in Britain between the Second World War and the early 1970s.

2.1.5 The 1940 British classification

Because Britain was relatively behind both Russia and the United States in starting a programme of intense field-mapping it was not until the post-war period that any considerable advance was made in systematic soil mapping. Obviously a working classification had to be in use to facilitate rapid mapping. In the light of field experience the deficiencies of the classification have become apparent and the production of a new definitional classification is in an advanced stage. This will be examined in more detail in the next section.

The semi-typological classification used on British soil maps was devised from limited experience by British pedologists in the 1920s and 1930s. Details of the original scheme are given by Clarke (1940) and it was modified as a result of further field experience. The final version closely resembles European soil classifications. (Full details are given in Appendix 2.) The scheme proposed six main soil groups:

I   Brown earths
II   Podzols
III   Gley soils
IV   Calcareous soils
V   Organic soils
VI   Undifferentiated alluvium

2.1.6 Definitional classifications

Recent soil classifications have been based on reasonably easily defined soil characteristics measurable either in the field or the laboratory. Examples of the new systems are that proposed for the United States (Soil Survey Staff, 1960; Smith, 1965), the Netherlands system (De Bakker and Schelling, 1966) and the British system (Avery, 1973). In the American and Dutch systems the soils are grouped according to the presence or absence of a specific 'diagnostic horizon' together with certain other specific criteria, such as morphological features and physical, chemical or mineralogical characteristics of specific reference layers.

    The United States Department of Agriculture (USDA) system is well known and has been explained elsewhere (see for example Bridges, 1970). Appendix 3 summarises the main groupings in this scheme, which is also known as the '7th Approximation'. Although of wide applicability, some British workers had serious misgivings about some of the premises upon which the classification was based (Webster, 1968). Attempts have been made to relate British soils to the USDA classification (Clayden, 1971; Ragg and Clayden, 1973) but it became generally apparent that a different system was necessary for Britain. By the late 1960s a great deal of information about British soils had been amassed by field-workers and Avery proposed a new system of classification (Avery, 1973), outlined below. This system has now been adopted by the Soil Survey of England and Wales.

2.1.7 The new British classification

Avery (1973) states the aim of the new classification:

> is to organise and communicate information about soil profiles by allocating them to named classes which subdivide the known variation in a useful and comprehensible way.

This then is a classification of soil *profiles* (vertical soil sections) which is itself an important difference from the USDA system. In the latter the basic unit of soil classification is the soil *pedon*, an artificial cuboid unit with a cross-section area depending on the lateral variability of properties which define classes.

Having set an arbitrary limit on the basic unit of classification (i.e. by defining it as the soil profile) limits were also set on the vertical definition of soil. It was defined as any unconsolidated material or organic layer thicker than 3.94 in (10 cm) occurring at the earth's surface: profiles shallower than 10 cm are not considered.

Another arbitrary limit is that soil is treated as a static, not dynamic, body. This obviously simplifies the problem of soil 'dimensions' which we referred to earlier. The use of this premise meant that properties used as diagnostic in the classification must be those which are relatively stable — in other words properties which for example alter daily (like moisture content) would be totally unacceptable as diagnostic features. In addition to the overriding objective that properties should be chosen as objectively as possible other considerations in the choice of differentiating criteria were:

(*a*) The criteria are usually recognisable in the field. The field-surveyor could there-fore put a particular profile in the correct class without reference to laboratory findings.

(*b*) The criteria are those which can usually be related to other environmental attri-butes such as land-form, geology, vegetation and climatic conditions. This is important in terms of field-mapping (see §2.2).

(*c*) Since the classification was conceived primarily for helping land-use decisions soils should be classified as far as possible on properties that affect land-use capability.

In the classification there is a twofold primary division into organic (peaty) soils and mineral soils. At a secondary level there are six groups of mineral soils and one group of organic (peaty) soils. Each group is further sub-divided into a total of 108 *sub-groups*. Full details of the various groups and sub-groups are given in Appendix 4.

§2.2 SOIL MAPPING

2.2.1 Maps for whom?

It is usually easier to construct reliable *special-purpose* maps but rather more difficult to produce useful *general-purpose* maps. With the former the map-maker is clear as to his specific aims and objectives. To achieve these he has to measure a relatively small number of soil attributes and plot their distribution on a map. An example of a fairly sophisticated special-purpose map is the hydrological classification of soils of northern England produced by Edmonds *et al.* (1970) (Fig. 2.1). The map was designed to help with flood prediction and it has been shown that the average minimum infiltration rate of the soil is a significant factor in determining the characteristics of river flood peaks. Soils are therefore grouped according to their minimum infiltration rate and it is this that is plotted on the map.

General-purpose maps are produced for a number of users. In Britain the most important users are agricultural advisers and soil maps are used by them in the

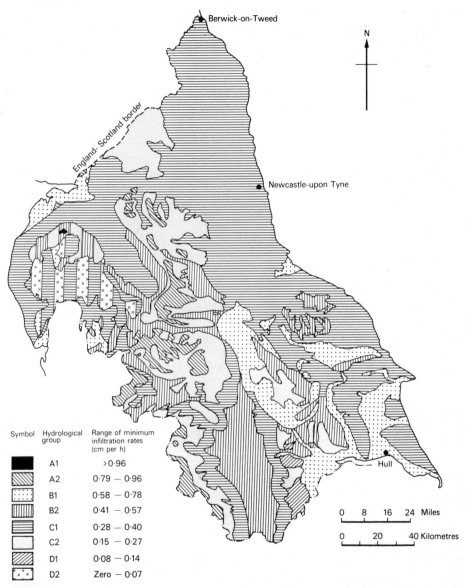

Fig. 2.1 Hydrological classification of the soils of northern England. (Source: Edmonds *et al.*, 1970)

assessment of land capability. The Ministry of Agriculture are producing their own *Land Classification* maps, based on soil and climate considerations. These maps are, in turn, used in making planning decisions. The close relationship between soil and land capability is explained in more detail in Chapter 14. Agricultural advisers also use soil maps in assessment of fertiliser need and drainage requirements.

Other users of general-purpose soil maps include civil engineers, interested in the potential strength of soil, and hydrologists, interested in water-retention properties. A large number of soil maps are sold to colleges and universities where they both aid students in understanding the surface of the physical landscape and also help researchers in explaining the processes that have resulted in the landscape as we know it.

2.2.2 Making general-purpose soil maps

The difficulty in producing good general-purpose soil maps lies partly in achieving an objective combination of measured variables and partly in selecting meaningful attributes which can be readily measured in the field. In more simple terms, firstly, how do we combine our observations in the most meaningful way and, secondly, and perhaps more fundamentally, what observations is it most useful to make? Generally, soil surveyors make subjective decisions based on field experience. With more experience decisions are taken about more important criteria and, also, observations become more accurate.

The new classification (see §2.1.7) is an attempt to relieve the soil surveyor of the problem of *what* to look for. With the classification key available, in the same way that the botanist has his key to species of plants, the soil surveyor can more easily 'pigeon-hole' the soil type he is examining. There is, however, an obvious danger in this approach, which is why soil surveyors will have to continue to take extensive field notes about profiles under examination. While (at least in Britain) the botanist can be fairly certain that he is unlikely to discover a new species of plant, the soil surveyor cannot be certain that any new profile he is examining will correspond to any that has been examined before or, indeed, that it will necessarily readily fit into the scheme of classification that he has available. (This point is discussed in more detail by Courtney, 1974.)

In Britain profiles are grouped together in *soil series*, each series consisting of a group of profiles with similar successions of horizons developed in lithologically similar parent materials. The soil series is named from a place where it is commonly found and usually where it was first described. The soil series name is also used to describe the *soil mapping unit*. The mapping unit is a geographic area where the majority of soil profiles (usually at least 60 per cent) conform to the general concept of the dominant soil series. So, for example, at least 60 per cent of the profiles in the area mapped as *Denchworth* mapping unit should lie within the conceptual limits set for the *Denchworth* series. Soil series are sub-divided where necessary into *soil phases* based on properties such as difference in stoniness or thickness of horizons. These phases may or may not be shown on the published map, depending on considerations of scale.

On Scottish soil maps, and on some English and Welsh county soil maps, the soil series are grouped into *soil associations*, the latter being shown on the map rather than the series mapping units. A soil association is defined as a group of topographically related soils developed on one geological parent material. The system of soil associations is explained in more detail by Glentworth (1954). In the original system the soil association is given a name and each soil series within the association received the association name together with an indication of drainage class. So, for example, the *Insch* freely drained and *Insch* very poorly drained soils were both soil series within the *Insch* association. In recent publications of the Soil Survey of Scotland, however, the English and Welsh practice of giving individual series separate names has been followed.

The production of a soil map is undertaken in a number of stages:

1. Field reconnaissance. The surveyor makes a general examination of the field area and looks at a selected number of profile pits in detail. The aim of this preliminary exercise is to establish the broad relationships that exist in the area between parent materials, landscape, vegetation and soils.

2. Soil identification and legend construction. Using the information from the reconnaissance the surveyor constructs a legend which can be used in the later detailed mapping. The legend must be constructed in the light of the existing national soil classification system. Figure 2.2 shows an example of a field mapping legend.

3. Surveying. Boundaries are drawn on field maps (see Fig. 2.3) between areas which correspond to the units in the mapping legend. The surveyor uses changes in slope and vegetation to guide him in drawing the boundaries. In order to check and substantiate the boundaries the surveyor makes a number of auger borings (usually at least $50/km^2$). In *free-mapping* the boreholes are sited according to the surveyor's judgement, while in *grid-mapping* the borings are sited according to a previously designed sample scheme. Grid-mapping is used either if there are few landscape or vegetation 'clues' to soil variation (as for example in the broad flat Fenland areas of eastern England) or where the survey is concerned with a less subjective statistical variability statement (see § 2.3.2).

4. Soil sampling and laboratory analysis. Profiles representative of the series present in the survey area are examined in detail in profile pits and samples from the various horizons are analysed in the laboratory. The most common analyses are for particle-size distribution, organic content, calcium carbonate content and pH. Some samples are also analysed for micromorphological and soil physical attributes.

5. Final map preparation. At this stage further detailed discussions take place between the surveyor and surveyors from adjacent or similar areas to solve problems of soil correlation. The final map is drawn up from the field maps and a record or memoir written and published.

2.2.3 Different map scales: the British system

In England and Wales, until 1966, maps were published on a scale of one inch to one mile or 1:63 360. The maps were based on the Ordnance Survey fifth-edition, this being the same as that used by the Geological Survey. Field-work for these maps was undertaken at a scale of 1:25 000. Because of the large areas involved in the production of these maps and the small staff available, map production was extremely slow;

Unit Symbol	Horizon	Colour	Depth	Texture	Stones	Carbonates		Ref. profile
10	A/R	10 YR-7·5 YR 4/2	0-20	L	s-vs, g-vg	ca	**Rendzina** May be on disturbed, quarried ground	
11	A/Bbc	7·5 YR 4/4	0-35	(Z)CL	s-vs, g-vg	ca	**Brown rendzina** Arable	Pit 1 (1192) Pit 13 (1547)
P11	A	7·5 YR 4/2	0-15	"	"	Maybe non ca	Pasture phase	
	Bbc	7·5 YR 4/4	15-35	"	"	ca		
W11							Woodland phase. To be defined	W11C
11	A/Bbc	"	<20	"	"	ca		Pit 2 (0898)
11	A/Bbc	"	0-35+	"	"	ca		
11C	A/Bbc	"	0-35	C	"	ca		Pit 2 (0898)
11L	A/Bbc	"	"	(fs)L	s?	ca or -?		
1̄1̄	A/Bbc	"	"	L,(Z)CLC	s?	-		
Coll	A/Bbc	"	0-50+	(Z)CL-C	s-vs		Valley bottom soils to be defined	
12	A	7·5 Y-10 YR 4/2-5/2	0-15	(Z)CL	ss	-	**Brown earth**	Pit 3 (1089)
	Bb	7·5-10 YR 4/4	15-35+	"	ss	-	or A/Bb deeper than 35	5 (1352)
12L	Bb	"	"	(fs)L	ss or t (sandy tiles)		Weak structure on plough	10 (3489)
1̄2̄L	Bb		15-90	(fs)L	ss	-		?
13	A	10 YR 4/2	0-15	C-(Z)CL	s	ca	**Gleyed brown rendzina** May have thin unmot. Bbc 15-25	Pit 15 (3083)
	B/Cgc	10 YR-2·5 Y 6/4	15-50	"	s	ca		
14	A	10 YR-4/2	0-15	C-(Z)CL	ss		**Gleyed brown earth**	Pit 11 (3795)
	Bb	10 YR-7·5 YR4/4-5/4	15-45	C-(Z)CL	ss			
14	B(t)/Cg(c)	2·5 Y 6/6-6/7	45+ <50		(s)	(ca)		
16	A	10 YR 4/2	0-15	(fs)L	ss	-	**Brown earth (argillic)**	Pit 4 4 (2498)
	E	7·5 Y-10 YR 4/4-5/4	15-25	L-CL	ss	-		
	Bt	7·5 YR-5 YR 5/4-5/6	25-50	CL-C	s		Weak structure on plough	

Fig. 2.2 Example of soil mapping legend. This table shows an early draft of the descriptive mapping legend in use during the mapping of 1:25 000 sheet SP12 (Stow-on-the-Wold), which is further referred to in Chapter 11, §6, p. 217. It can be seen that the legend is far more detailed than the legend provided on the final soil map. The coded unit symbols approximate to series divisions in that each group is equivalent to one series. For example, 10 = *Yatton* series, 11 = *Sherborne*, 12 = *Waltham*. (Source: Soil Survey of England and Wales Mapping Legend, unpublished)

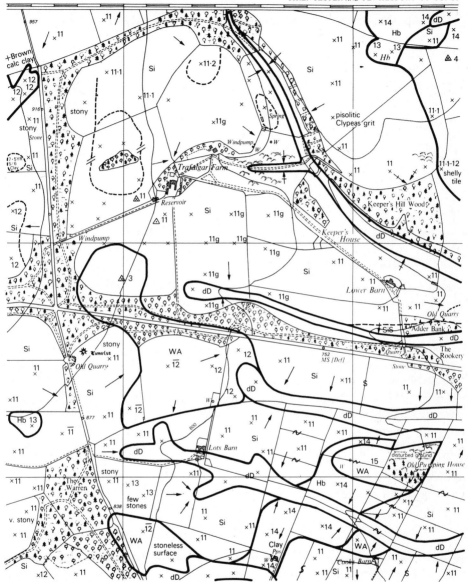

Fig. 2.3 Example of soil survey field map. The map is an extract of one of the 1 : 10 560 (6 in) field sheets used in the compilation of the 1 : 25 000 published soil maps. In this case the area shown is part of that eventually published as sheet SP12 (Stow-on-the-Wold), further referred to in Chapter 11, §6, p. 217. The small crosses on the map indicate points at which auger borings were made or pits were dug, while the numbers are coded references to particular soils series (11 = *Sherborne*, 12 = *Waltham*, 13 = *Haselor*, etc.) The heavy lines separate the different mapping unit areas, which are identified by letters (Si = Sherborne, WA = Waltham, Hb = Haselor, dD = Didmarton). Arrows point downslope and indicate the dominant direction of run-off. (Source: Soil Survey of England and Wales Field Map, unpublished)

only about twenty maps being produced between 1950 and 1965. For this reason policy was changed in 1966 from an aim of providing a national cover of 1 : 63 360 maps to one of producing 1 : 25 000 maps for selected areas in each administrative county. These new maps are chosen to cover selected areas of geomorphological and agricultural interest, and up to the end of 1975 about thirty maps had been published. the 1 : 25 000 maps are intended to serve as sample areas in the later construction of county, regional and national maps; production of some of these is already in progress[1].

The situation in Scotland and the Republic of Eire is rather different. In both these areas soil maps are still being published at a scale of 1 : 63 360 though the scale of mapping units is slightly courser than those in use in England and Wales. Reference to Scottish soil maps has already been made.

§2.3 NUMERICAL AND STATISTICAL TECHNIQUES

2.3.1 Development of multivariate techniques in soil studies

One of the main objects of statistical studies of soils is to examine classificatory relationships. Much of the earlier work in this sphere developed from that of ecologists studying plant communities. A study by Rayner (1966) followed the methods of numerical taxonomy taking twenty-three profile descriptions of soils from Glamorgan-shire and the results of laboratory measurements on soils samples of the ninety-one horizons into which they were divided by the surveyor. Calculations were made of similarity co-efficients and this matrix was then sorted to give a dendrogram or 'tree-diagram' (Fig. 2.4). The classification of soils produced by the similarity index corresponded closely to that delineated by the surveyor, although better distinction was obtained between different brown earths than between different gley soils.

Since Rayner's work there have been a number of studies in which numerical techniques have been applied to the study of British soil distribution. It is, however, important to realise the limitations of these statistical techniques. These have been succinctly summarised by J. M. Norris (1970):

> . . . someone must initially fix all the decisions and concepts — but once this is done the rules are defined and can be explicitly stated, and they are uniform and consistently applied throughout the analysis.

In the numerical classification of soils Norris considers that there are two basic initial decisions to make. The first is the selection of a similarity co-efficient or distance measure between the individuals or variables, depending on which is being classified. The second decision concerns the strategy to be used: this is the method by which the individuals are joined together to make groups.

2.3.2 The assessment of soil variability

Beckett and Webster (1971) have shown how variances and co-efficients of variation

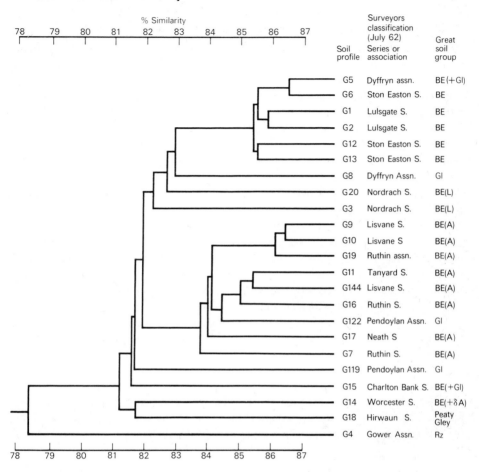

Fig. 2.4 Dendrogramme of soil-profile similarities based on average horizon similarities. (Source: Rayner, 1966)

increase with the size of the area samples. They show that up to half the amount of variation within a field may be present within any m² in it. Elsewhere, it has been shown that up to 40 per cent of the variation in a soil-mapping unit covering a wide area in Gloucestershire was present in 20 m² areas (Courtney, 1972). In the same study Courtney (1973) was concerned with confirming the surveyor's classification of some Cotswold soils (*Sherborne* series). One hundred and eighty-four sites within the area of the mapping unit delineated by the surveyor were reinvestigated and the data subjected to principal components analysis. The result showed the essential modality of the mapping unit and also that the original surveyor had been very successful in including mostly profiles within the limits of the original definition of the mapping unit.

2.3.3 Particular problems in the application of statistical techniques to soil data

A complication in the use of any numerical technique with soil data is that with a large number of soil individuals the data is very heterogenous. (For example, one or two principal components often only account for a small amount of the variation. Often in social science they may account for up to 70 per cent.) Sarkar *et al.* (1966) developed a hierarchical model designed to achieve maximum efficiency of discrimination with the incorporation of fewest characteristics. They concluded that a small group of carefully selected variables could be as effective in discriminating between soil types as a large number of unselected characters. However, in order to check which criteria should be discarded it is necessary to have measurements of all characteristics in most of the profiles, or at least a large sample. This particular approach therefore demands a two-stage sampling programme.

A further problem is the question of how to deal with different horizons. There are a number of approaches available. Rayner (1966) treated all horizons as separate units and compared all horizons with all other horizons. An alternative to this is to compare all properties at specific depths in all profiles. Grigal and Arneman (1969) used a variation of this method, comparing horizons at the same depth and also those above and below the two which are being directly compared.

In many ways the results of the application of numerical techniques to the production of soil maps and classifications has proved rather disappointing. This is perhaps because the limitations of statistical methods were only dimly understood. Work is at present being conducted by the Soil Survey of England and Wales to see how far it is possible to produce data for automatic plotting by digital computer from grid-mapping. It has already been shown that it is possible to automate the carto-graphic stage of the operation. Perhaps the best future for numerical techniques lies in the checking and confirmation of decisions produced by the surveyor.

FURTHER READING

Soil classification

Avery (1969) summarises some of the problems of typological classifications. This first paper forms a useful introduction to his later British classification (Avery, 1973) which should be read by all students of British soils. Simonson (1968) provides a thorough, if lengthy, introduction to American ideas.

Soil mapping

The annual reports of the Soil Survey of England and Wales[2] provide a useful back-ground to a study of the development of soil mapping in Britain. Findlay (1970)[2] provides a very concise summary of the details of soil-mapping techniques. The application of photography to soil mapping from the air is discussed by Curtis (1973).

Numerical techniques

There have been a number of recent reviews of numerical soil techniques, though these are often difficult to get to grips with if you lack a basic statistical background. Among the more easily assimilated is that of Norris (1970) which, although it is rapidly becoming outdated, covers a large field in a very clear manner. The paper by Webster and Beckett (1968) is of particular interest as it can now be seen as something of a milestone in this field. Summaries by Beckett and Webster (1971) on soil variability and by Bie and Beckett (1970) on costs of soil survey are also useful.

NOTES

1. In January 1976 this policy was under review.

2. These items are available in booklet form from the Soil Survey of England and Wales, Rothamsted Experimental Station, Harpenden, Herts.

3   THE SOILS OF BRITAIN IN A WORLD SETTING

§3.1 INTRODUCTION

As is seen in Chapter 2, world soils can be classified into three soil orders: Zonal, Intrazonal and Azonal. Zonal soils are those with characteristics dominated by the influence of climate and vegetation. Intrazonal soils are those that show the influence of some local factor and can be sub-divided as follows:

Intrazonal soils
- hydromorphic soils — formed where excess of water is present and periodic waterlogging takes place.
- halomorphic soils — formed where salts have accumulated.
- calcimorphic soils — formed where calcium has accumulated.

The Azonal soils are the immature soils without developed profiles occurring on recent deposits. These soils can be sub-divided according to the nature of the initial deposit as follows:

Azonal soils
- Lithosols — immature soils forming on rock material.
- Regosols — immature soils forming on particulate deposits such as loess or sand.
- Alluvial soils — immature soils forming on recent alluvial deposits.

The general pattern of zonal soils in relation to climatic conditions is shown in Fig. 3.1. Britain is situated in a humid temperate region and the zonal soil groups characteristic of such areas are the Podzol group and the Brown Earth (Brown Forest) group. In addition to these two major soil groups of the zonal order the soils of Britain include substantial areas of intrazonal soils. These are mainly represented by the hydromorphic soils of the Gley group and Organic group. Smaller areas carry soils of the Brown Calcareous and Calcareous Gley soils where calcimorphic parent materials occur.

The place of British soils within a world pattern of soil distribution is shown in Fig. 3.2. It will be apparent that the soils of Britain have similar profiles to many of those in North America and Europe. These areas are those in which agricultural technology is highly developed and where agricultural productivity is highest (Fig.

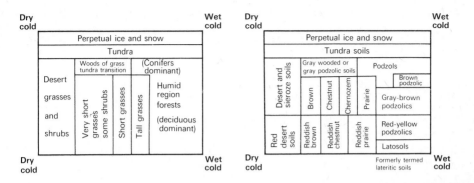

Fig. 3.1 Left: a simplified diagram showing how the natural vegetation varies with climate. Right: A simple diagram indicating the relationship of zonal soils to climate. Note that soil boundaries correspond rather closely to those of the natural vegetation. The right to left successions of soils and vegetation shown in the diagrams approximate those that will be encountered when travelling from the east coast of North America to Sierra Nevada and Cascade Mountains. Typical desert conditions occur in the inter-mountain regions west of the Rockies. (Source: Buckman and Brady, 1974)

3.3). Therefore the soils of Britain must be considered not only in relation to British needs but also as productive elements in the world pattern of natural fertility of soils (Fig. 3.4). The principal pedological features of the soil groups are given in the following sections.

The horizon nomenclature (L, F, H, Ea, B, C, etc.) is that used in many British publications at present available to the reader. Appendix 1 provides the reader with explanatory notes concerning the lettering employed for different horizons.

§3.2 PODZOLS

As a result of the translocation of humus and/or sesquioxides certain horizons are formed which typify the Podzol group of soils. Below the raw humus (H layer) there is a grey and sometimes structureless Ea horizon from which virtually all free iron has been removed. The formation of a humus—iron podzol requires leaching by water and the presence of raw humus or mor to provide soluble organic substances capable of mobilising ferric and aluminium oxides. The B horizon of illuviation includes a dark humus-enriched layer but in many areas the most prominent feature is a strong brown or rusty coloured Bs horizon of iron and aluminium enrichment.

Podzols may be sub-divided into sub-groups consisting of humus—iron podzols, peaty gleyed podzols, gleyed podzols and indurated podzols (Fig. 3.5).

Humus—iron podzols have Ea and Bh horizons, with or without Bs horizons, and they lack the mottled Btg horizons of the gleyed podzols described below.

Peaty gleyed podzols have acid organic O horizons or peat overlying Eag eluvial

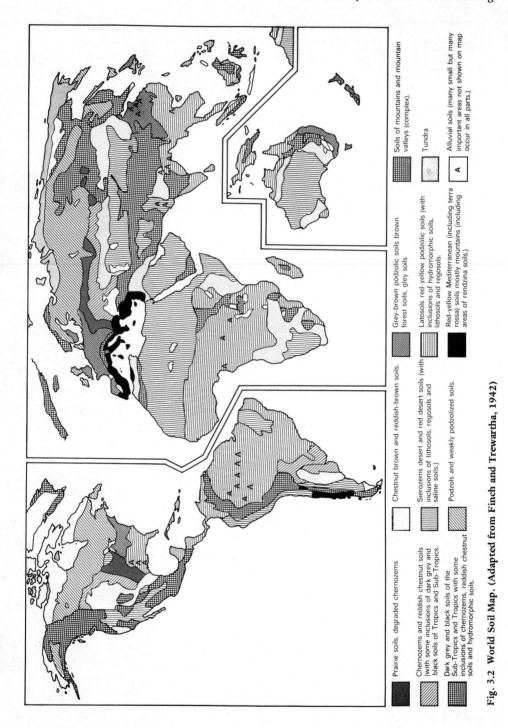

Fig. 3.2 World Soil Map. (Adapted from Finch and Trewartha, 1942)

Prairie soils, degraded chernozems

Chernozems and reddish chestnut soils (with some inclusions of dark grey and black soils of Tropics and Sub-Tropics.

Dark grey and black soils of the Sub-Tropics and Tropics with some inclusions of chernozems, reddish chestnut soils and hydromorphic soils.

Chestnut brown and reddish-brown soils.

Sierozems desert and red desert soils (with inclusions of lithosols, regosols and saline soils.

Podzols and weakly podzolized soils.

Grey-brown podzolic soils brown forest soils, gley soils

Latosols red-yellow podzolic soils (with inclusions of hydromorphic soils, lithosols and regosols.

Red-yellow Mediterranean (including terra rossa) soils mostly mountains (including areas of rendzina soils).

Soils of mountains and mountain valleys (complex).

Tundra

Alluvial soils (many small but many important areas not shown on map occur in all parts.)

A

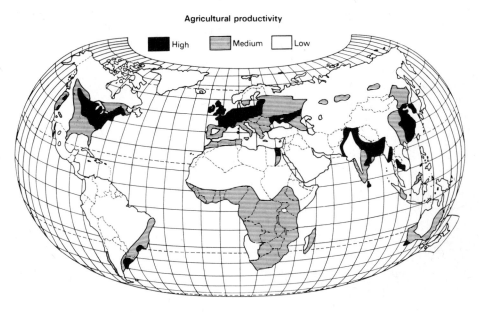

Fig. 3.3 Geographic distribution of land in each continent showing approximate classes of agricultural productivity of cultivated land. (Source: Woytinsky, 1955)

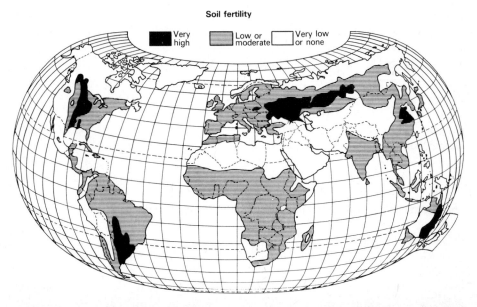

Fig. 3.4 Geographic distribution of land in each continent showing approximate levels of soil fertility. Note that some areas of high agricultural productivity have only moderate soil fertility. (Source: Woytinsky, 1955)

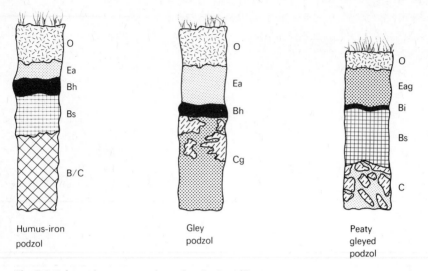

Fig. 3.5 **Schematic representations of podzol profiles.**

layers. The lower side of the Eag-layer is sharply bounded by a thin iron pan (Bi) horizon which is fairly impermeable to water and constitutes a barrier to roots so that a root mat normally forms on the upper surface. Below the Bi horizon diffuse Bs horizons commonly occur. The humus accumulation characteristic of Bh horizons is normally absent or weakly developed in these soils.

Two principal modes of formation of the iron pan have been advanced in the case of peaty gleyed podzols (Muir, 1934). In the case of some Scottish soils where the iron pan has developed in association with Bh and Bs horizons it is suggested that the impeding effect of the Bs horizon on downward movement of water may have led to iron precipitation. In the more common case, where the Bh horizon is absent, Muir suggested that reduction and solution of iron occurred through the action of overlying acid peat. Subsequent re-precipitation was thought to occur in the better drained subsoil horizons as a result of oxidation. Bloomfield (1951, 1954) has shown that sorption of ferrous iron on ferric oxide in Bs horizons may provide a mechanism for precipitation of iron. Stobbe and Wright (1959) also review work indicating the inter-action of organic acids (e.g. humic and crenic) with sesquioxides to form soluble organo—mineral complexes which precipitate in the B horizons.

Coulson *et al.* (1960), have shown that polyphenols derived from beech leaves can lead to the reduction of iron and movement of iron in soil profiles. In the case of the podzol a stage in the formation must be precipitation of iron from such an iron—polyphenol complex. Studies of the mechanisms of precipitation of iron include those of Schnitzer (1969) who prepared complexes of organic compounds (fulvic acid) and inorganic soil constituents (Fe^{+++} and Al^{+++}). He found that metals—fulvic acid complexes were water soluble in the ratio 1:1, but that when further metals were added so that the ratio became 6:1 the complexes became insoluble. He has suggested, therefore, that after being produced in the organic surface horizon, fulvic acid makes complexes with metals as it passes downwards through the soil profile. The water soluble metals—fulvic acid complexes gradually react more and more with the metals and solubility decreases. Eventually the complexes become completely insoluble and precipitate in the B horizon. An examination of an iron pan in a Humus Podzol has shown it to be essentially a $6:1$ molar Fe^{+++}—fulvic acid complex. This supports the view that iron—organic compound complexes are involved in the formation of precipitation layers in podzols. How far micro-organisms affect such precipitation is difficult to assess but Ten Khak-Mun (1973) has provided evidence of the binding action of iron—manganese bacteria.

Gleyed podzols formerly described as 'podzols with gleyed B and C horizons' (Clarke, 1957) differ from humus—iron podzols in having mottled Btg horizons below the podzol B horizons. They mainly occur in composite parent materials, the upper part being in loamy or gravelly deposits, while the lower horizons are formed in a finer textured material.

In addition to these established sub-groups in the classification of podzols in Britain there is a further group of podzolised soils termed the 'podzol-with-gleying' type by Crampton (1963). Although this sub-group has not been formally adopted in British classificatory systems it is used in discussion later in this book because of its clear association with site and aspect. In the podzol-with-gleying profile peaty organic horizons (O) overlie a leached horizon with well developed prismatic structure. Within the prisms there are gleyed cores which are bounded by a very thin iron pan. These Eag horizons are underlain by Bs horizons some of which also contain translocated clay (Bst horizon).

§3.3 BROWN EARTHS

This group as defined in earlier editions of the *Handbook of the Soil Survey of England and Wales* (in Clarke, 1957) includes a wide variety of soils. Brown earths (or Brown Forest soils as they are sometimes termed) are freely or moderately well drained and contain no free calcium carbonate, or redistributed iron, aluminium and organic matter. The SiO_2/R_2O_3 ratio in the clay fraction is fairly constant throughout the profile. The solum generally has an overall brownish colour due to hydrated iron oxides formed by weathering.

Following Duchaufour (1970), the brown earths may be sub-divided into soils with or without a Bt horizon of clay accumulation. Brown earths *sensu stricto* (sols bruns; Duchaufour, 1965) have no accumulations of translocated clay, and the B horizon is normally distinguishable only by colour and structure, or by structure alone. There can, however, be clay skins (cutans) below the B horizon in layers designated B/C (Fig. 3.6).

Sols lessivés have Bt horizons with or without clay-depleted Eb horizons above. Such horizons are relatively rare in south-west Britain possibly because the soils dry out less regularly than those in lower rainfall areas further east (Clayden, 1971). Thus

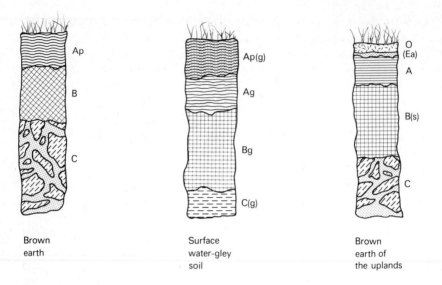

<table>
<tr><td>Brown
earth</td><td>Surface
water-gley
soil</td><td>Brown
earth of
the uplands</td></tr>
</table>

Fig. 3.6 Schematic representation of brown earth and gley profiles.

fissures down which fine material can be washed do not form so readily.

Brown earths may also be sub-divided according to their percentage base saturation under semi-natural conditions into brown earths of high and low base status. For
the most part upland brown earths are brown earths of low base status (*sols bruns
acides*; Duchaufour, 1970) (Fig. 3.6).

§3.4 GLEYED BROWN EARTHS

This sub-group includes those brown earths with mottled B and C horizons, transitional to surface-water gley soils in morphology. The extent of gleying depends on the
length of the period of waterlogging and is usually characterised by grey ped faces,
mottling of grey and reddish-brown and iron-manganese concretions in the B and C
horizons.

§3.5 GLEY SOILS

Gley soils are those in which the profile morphology reflects periodic waterlogging.
Gley morphology develops where pore spaces are filled by water containing dissolved
organic substances, which, with anaerobic bacteria, bring about the reduction and solution of iron compounds. Gley horizons are grey with ochreous mottling (Crompton,
1952). A convenient and practical division into surface-water and ground-water gley
soils derives from their two main modes of formation. In the surface-water group
drainage is impeded by an impermeable layer within the soil so that waterlogging

occurs mainly in the upper horizons of the profile (Fig. 3.6). Ground-water gley soils develop where a ground-water table approaches the surface, waterlogging the lower part of the profile.

Humic or peaty gley soils include gley soils with either a surface peaty layer (O horizon) or a thick, very dark coloured A horizon.

§3.6 CALCAREOUS SOILS

This group includes well drained and imperfectly drained soils developed on highly calcareous sediments. Although undergoing leaching in the British climatic conditions these soils, unlike the brown earths, retain calcium carbonate throughout the profile. The calcareous soils occur in three broad categories: Rendzinas, Brown Calcareous soils and Gleyed Calcareous soils. The Rendzinas are shallow soils formed over limestone. They usually show a dark-coloured organic horizon (A) resting directly on weathering limestone (C). Thus they are sometimes termed A/C or humus—carbonate soils. The Brown Calcareous soils are far more extensive. In these soils the A horizon grades into a brown or reddish-brown calcareous B horizon. The brown or reddish-brown colour characteristic of these soils is due to ferric oxides occurring mainly in the clay particles. The colour tends to conceal the presence of organic matter but the A horizon is normally very well structured and is usually a granular mull. The brighter coloured B and B/C horizons normally have markedly blocky structures and deposits of secondary carbonate are often seen. The soils are mainly neutral or somewhat alkaline in reaction.

In the case of the Gleyed Calcareous soil sub-group the soils are mostly derived from highly calcareous clay and shale. The soils have impeded drainage and are normally deep, coarse structured and have zones of secondary carbonate deposition. The A horizons are usually dark in colour with a fine blocky structure. The B horizons are often faintly mottled with yellow-brown and grey, below which grey gleyed (C(g)ca) horizons occur in which carbonate concretions are abundant. The subsoil structures are typically prismatic but with increasing depth the prisms are replaced by the laminations of the clay shales, and concretions and mottles disappear.

§3.7 ORGANIC SOILS

This group includes soils derived from deposits of peat which have accumulated in bogs. Many different schemes of classification have been devised for organic soils of the British Isles (Barry, 1954; Fraser, 1943, 1954; Godwin, 1941).

Godwin (1941) proposed the following scheme of classification:

I. Topogenous Mires or Fens.
    A. Eutrophic fen.
    B. Oligotrophic fen.

II. Ombrogenous mires.
    A. Blanket bogs.
    B. Raised bogs.

Fraser (1943, 1954) uses the following categories in the upper levels of his classification:

I. Climatic or Zonal Bogs.
    A. Bogs of cool temperate regions formed under maritime rainfall at lower elevations — blanket bogs.
    B. Peat bogs of hill and mountain masses developed under high rainfall and low temperature, particularly on high plateaus — hill peat.
    C. Sub-arctic climatic bogs of tundra regions.
    D. Arctic—Alpine climatic bogs of some alpine plateaus.

II. Intrazonal Bogs.
    A. Peat developing in or on free water.
        1. Lake basin peat, basal deposition.
        2. Shallow lakes with swamp vegetation, basal deposits.
    B. Peat developing on waterlogged or intermittently flooded mineral soil and vegetation.
        1. Valley bog.
        2. Flush bog.

Moore and Bellamy (1973) have pointed out that peat growth in water continues up to a point at which the surface of the peat reaches the level at which water drains from the reservoir. Beyond this point the peat no longer acts as an inert mass within the water body simply displacing some water volume. It now acts as an active reservoir of water itself holding a volume of water against drainage. In the early literature this stage was referred to as raised bog. However, the distinctions between different peat phenomena are somewhat ill defined and Moore and Bellamy (1973) have proposed the terms 'primary', 'secondary' and 'tertiary' as defined below.

1. Primary peats are those formed in basins or depressions and as the peat grows it reduces the depression storage of water in the basin. These are the valley bogs (Fraser, 1943, 1954) or fens (Godwin, 1941).

2. Secondary peats are those formed beyond the physical confines of the basin or depression. The peat itself now acts as a water reservoir and increases the water storage capacity of the landscape.

3. Tertiary peats are those which develop above the physical limits of the groundwater. The peat itself acts as a reservoir holding a volume of water up above the main ground water-table of the landscape. In other words the water held in the Tertiary peats is in the form of a perched water-table and is fed by the precipitation it receives. These peats include those termed ombrogenous mires by Godwin (1941).

Oh
Om
Oh

Ea(g)

Bi

Bs

B/C

1a
Peaty gleyed podzol in Head derived from Devonian slates and sandstones, Exmoor. (Curtis)

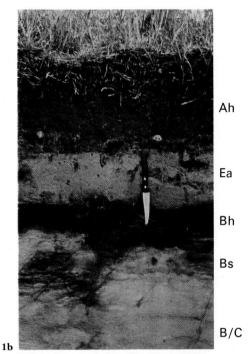

Ah

Ea

Bh

Bs

B/C

1b
Humus—iron podzol on Keuper sandstone, Lancashire. (Soil Survey of England and Wales)

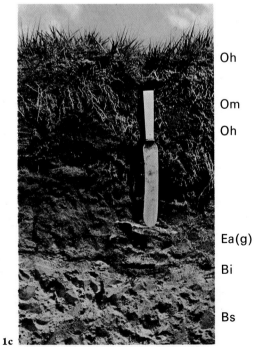

Oh

Om

Oh

Ea(g)

Bi

Bs

1c
Peaty gleyed podzol on Carboniferous sandstone, Pennines, Lancashire. (Soil Survey of England and Wales)

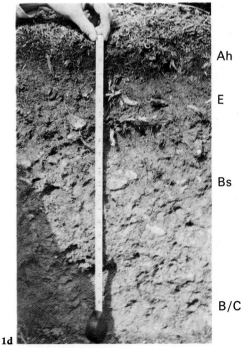

Ah

E

Bs

B/C

1d
Acid brown earth (Sol brun acide) on Jurassic grit, Saltersgate, North York Moors. (Curtis)

H
A
Eg

Bg

2a Peaty gley soil over clayey Bovey beds, south Devon. *(Curtis)*

A
Eg

B/Cg

Cg

2b Humic gley soil on Lias clay, south Gloucestershire. *(Curtis)*

Ag

Bg

B/Cg

2c Humic gley soil on Triassic drifts on lowlands of Lancashire. *(Soil Survey of England and Wales)*

A

Eb(g)

Bg

2d Gleyed brown earth on Glacial Drifts, Derbyshire. *(Curtis)*

FURTHER READING

Farnham, R. S. and Finney, H. R. (1965). 'Classification and properties of organic soils', *Adv. Agron.*, **17**, 115–62.

Hodgson, J. M. (1974). *Soil Survey Field Handbook*, Technical Monograph No. 5, Soil Survey, Harpenden, England.

Kubiena, W. L. (1953). *The Soils of Europe*, Murby, London.

Strahler, A. N. (1973). *Introduction to Physical Geography*, J. Wiley.

4 THE INFLUENCE OF MESOLITHIC AND NEOLITHIC MAN ON SOIL DEVELOPMENT IN THE UPLANDS

§4.1 POLLEN ANALYSES IN RELATION TO UPLAND SOIL DEVELOPMENT

Standing on heather-clad moorland hills one can often see a sharp break between moorland vegetation and the green fields of the farms below (Fig. 4.1). It is tempting to think that the heather represents the natural and untouched parts of the landscape. However, the development of the technique of pollen analysis has led to studies which have shown that this is not so. This chapter seeks to illustrate the evolution of upland moorlands and upland soils by means of case studies drawn from the North York Moors, Dartmoor and the Welsh uplands.

Pollen grains are produced in great profusion (400 million grains from a single plant of Sheeps' Sorrel) and they are so small (10—50 μm) that they can be carried over vast distances. When the pollen rain reaches the ground the grains are incorporated in the surface layers of soils. The external walls of the pollen grains are composed of a non-nitrogenous substance termed sporopolenin which is extraordinarily resistant to chemical attack and provides a highly protective cover which resists decomposition. Preservation is particularly good in anaerobic and acid environments; it is less good in calcareous, well-drained soils.

When the external surfaces of pollen grains are examined under a microscope they are seen to have distinctive features (Fig. 4.2). These features allow the pollen from different plant genera to be recognised (Godwin, 1956; Faegri and Iversen, 1964). Thus it is possible to identify the composition of the vegetation cover which contributed to the pollen rain at particular periods of time in the past. Furthermore, by counting the numbers of different types of grains it is possible to compile pollen diagrams (see Fig. 4.9) showing the percentages of different types of pollen. In this manner the percentages of different plant genera forming the vegetation covers of earlier periods can be interpreted (Crabtree, 1968). The interpretation of a pollen diagram requires considerable skill and experience and must take into account such variables as different quantities of pollen produced by different plants, variations in preservation, the effects of topography on wind dispersal and local environmental conditions. Most palynological studies have been carried out on peat deposits and as a result of many studies a general sequence of pollen zones characterised by different

Fig. 4.1 Edge of Long Mynd. (Cambridge University Collection − copyright reserved)

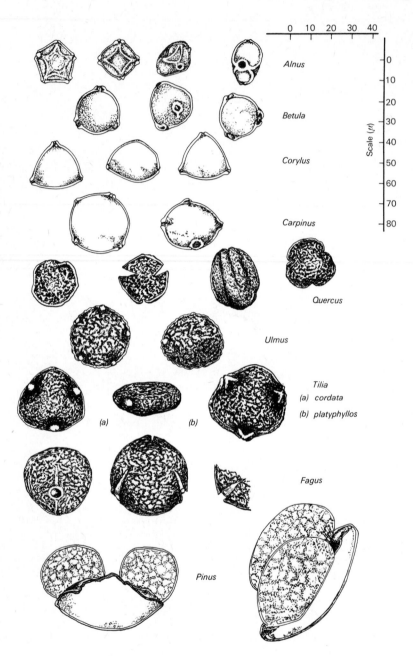

Fig. 4.2 Drawings of the chief types of tree pollen found in British Post-Glacial deposits, drawn to a common scale of size. (Source: Godwin, 1956)

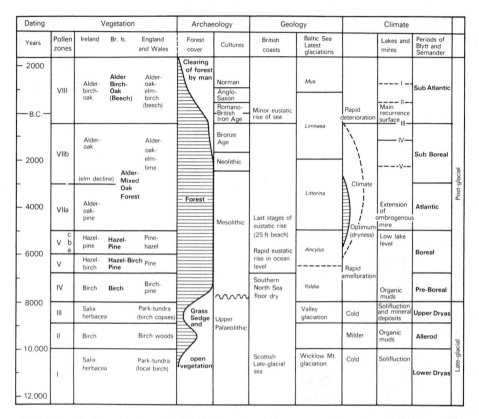

Dating		Vegetation			Archaeology		Geology		Climate			
Years	Pollen zones	Ireland	Br. Is.	England and Wales	Forest cover	Cultures	British coasts	Baltic Sea Latest glaciations		Lakes and mires	Periods of Blytt and Sernander	
2000	VIII	Alder-birch-oak	Alder-Birch-Oak (Beech)	Alder-oak-elm-birch (beech)	Clearing of forest by man	Norman / Anglo-Saxon / Romano-British Iron Age	Minor eustatic rise of sea	Mya	Rapid deterioration	Main recurrence surface — I — II — III	Sub Atlantic	Post-glacial
B.C.	VIIb	Alder-oak (elm decline)	Alder-Mixed Oak Forest	Alder-oak-elm-lime		Bronze Age / Neolithic		Limnaea	Climate	— IV — — V —	Sub Boreal	
2000 4000	VIIa	Alder-oak-pine			Forest	Mesolithic	Last stages of eustatic rise (25 ft beach)	Littorina	Optimum (dryness)	Extension of ombrogenous mire	Atlantic	
6000	V c b a	Hazel-pine	Hazel-Pine	Pine-hazel			Rapid eustatic rise in ocean level	Ancylus		Low lake level	Boreal	
	V	Hazel-birch	Hazel-Birch	Pine					Rapid amelioration			
8000	IV	Birch	Birch	Birch-pine			Southern North Sea floor dry	Yoldia		Organic muds	Pre-Boreal	
	III	Salix herbacea		Park-tundra (birch copses)	Grass Sedge and	Upper Palaeolithic		Valley glaciation	Cold	Solifluction and mineral deposits	Upper Dryas	
10,000	II	Birch		Birch woods					Milder	Organic muds	Allerod	Late-glacial
12,000	I	Salix herbacea		Park-tundra (local birch)	open vegetation		Scottish Late-glacial sea	Wicklow Mt. glaciation	Cold	Solifluction	Lower Dryas	

Fig. 4.3 Correlation table showing the main events of the Late-Glacial and Post-Glacial periods in the British Isles. (Source: Godwin, 1956)

vegetation covers in the Post-Glacial period has been established (Fig. 4.3).

Dimbleby (1961*b*) developed a further technique by which pollen contained in the different horizons of soil profiles could be extracted by hydrofluoric acid and then subjected to microscopic examination and counting. Thus pollen diagrams indicating the nature of the vegetation cover at different periods in the evolution of the soil profile can be constructed. Normally the lower soil horizons give pollen diagrams characteristic of the earlier vegetation communities whereas the upper horizons reflect the more recent plant cover.

§4.2 POLLEN AND CHEMICAL ANALYSES OF SEDIMENTS OF NORTHERN ENGLAND IN RELATION TO SOIL DEVELOPMENT

In early studies by means of soil pollen analysis Dimbleby (1952) examined various sites on the North York Moors. In particular he made comparisons between the pollen sequences in soils preserved beneath archaeological sites formed by Bronze Age barrows and those of the soils occurring on open moorland. It was found that the soil beneath a barrow at Hackness was *not* podzolised and that the bulk of the pollen came from deciduous trees. In particular, hazel (*Corylus avellana*), a species which does not grow on developed podzol soils, was well represented in the pollen. This indicated that prior to Bronze Age times the soils on the uplands were essentially brown earths or gleyed brown earths.

On the open moorland at sites such as Silpho Moor and Ralph's Cross on the moors near Whitby, the present soils consist of peaty gleyed podzols (i.e. thin iron pan soils). These soils have been found to contain two distinct pollen categories. The upper horizons are dominated by heather pollen whereas the lower layers are typified by deciduous tree pollen. This distribution has led Dimbleby to suggest that the tree pollen is characteristic of pre-Bronze Age conditions and the heather developed in post-Bronze Age times.

Pollen frequencies can be expressed as percentages of total pollen (T.P.), total arboreal pollen (T.A.P.) or non-arboreal pollen (N.A.P.). The N.A.P. percentages are often of interest because they provide indications of ground flora and species associ- ated with glades or clearances within a forest. In the Hackness area of the North York Moors, Dimbleby found that Plantain (*Plantago* spp.), Goosefoot (*Chenopodium* spp.) and Knawel (*Scleranthus* spp.) were among the non-arboreal pollen types. These species are normally regarded as typical of cleared areas in forests. Thus the pollen record as a whole for the North York Moors led Dimbleby to conclude that the development of heathland was primarily due to clearance of the upland deciduous forest by Neolithic man.

One may ask, however, why did the original forest disappear so completely? It is possible to argue that gradual leaching of parent materials since the glacial epoch led to gradual change in soil conditions. Pearsall (1950) suggested that before Neolithic man came on the scene the upland soils were relatively unchanged by man. He put forward the view that soil impoverishment by the continuous process of leaching throughout the Post Glacial period could explain the changes in forest cover. In his opinion it was not necessary to invoke climatic changes as a causative factor of changes in the vegeta- tion. Mackereth (1965) has provided interesting evidence from the Lake District to support this argument. By chemical analysis of cores through the bottom sediments of valley lakes with large catchment areas Mackereth showed that the history of soil development in the catchment was recorded in chemical form in the sediments. Carbon was deemed to have been highest in periods of stability of the land surface and lowest when erosion mixed particles of mineral matter with the organic debris. The periods at which a fall in carbon content of the sediments indicates increased erosion are dated as 5 000 and 2 000 BP (BP = Before Present). The former date is contem- porary with the elm (*Ulmus* spp.) decline and the presence of Neolithic cultures. Mackereth argues that the increasing content of elements capable of forming salts by combining with metals (halogens) found in the sediments indicates progressive removal of soluble anions from the mineral matter of the soil (Fig. 4.4). The general form of the graph showing analyses of the sediments reveals an increase in halogen content

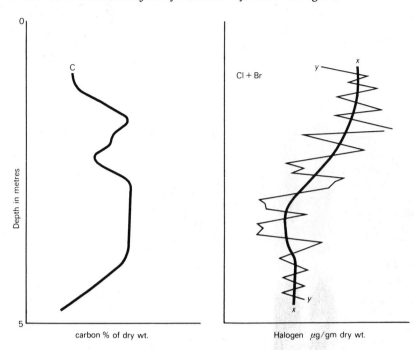

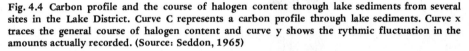

Fig. 4.4 Carbon profile and the course of halogen content through lake sediments from several sites in the Lake District. Curve C represents a carbon profile through lake sediments. Curve x traces the general course of halogen content and curve y shows the rythmic fluctuation in the amounts actually recorded. (Source: Seddon, 1965)

through time. However, Seddon (1965) has pointed out that the general form of the graph hides marked fluctuations in the figures for halogens recorded by Mackereth. These fluctuations may reflect a climatic factor of a cyclical nature with a periodicity of 50—100 years. Thus climatic variation should not be dismissed entirely as a factor in the development of soils in the Post Glacial period.

It is questionable whether soil changes alone were sufficiently drastic to eliminate the deciduous tree on the uplands. Likewise the climatic deterioration in the Sub-Atlantic and Sub-Boreal periods may not have been solely responsible for the decline in tree cover. Therefore the evidence put forward by Dimbleby from sites at Oak-hanger, near Farnham, Surrey, for the use of fire by Mesolithic man is of considerable interest. On the basis of this evidence it is possible to envisage the use of fire by Meso-lithic peoples to provide themselves with larger grazing grounds and to increase the game supply. Such fires might well go out of control periodically and burn large areas. The increase in game supply would also lead to intensified grazing by deer and other wild animals. Thus grazing would be added to fire as a pressure against the forest. The importance of the possible Mesolithic influence on vegetation is that it may explain some earlier modification of the forest cover which rendered it more prone to wide-spread disruption in the ensuing periods.

The main impact of prehistoric man on soil development on the uplands began

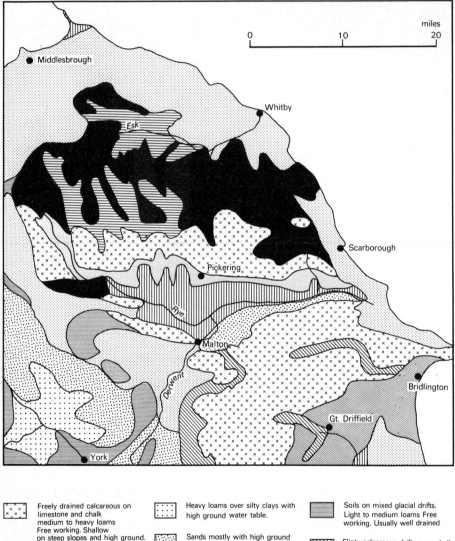

Fig. 4.5 **Sketch map of the soils of Yorkshire. (Adapted from Crompton, 1961)**

however, in the Bronze Age. It has been suggested that there may have been a clear division between Bronze Age activities on the uplands and the lowlands. On the uplands there is little evidence of agriculture and the activity seems to have been essentially pastoral aided by fire clearance. On the lower hills, for example, at Springwood Barrow, Tabular Hills, North Yorkshire, there is preserved material indicating cultivation of crops by Bronze Age cultures. Their economy may have been based on a form of agriculture of the shifting cultivation type.

In general a picture of declining soil fertility can be drawn for the prehistoric period in Britain. This decline largely resulted from the switch from a nutrient cycling system under deciduous woodland to an acidifying process under heather. Contributory factors to the decline were loss of soil structure and destruction of the nutrient capital in organic horizons as a result of burning. Under these circumstances the fertility of upland soils may have fallen substantially by the Iron Age. Iron-Age communities may, therefore, have had an increasing incentive to attack the woodland at the forest margins at the lowland—upland interface in search of more productive soils.

Many of the sites used in early studies by Dimbleby lay on moorland in the North York Moors north of Pickering and inland from Whitby and Scarborough (Fig. 4.5). These areas carry soil groups dominated by organic soils (hill peat), peaty gley soils, peaty gleyed podzols and acid brown earths. Similar soil groups occur on the uplands on both sides of the Pennines and can be found in Lancashire (Fig. 4.6). The peaty gleyed podzols (e.g. *Belmont* series) and acid brown earths (e.g. *Anglezarke* series) are widespread in the Pennine areas of Lancashire and Yorkshire in sites where traces of prehistoric settlement are found.

§4.3 POLLEN ANALYSES OF DARTMOOR SOILS IN RELATION TO SOIL DEVELOPMENT

A similar range of soil groups occurs on Dartmoor, but here the parent material consists of a relatively uniform mantle of rock waste derived from granite (Fig. 4.7). Blanket bog, with associated peaty gley soils occupies the gently sloping summits of the northern and southern plateaus, which are separated around Two Bridges (Clayden and Manley, 1964). The uncultivated moorlands of Purple Moor Grass (*Molinia Caerulea*) or wet heath (*Erica tetralix*) that surround the plateaus are characterised by soils with a thin iron pan termed the *Hexworthy* series. These soils belong to the sub-group of peaty gleyed podzols and may be correlated with areas described by Vancouver (1808) in his *General View of the Agriculture of Devon* as 'black peaty moor' In the north-east the soils generally lack the surface layer of peat which is typical of the other areas. The soils in this region are acid, free draining, brown earths named the *Moretonhampstead* series (Fig. 4.8). The acid brown earths occupy approximately the Granite Gravel District described by Vancouver and a high proportion of these soils comprise acid grassland in enclosed farmland.

Pollen diagrams by Simmons (1964) show a number of phases in the vegetational history of Dartmoor in the Post Glacial period (Fig. 4.9). The first phase (A) is characterised by sparse tree pollen of Birch (*Betula* spp.) and Pine (*Pinus sylvestris*) only with

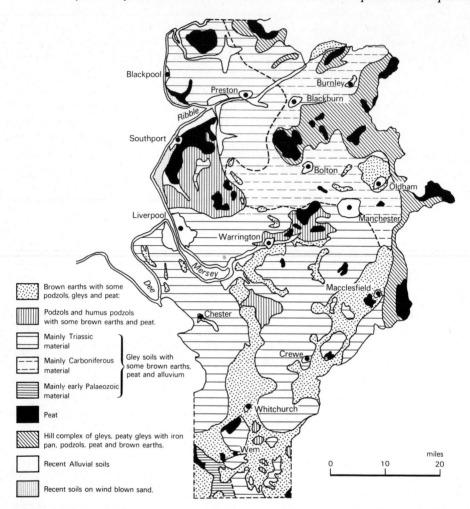

Brown earths with some
podzols, gleys and peat:

Podzols and humus podzols
with some brown earths and peat.

Mainly Triassic
material

Mainly Carboniferous   Gley soils with
material   some brown earths,
                        peat and alluvium
Mainly early Palaeozoic
material

Peat

Hill complex of gleys, peaty gleys with iron
pan, podzols, peat and brown earths.

Recent Alluvial soils

Recent soils on wind blown sand.

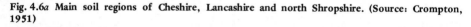

Fig. 4.6a Main soil regions of Cheshire, Lancashire and north Shropshire. (Source: Crompton, 1951)

extensive shrub growth. The latter consists of Willows (*Salix* spp.) with some Dwarf Birch (*Betula nana*). During this phase very high frequencies of pollen of the heath plants Crowberry (*Empetrum nigrum*) and heather (*Calluna vulgaris*) are recorded. The peaty areas that preserved the pollen were dominated by sedges (*Carex* spp.). Thus during the earliest part of the Post Glacial the upland must have been largely clothed in heath, with willows and birches in small thickets and sedge swamps in hollows.

The second phase (B) was marked by an immigration of deciduous tree species. At first Hazel (*Coryllus avellana*) occupied large areas followed by Oak (*Quercus* spp.) and Elm (*Ulmus* spp.). The forest eventually consisted mainly of oak and the subsidiary species was generally hazel with smaller amounts of elm.

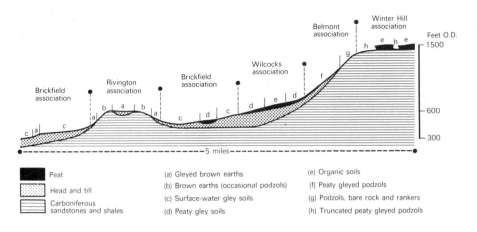

Fig. 4.6b Distribution of soils in the *Winter Hill, Belmont, Rivington, Wilcocks* **and** *Brickfield* **Associations. (Source: Hall and Folland, 1970)**

Phase C (Zones V and VI) shows the onset of disforestation. The pollen rain reveals the sudden entry of Bracken (*Pteridium* spp.) and an increase in grasses and weed species such as Plantain (*Plantago lanceolata*) and Sheeps' Sorrel (*Rumex* spp.). During this period the proportion of tree pollen declines but Ash (*Fraxinus*) with Rowan (*Sorbus*), both light-demanding trees, are increasingly represented. Thus there is evidence of disforestation at high altitudes, probably near the tree line. Since the Neolithic period begins at Zone VIIb the signs of declining tree cover in Zones V and VI point to Mesolithic influences. Abundant archaeological finds of Mesolithic date have been recorded by Radford (1952) and others.

The decline of Elm (*Ulmus*) which occurs in the middle of phase C has been found to be typical of conditions at the beginning of the Neolithic period in many parts of Europe. Soon afterwards clearance of the forest is apparent in the pollen diagrams although on Dartmoor it would appear that the clearings subsequently reverted to forest for a time.

In phase D Bronze Age settlements are associated with further opening up of the forest and at this time cereals were grown in some of the cleared areas. However in the ensuing Iron Age around 500 B.C. wetter and cooler climatic conditions appear to have led to an increasing development of bog moss. At this time there appears to have been a withdrawal of Iron Age settlements towards the better drained margins of the uplands.

Simmons has suggested a possible mechanism for the sequence of clearing by Mesolithic man (Fig. 4.10) drawing upon evidence from Blackland Brook, Dartmoor (1 499 ft; 457 m), Egton High Moor, North York Moors (1 215 ft; 370 m) and Stump Cross, Grassington, Yorkshire (1 000 ft; 305 m). The indications are that only in upland areas where the soil was changed rapidly after clearance by pedogenic processes such as podzolisation did man exert a strong or lasting effect on his environment. This

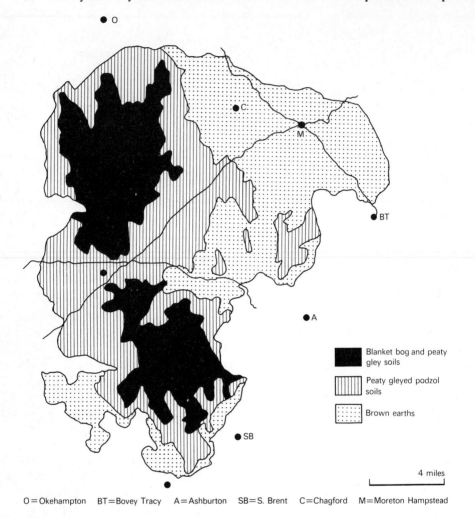

O = Okehampton   BT=Bovey Tracy   A=Ashburton   SB=S. Brent   C=Chagford   M=Moreton Hampstead

Fig. 4.7 Sketch map showing the distribution of the main soil groups on the Dartmoor granite: 1 = blanket bog and peaty gley soils: 2 = peaty gleyed podzol soils; 3 = brown earths. (Source: Clayden and Manley, 1964)

implies that rapid leaching of low base status parent material promoted permanent changes in soil dynamics. The generally low base status of upland parent materials and the effective leaching power of upland climates have been discussed earlier (see Chapter 1). However, the extent and the speed of the changes wrought in the soils would be greatly affected by soil permeability. In turn the permeability, particularly in the early stages of soil development, depends on the texture of the soil (or parent material). Parent materials such as shales which weather to form clay soils are likely to

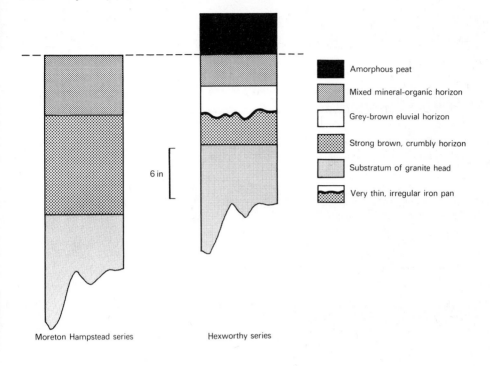

Fig. 4.8 Diagrammatic representation of soil profiles of the *Mortonhampstead* and *Hexworthy* series: 1 = amorphous peat; 2 = mixed mineral—organic horizon; 3 = grey-brown eluvial horizon; 4 = strong brown, crumbly horizon; 5 = substrata of granite head; 6 = very thin, irregular iron pan. (Source: Clayden and Manley, 1964)

have been more strongly buffered against change than the coarse-textured parent materials such as sandstones and granites.

§4.4 POLLEN ANALYSES OF SOILS IN WALES IN RELATION TO SOIL DEVELOPMENT

When the soils of the uplands of North Wales are examined (Fig. 4.11), it is evident that they include substantial areas of acid brown earths of low base status (e.g. *Denbigh* series), podzolised soils (e.g. *Hiraethog, Manod* and *Cymmer* series) together with peats and gleys (Ball, 1960, 1963). The *Denbigh* series is probably the most frequently occurring soil in much of the less mountainous parts of Wales. It is based on drift derived from Silurian shales and the normal phase consists of 20 cm of moderately stony silty loam or dark brown to dark yellowish-brown colour passing into a yellowish-brown subsoil and a paler B/C horizon at about 70 cm. The pH values normally range from about 4.9 at the surface to 4.6 at depth. The *Denbigh* series is often found in association with the *Manod* series which is transitional between the

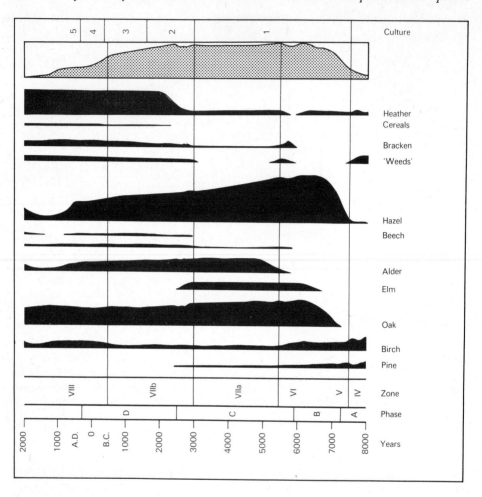

Fig. 4.9 Generalised pollen diagram for the Dartmoor region. The curve at the extreme right attempts to show the amount of tree cover on the upland. Culture periods: 1 = Mesolithic; 2 = Neolithic; 3 = Bronze Age; 4 = Iron Age; 5 = 'Historical'. (Source: Simmons, 1964)

brown earths and podzols. The *Manod* soils are strongly leached and show an accumulation of iron in the subsoils. This led Robinson (1949) to describe them as Crypto-podzols (hidden podzols). The *Hiraethog* series includes soils of the peaty gleyed podzol group (i.e. thin iron pan soils). It is based on drifts derived from Silurian sedimentary rocks. The profile characteristics are similar to those of the *Hexworthy* series on Dartmoor, the *Belmont* series in the Pennines and the *Burcombe* series of Exmoor. Each series is, however, based on different parent materials, and so textures and structures vary somewhat. The common feature of these soil series is the thin iron pan which occurs in the profile. Also one may note that they generally occupy similar topographic positions at the margins of the upland plateaus.

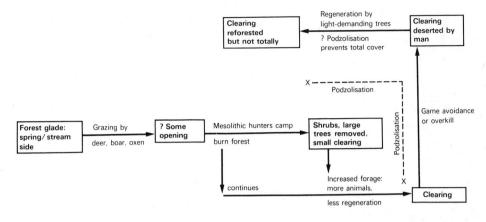

Fig. 4.10 A possible mechanism for the sequence of clearing by Mesolithic man in upland areas. (Source: Simmons, 1964)

In a discussion of the changing vegetation of west-central Wales in the light of human history, Moore and Chater (1969) take the view that in Wales Mesolithic man did not interfere in any substantial way with the vegetation of the area. The first evidence of human interference with the vegetation cover occurs in deposits dated to the Neolithic period. The advent of Neolithic man is accompanied by a rapid decline in Elm (*Ulmus* spp.) pollen. Further clearance in the area was associated with the Bronze Age cultures (Turner, 1964), but as the climate deteriorated these peoples seem to have moved towards the margins of the uplands. However, in the ensuing Iron Age there continues to be evidence of pastoral activity on Plynlimon and the establishment of hill forts appears to have had a considerable effect on the local forest.

The main impact of man on the woodlands of Wales seems to have taken place in Roman and mediaeval times when particularly severe clearance occurred. Apart from the increase in agriculture and clearance for roads in Roman times it may be noted (Davies, 1961) that Julius Agricola ordered the destruction of the Ordovices territory following their attacks on Roman cavalry regiments. It is clear, however, that intensive clearance of woodland in Wales began around A.D. 1300. This mediaeval clearance was linked to the growth of the cattle industry and with the smelting of lead in the uplands. This destruction of woodland was widespread in upland areas outside Wales in the mediaeval period. It was dominantly due to the desire for extension of the capabilities of lowland settlements and to provide fuel for mineral smelting, e.g. lead and iron production. For example, the monastic holdings in Lancashire and Yorkshire developed vaccaries (cattle breeding stations) in the thirteenth and fourteenth centuries. Although most of these stations had a streamside location, some were sited on valley slopes and on the moorland edge where they doubtless played a part in modifying the vegetation cover. The sixteenth century witnessed a rising population and a quickening of economic activity brought new farms in appreciable numbers. It is

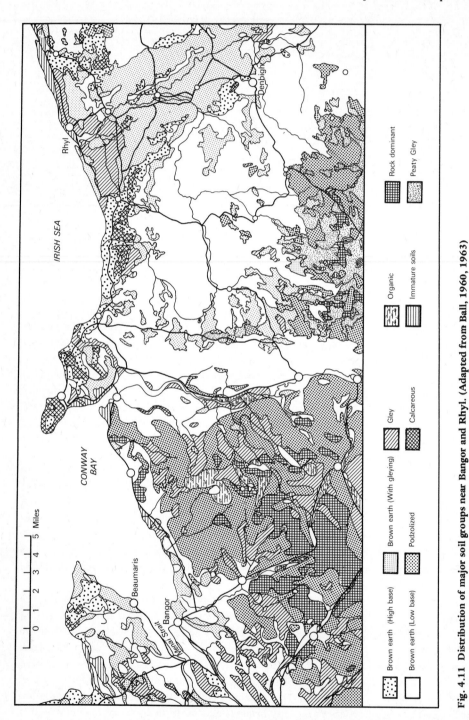

Fig. 4.11 Distribution of major soil groups near Bangor and Rhyl. (Adapted from Ball, 1960, 1963)

noteworthy, for example, that Leland (*c.* 1540) reported that 'Weredale is wel woodid' whereas by 1652 a survey showed 'no timber . . . in Weardale and no coppice or under-wood near Wearhead'. Therefore the impact of man on the woodland and vegetation at the margins of the uplands was considerable and probably triggered off local patches of erosion and deposition. Some of the soil profiles were truncated and lower horizons were exposed on which new soils developed; other soils were buried beneath eroded material and the old ground surfaces overlain with new deposits. Such buried levels are not always distinguishable by eye but can be recognised by discontinuities in the pollen distribution. Dimbleby (1961a) examined a number of soil profiles and found that about one-third of them contained old surfaces buried beneath later deposits. The buried surfaces occurred in various horizons at different levels within the soils but were mostly within 50 cm of the surface. These buried levels are evidence of unstable phases in the history of the landscape and can be considered within the K cycle concept put forward by Butler (see p. 16).

FURTHER READING

Dimbleby, G. W. (1962). 'The development of British Heathlands and their soils', *Oxford Forestry Memoir*, No. 23.

Simmons, I. G. (ed.) (1964). *Dartmoor Essays*, Devon Association for the Advancement of Science.

5 RECLAMATION AND FORESTRY IN UPLAND AREAS

§5.1 INTRODUCTION

In the last chapter we discussed the part played by prehistoric man in changing the vegetation and soils of the uplands. In so doing early man induced an impoverishment of the soil resource. However, there was plenty of untouched land in the valleys and lowlands which could be cleared for agriculture. The uplands were, therefore, gradually abandoned for the less exposed and potentially richer soils of the lowlands. In the ensuing centuries most of the favourable lowland was enclosed and brought into agricultural production. By the nineteenth century the pressures of increasing population accompanied by expanding urban communities created additional demands for farm produce. This coincided with a period of invention in industry and innovation in agricultural techniques. Thus it is understandable that nineteenth-century man looked towards the relatively unused upland areas and felt that he could master the environmental obstacles presented by them by applying his new-found technology and agricultural science. So it was that some 2 000 years after his forebears had initiated change on the uplands that nineteenth-century man sought to undo the effects of these changes and tried to create yet another soil environment in the hills.

Before considering examples of upland reclamation it is desirable to review the environmental problems which had to be overcome. The range of soils in upland sites consisted mainly of peats (organic accumulations more than 30 cm deep), peaty gleys, gleys, podzols, peaty gleyed podzols and acid brown earths (Fig. 5.1). Some areas were also characterised by shallow soils in which acid organic matter lay directly on base poor parent rocks. These soils are termed 'ranker' soils and are typical of glaciated areas in the Lake District (Fig. 5.2) and elsewhere. The principal problems requiring attention can be summarised as follows. First, the soils were generally acid except where base-rich parent materials such as limestone provided anomalous and favoured areas. The pH values generally ranged from pH 3—5 at the surface to pH 4.5—6 in the subsoil. Since the desirable pH for plant growth is pH 6.5 it was necessary to add lime in order to raise the soil pH to reasonable levels. Second, leaching had removed most of the available plant nutrients so that calcium, magnesium, potassium and phosphate were generally reduced to very low levels. Third, the soils were often poorly drained

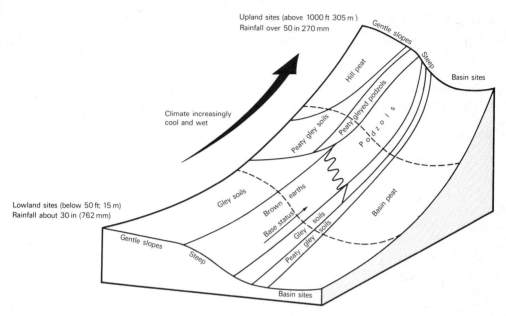

Fig. 5.1 Relationships between soils, climate and relief. (Source: Crompton, 1966)

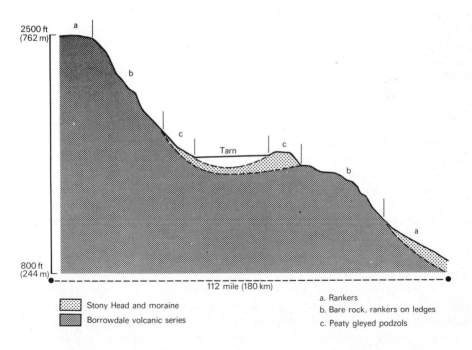

Stony Head and moraine
Borrowdale volcanic series

a. Rankers
b. Bare rock, rankers on ledges
c. Peaty gleyed podzols

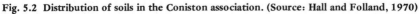

Fig. 5.2 Distribution of soils in the Coniston association. (Source: Hall and Folland, 1970)

with peaty surface horizons. In the case of the peaty gleyed podzols the presence of a thin iron pan restricted downward penetration of moisture and limited the depth of soil in which moisture was stored. Therefore these soils were often waterlogged in winter months but suffered from drought in dry summer periods. Fourth, the uplands were largely open and unprotected from adverse weather. The provision of shelter for crops and animals was an essential part of any attempt to reclaim the uplands.

§5.2 THE RECLAMATION OF EXMOOR FOREST: A CASE STUDY

The reclamation of Exmoor Forest provides a good example of the upland reclamation practice adopted in many areas fringing the highland zone of Britain in the nineteenth century. The Forest had been protected from drastic change in mediaeval times because it was one of the sixty-seven royal forests of England, in which the deer and certain other wild animals were reserved to the King and protected by Forest Law. The protection and administration afforded by the Crown virtually excluded agricultural activities for many centuries so that the general character of the Forest remained almost the same for nearly 1 000 years (MacDermot, 1911, reprinted 1973). A survey of Exmoor Chase in 1651 described it as 'mountainous' and cold ground much affected by fog and mist and mainly composed of heath of little value. The main use of the upland at this time was for summer grazing (agistment) and records show that in 1736 some 30 136 sheep were agisted on the moor. In 1814 records show 32 000 sheep and 640 horses were grazing the moorland during the summer months.

Apart from turf cutting little disturbance of the soils took place prior to the works of reclamation embarked on by John Knight and his son Frederic following the inclosure and sale of Exmoor Forest in 1815 (Orwin and Sellick, 1970). Immediately before the sale the Commissioners of Woods, Forests and Land Revenues ordered a survey of the property to be made. In the report of June 1814 dealing with aspect and soil both surveyors agree

> in reporting the general character of the Forest to be mountainous, that the hills rise very high towards the boundaries, sloping in the interior into deep ravines, that there is little extent of level ground in the valleys, but considerable plains on the summit of the hills; that the soil consists, for the most part, of a black peaty earth, of good depth, but on a rocky bottom; that there are many tracts of superior land, containing from one to four hundred acres; that there are also large tracts of swampy land in the plains, but which are capable of being easily drained, and made sound and useful.

The progress of farming and reclamation by John Knight and his son Frederic can be divided into two periods in which different methods were adopted. First, there was the period of desmesne farming in which John Knight attempted to farm his property as a single unit on a large scale. The earliest work was the construction of a boundary wall which had a total length of 29 miles (46 km) around the estate. The construction of roads was also begun at an early stage and some 22 miles (35 km) of public roads were made by John Knight within the Forest.

Recognising that acid brown earths on the valley slopes were potentially useful for pasture, attempts were made to increase their grazing properties by means of primitive forms of irrigation. Water was carried from higher ground along a horizontal gutter, or by a series at distances 20–39 ft (6–12 m) below each other. To irrigate, a square sod of turf was placed in the gutter to block it so that water overflowed on the ground below the gutter. Each lower gutter caught water from above and redistributed it so that the tendency of water to run in separate channels in depressions on the surface was corrected and the slope was watered evenly (Barker, 1858).

Within Exmoor Forest this method, known locally as 'catch meadows', was often employed on steeply sloping hillsides. Robert Smith, farm bailiff to Frederick Knight, exploited this irrigation technique and was well aware that the principal benefit that accrued was due to the warming effect of the water on the soil. Smith, therefore, recommended the use of warm springs and made temperature measurements in 1854–55 of spring waters at several parts of the moor and compared these with the temperature of the air and the river water in the Barle. Irrigation gutters remain on several hillsides within Exmoor Forest and their distribution is shown in Fig. 5.3. They are now mostly inoperative but some were used in the Sherdon district of Exmoor in the late 1950s.

The peaty and ill-drained nature of much of the land indicated the need for drainage. John Knight, and more particularly his son, attempted to drain large tracts by cutting surface drains. These are largely useless today, having become overgrown and partially infilled. Nevertheless they can be traced over wide areas of Exmoor (Fig. 5.3), and they doubtless had some effect, but satisfactory soil drainage only became possible at a later stage when subsoil ploughing was carried out.

John Knight proposed to bring the land into rotation farming and the warmer and dryer south-facing slopes appeared to be the most promising for this purpose. His first operations were on slopes above the River Barle and bullock teams of six were used in the ploughing. The cultivations comprised:

1. Spading, in which 2–3 in (5–7.5 cm) of the top layer of turf were pared off;

2. Burning, in which the top layers were gathered in heaps, burned and then spread;

3. Liming, in which lime was spread at the rate of 3 tons per acre (6 tonnes per hectare) in some areas;

4. Halving, in which the bullock teams ploughed every alternate furrow width.

The furrow slice was turned over on to the unploughed strip. Where there was no iron pan the depth of ploughing was such that the soil turned up yellow, that is deep enough to reach the B horizon. When there was iron pan every alternate furrow was ripped with a subsoiler to break the pan. The land was left in low ridges until the following spring when it was cultivated for spring sowing.

In addition to the reclamation of the soils Knight actively sought to enclose large areas for grazing purposes. These areas were chiefly those judged difficult to reclaim for arable purposes. Both cattle and sheep were brought on to the moor in considerable numbers so that grazing intensity may have been increased in certain areas. Long-

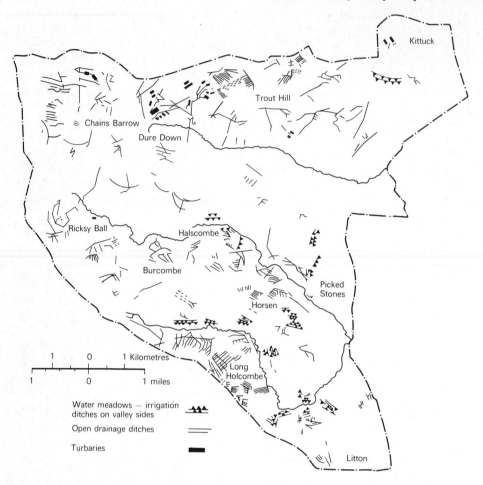

Fig. 5.3 Water meadows, drainage ditches and turbaries on Exmoor. (Source: Curtis, 1971).

continued grazing by cattle can destroy heather and this may explain why there is now
little heather within the Forest. Where the Forest boundary adjoins the common land
it displays a striking change in vegetation pattern (Fig. 5.4), which probably reflects
the effect of grazing intensity within the Forest compared with that on the commons.

The attempts by the elder Knight to carry out arable farming at altitudes of
1 312 ft (400 m) were not successful and the policy of desmesne farming meant that
there was little extension of settlement apart from the home farm. Thus his workers
suffered from social isolation. When his son took control he introduced a new policy
of building farmsteads and letting them to tenants and a new period of reclamation
ensued based on a different approach. The new farms were let on long leases at low
rents. Much of the land was unimproved by Knight except for the buildings and a

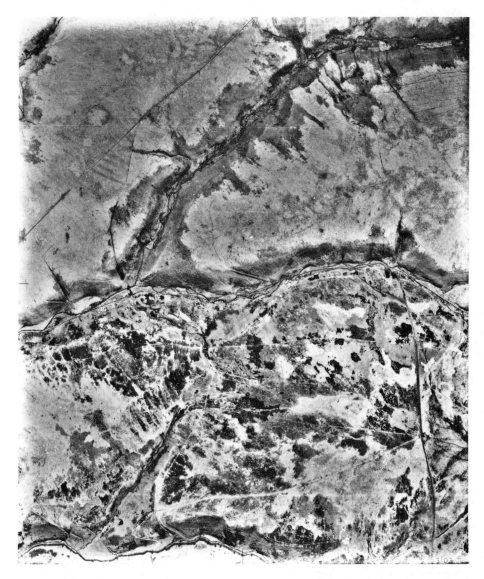

Fig. 5.4 Air photo of northern boundary of Exmoor Forest between Lannacombe and Brendon Common. (Ministry of Defence (Air Force Department) Crown Copyright Reserved)

certain amount of hedging. It was at this stage that the beech hedges were introduced into the landscape. The general policy adopted by the tenants in reclamation was:

1. Cutting of open drains on the summits to allow grazing;

2. Reclamation of land at the margins of the summit plains, preferably those with southerly aspect;

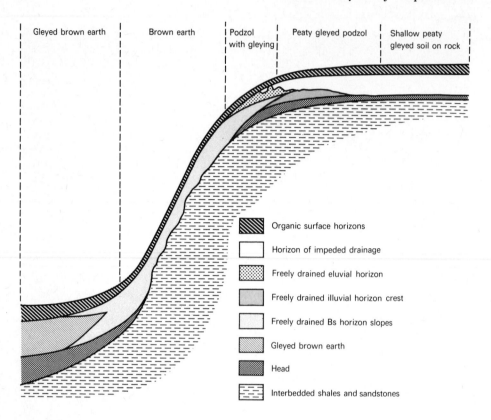

| Gleyed brown earth | Brown earth | Podzol with gleying | Peaty gleyed podzol | Shallow peaty gleyed soil on rock |

Organic surface horizons

Horizon of impeded drainage

Freely drained eluvial horizon

Freely drained illuvial horizon crest

Freely drained Bs horizon slopes

Gleyed brown earth

Head

Interbedded shales and sandstones

Fig. 5.5 Distribution of soils on south- and west-facing slopes on Exmoor. (Source: Curtis, 1971)

3.   The creation of catch meadows to irrigate steeply sloping areas.

Until the 1870s ploughing and reclamation was mainly confined to the drier soils (acid brown earths and some podzols). However, the demand for permanent pasture created by the extensive sheep farming gave further impetus to reclamation. Tracts of wet peaty land (peaty gleys) were now considered for conversion to permanent pasture. Heavy machinery using steam power was now brought in. Thus the Sutherland steam plough was given a trial on Exmoor in 1876 and by 1877 steam engines were operating the plough to reclaim some 400 acres (160 ha) of Knight's land in the vicinity of Titchcombe. The work was done at the rate of 5 acres (2 ha) per day. After breaking the land it was sown for 3 years in succession with rape, which was eaten off by sheep. By this time the peaty turf was well rotted and grass seeds were then sown. In this manner several hundred acres were broken up on Duredown, Titchcombe, Prayway Head and Ashcombe. In addition to summer grazing of cattle some 9 000 ewes were kept on the Forest by 1879. Thus the general pattern of soils, relief and land-use on

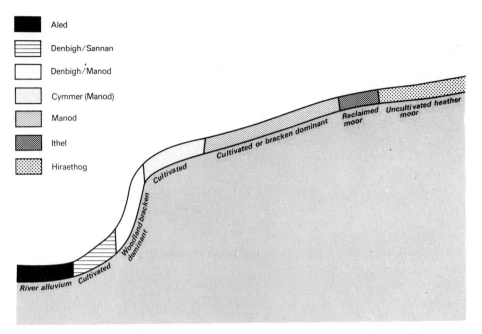

Fig. 5.6 Relationship between soils and relief on the moorland fringe in north Wales. (Source: Ball, 1960)

the moorland fringe was as shown in Fig. 5.5. Similar patterns of soils and land-use occurred elsewhere, as in the Welsh uplands (Fig. 5.6).

§5.3 EXMOOR SOILS AND RECENT RECLAMATION

The present soil pattern of Exmoor Forest (Fig. 5.7) consists of podzols, brown earths, gleys and organic soils. The most useful agricultural land occurs where acid brown earths (*Cornham* and *Sherdon* series) are present and these soils were mostly reclaimed in the early stages of settlement. In the podzol group the soils most favoured for reclamation were the peaty gleyed podzols (*Burcombe* series). The iron pan in these soils frequently occurs at a depth of 10 in (*c.* 25 cm) and Curtis (1971) has identified considerable areas where the pan has been broken by reclamation works. In the Gley group the *Ashcombe* series has also been favoured for reclamation by drainage and ploughing (Table 5.1).

The reclamation of moorland areas for agriculture in recent years has attracted the attention of those concerned with amenity values of the Exmoor National Park. For example, the Exmoor Society sponsored land-use mapping in 1965—66 (Exmoor Society, 1966). It was claimed that comparison between the land-use map of 1966 with earlier records indicated that some 8 080 acres (3 264 ha) of the National Park, or 13.75 per cent of the total moorland area existing in 1957—58 had been lost. These

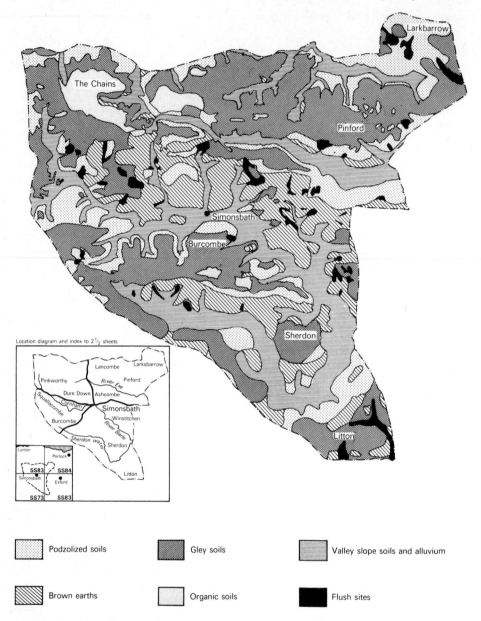

Fig. 5.7 The soils of Exmoor Forest. (Adapted from Curtis, 1974)

figures were disputed by the Country Landowners Association and National Farmers Union (1967) which asserted that in the period 1955–62 in Exmoor Forest only about 1 001 acres (405 ha) were rehabilitated and 730 acres (295 ha) of worn-out pasture reseeded. Afforestation took place on 500 acres (202 ha). In the period

1955—65 agricultural statistics for Exmoor parish show the total for sheep and lambs increased from 19 839 to 23 876 (*cf.* 32 000 sheep in 1814 agisted) but cattle and calves remained fairly constant in number, ranging from 1 698 to 1 643 (*cf.* 640 horses agisted in 1814). The general picture for Exmoor Forest suggests gradual improvement rather than major intakes of land. Furthermore some areas reclaimed in the early stages have undoubtedly reverted to rough pasture and have required rehabilitation.

When the existing soils (Table 5.1) are considered in terms of their potential use for agriculture the following observations can be made. The Organic soils (*Chains* and *Pinkworthy* series) have low value for agricultural purposes, and since they often

Table 5.1 Soil classification of Exmoor soils. (Source: Curtis, 1971)

Major group	Sub-group	Units	Phases	Area hectares	Texture	Area (%)
Podzol	Humus—iron podzol	Larkbarrow series		433	loamy	4.8
	Peaty gleyed podzol	Burcombe series	(i) moorland phase	734	silty	8.2
			(ii) cultivated phase	440	silty	4.9
Brown earth	Acid brown earth (*sol brun acide*)	Cornham series	(i) deep phase	711	silty	8.0
			(ii) shallow phase	407	silty	4.6
		Sherdon series		192	silty	2.2
Gley soil	Humic or peaty gley soil	Aschombe series		1 746	silty	19.6
Organic soil	Blanket bog	Chains series		334	(peat)	3.75
	Half bog	Pinkworthy series		1 121	(peat)	12.5
Podzol-brown earth-gley soil complex		Exmoor complex		2 333	silty and loamy	26.15
		Undifferentiated alluvium		252	silty	2.8
		Flush sites (flush gleys and flush peats)		220	—	2.5

occupy skyline positions they should remain as open moorland for rough grazing and amenity use. The Gley soils (*Ashcombe* series) where they occur on north- and east-facing slopes have limited value and are best considered for rough grazing and amenity purposes. On south- and west-facing slopes these gleys have already been partly re-claimed and they could be further improved. The peaty gleyed podzols (*Burcombe* series) have been shown to be useful for reclamation and much of the existing farm-land has been carved out of these soils on south- and west-facing slopes. On north- and east-facing slopes the value of reclamation is more questionable and the *Burcombe* series in these sites might be better left for rough grazing and amenity use.

In the previous paragraph emphasis has been placed on the importance of aspect. When the contrasts in soil temperatures on opposing slopes on Exmoor are examined (Fig. 5.8) it will be apparent that south-facing soils may show approximately 190 per cent higher surface temperatures than similar soils on north-facing slopes. Studies have also shown that west-facing soils may show an advantage of approximately 50 per cent in surface soil temperatures. These differences are bound to assume considerable weight when land-use decisions are being made, since crop growth largely depends on soil temperature.

Table 5.2 Soil moisture contents. (Source: Curtis, 1971)

Year	Month	Location	Aspect	Slope	Angle (°)	Mean moisture content (per cent dry weight)	Standard deviation
1963	December	Drybridge Combe (SS 760380)	East	(075°)	22	86.8	5.8
			West	(225°)	22	66.2	4.7
1964	July	Drybridge Combe	East	(075°)	22	78.1	2.5
			West	(255°)	22	56.9	0.2
1964	August	Exe Valley (SS 795405)	North	(020°)	29	72.9	5.7
			South	(200°)	22	45.9	6.4
1966	June	Drybridge Combe	East	(075°)	22	91.1	19.4
			West	(255°)	22	72.5	13.4
1969	September	Exe Valley	North	(020°)	29	67.3	9.3
			South	(200°)	22	55.4	3.8
		North and east-facing slopes				79.2	8.7
		South and west-facing slopes				59.4	9.2

N.B. Mean bulk density values: north and east facing 1.35, north and west facing 1.40.

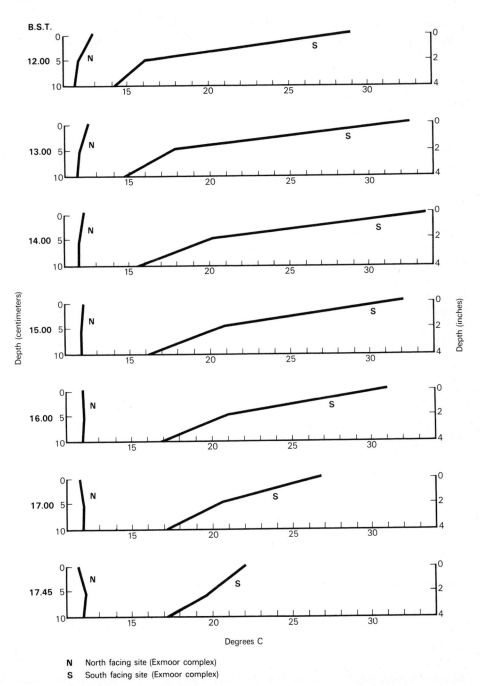

N   North facing site (Exmoor complex)
S   South facing site (Exmoor complex)

Fig. 5.8 Soil temperatures in the Exe valley, 30 August 1964. (Source: Curtis, 1971)

The effects of insolation on different slopes are also seen in contrasts in moisture status of opposing slopes (Table 5.2). Not only are the north- and east-facing slopes less attractive for agriculture through being cooler and wetter but the higher moisture contents render them more liable to waterlogging. As a result landslipping and erosion is more marked on these slopes. It is clear that on Exmoor and in other upland areas steeply sloping areas with shallow soils should be carefully monitored for signs of erosion particularly where the impact of grazing and amenity use becomes severe. In general, the north- and east-facing slopes are likely to be those particularly requiring conservation and management.

§5.4 FORESTRY ON UPLAND SOILS

The use of upland soils for forestry has been a subject of some interest and dispute for a considerable time. Tansley (1953) suggests that the seventeenth century may be said to mark the exhaustion of our forest reserves and that England then became dependent on imported timber. As timber became scarce planting schemes were adopted in Scotland in the seventeenth century and reached its zenith there in the eighteenth century. In England and Wales planting lagged behind but became important in the eighteenth and nineteenth centuries.

From the standpoint of Britain's soil resources the planting operations were not always well conceived. For example, in Scotland some areas were planted to trees which included brown earths (brown forest soils developed under deciduous forest) of considerable agricultural value. Notable amongst these 'Planting Lairds' as they came to be known was Sir Alexander Grant of Monymusk who is said to have planted 50 000 000 trees during his lifetime. Some of the trees he planted are still growing in Paradise Wood on the banks of the River Don. One of the reasons for this upsurge in tree planting lay in the return from exile of many of the Scottish lairds. They brought with them from the Continent new ideas on forestry and species of conifers such as European silver fir (*Abies alba*), European larch (*Larix decidua*) and Norway spruce (*Picea abies*).

In so far that brown earths were planted to native Scots pine (*Pinus sylvestris*), and hardwoods little damage was done because a natural undergrowth was developed. Where exotic species such as spruce were planted on these soils, however, the situation was different because the trees developed thick carpets of dead needles which promoted podzolisation and lowering of soil fertility.

Extensive areas had been planted by the middle of the nineteenth century when the introduction of hitherto unknown conifer species from north-west America by Scottish explorers such as Douglas and Menzies gave fresh impetus to the work. Principal among the species introduced were Douglas fir (*Pseudotsuga menziesii*), Sitka (or Menzies) spruce (*Picea sitchensis*), Western hemlock (*Tsuga heterophylla*), Western red cedar (*Thuja plicata*) and various silver firs. Thus the introduction of coniferous species continued and since they were fast growing when compared to hardwoods they seemed an obvious choice for the planting schemes.

Fig. 5.9 Plough development — Afforestation of peaty soils. The Parkgate Humpy with the deep line double mouldboard, ploughing for planting, Coed y Rhaiadr Forest (Powys). (Source: Toleman, 1974)

The freely drained brown earths were the most sought after soils by the forester because they required little or no preparation and could grow almost any tree species, both broadleaved and coniferous. If these were not available the imperfectly drained brown earths were favoured because they were intrinsically fertile and merely required drainage to a greater or less degree. On extensive sites where large-scale drainage schemes are economic the drainage channels are now created by large hill-drainage ploughs drawn by crawler tractors (Fig. 5.9). In the wetter sites the Norway and Sitka spruces are commonly planted because of their tolerance of wet soils.

The next most valuable soils for forestry are the freely drained podzols which again can often be planted without much preparation work. The chief species used in Scotland are Scots pine, European larch, Japanese larch and hybrid larch. Douglas fir may also be planted on these soils. Since these soils were already podzolised the use of conifers did not cause any further deterioration in soil condition; indeed, the moisture-retention properties of the soils may have been improved with resultant advantage to regional water storage.

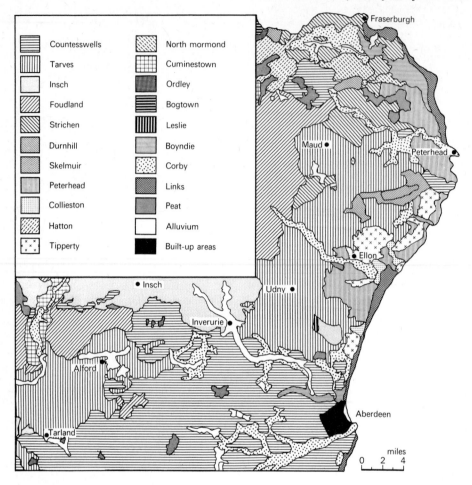
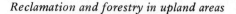

Fig. 5.10 Distribution of soil associations near Aberdeen. (Source: Glentworth and Muir, 1963)

Imperfectly drained iron podzols are much more difficult to afforest. Intensive preparation work is carried out and these soils are often planted to a nurse crop such as Japanese larch (*Larix kaempferi*), hybrid larch (*Larix curolepsis*) and contorta pine (*Pinus contorta*). The peaty podzols, gleys and peaty gleys all require drainage and in wetter areas of Scotland up to 120 chains of drains per acre are sometimes needed. Planting is nowadays carried out on the upturned turves which have been manured with ground mineral phosphate. Norway and Sitka spruce are often planted on better drained sites but Norway spruce only is put in frost hollows. As drainage gets poorer and site quality deteriorates an increasing proportion of contorta pine is included with the spruce until on the worst sites contorta pine is planted pure. The distribution of soil associations near Aberdeen provides an example of the soils under discussion (Fig. 5.10).

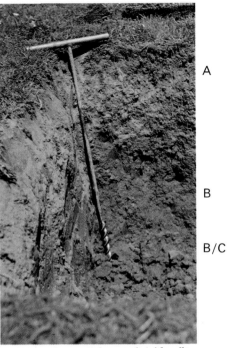

A

B

B/C

3a

Brown earth on Trias marl with silty surface horizon, south Derbyshire. *(Curtis)*

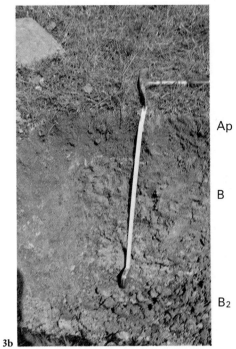

Ap

B

B₂

3b

Argillic brown earth on Jurassic limestone, south Cotswolds. *(Courtney)*

Ap
Eb

Bt

C

Brown earth (*Sol lessivé*) on drift over chalk in Breckland, East Anglia. *(Curtis)*

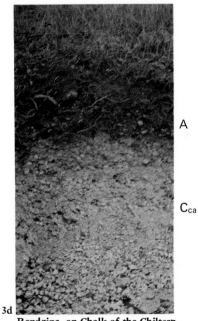

A

C_{ca}

3d

Rendzina, on Chalk of the Chiltern Hills. *(Courtney)*

4a
Buried podzol, Culbin Sands Scotland.
(Curtis)

4b
Buried peaty gleyed podzol in Pigtrough
Griff, Levisham Moor, North York Moors.
(Curtis)

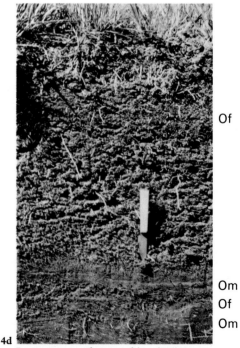

4c
Buried podzol in Glacial Drift. Ayrshire,
Scotland. *(Curtis)*

4d
Peat moss, south Lancashire. *(Soil Survey
of England and Wales)*

The foregoing summary of the relationships between soils and planting programmes in Scotland holds good for much of the upland of Britain although climatic conditions lead to somewhat different species combinations further south (Toleman, 1974).

§5.5 COMPETITION FOR UPLAND SOILS

The Forestry Commission is now the largest landowner in Britain and carries the responsibility for good management of vast areas of upland soils. The distribution of the larger commission forests are shown in Fig. 5.11. There were some 759 080 acres (307 196 ha) in England, 1 849 066 acres (748 307 ha) in Scotland and 394 329 acres (159 583 ha) in Wales under Forestry Commission management in 1972. It is therefore important to recall the conditions under which the Commission was brought into existence and the task it has sought to fulfil. The first Forestry Commissioners were appointed under the Forestry Act, 1919. This was immediately after the First World War, 1914–18, which had brought large-scale destruction of the scant forest resources of Britain. As a result of the dearth of timber in Britain the Acland report (1918) recommended afforestation of 8 million acres (3.5 million ha) with softwood species. By 1939 the Forestry Commission was already the largest landowner with over 1 million acres (0.5 million ha) in hand. Subsequent large-scale destruction of an already inadequate reserve of timber in the Second World War further underlined the need for greater timber resources. Thus the planting of softwood (coniferous) species was justifiable in view of the need for quick growth. There was, at this time, nothing in the Forestry Commission brief concerning amenity and emphasis was placed on conifers with little planning for public access to forests or for landscape presentation. As a result the Forestry Commission came into conflict with amenity groups in the post-war years. In the Lake District for example there was strong criticism of unthinking afforestation of sites of scenic beauty. The Hardknott National Forest Park Committee consequently recommended that forestry policy should:

1. Use different tree species to achieve as natural an effect as possible;

2. Follow the contours in the outlines of plantations;

3. Leave scenic viewpoints unplanted and enough land clear so that the views can still be enjoyed after trees become fully grown;

4. Protect skylines by not planting trees too close to them;

5. Keep planting back from certain becks and roads;

6. Leave notable rock formations and stream banks free of trees;

7. Preserve the sylvan element at the heads of dales.

Thus the indiscriminate use of large blocks of coniferous plantations was attacked and the needs of the community for recreation areas was emphasised. In the post-war period, therefore, there was competition for hill or marginal land, especially in

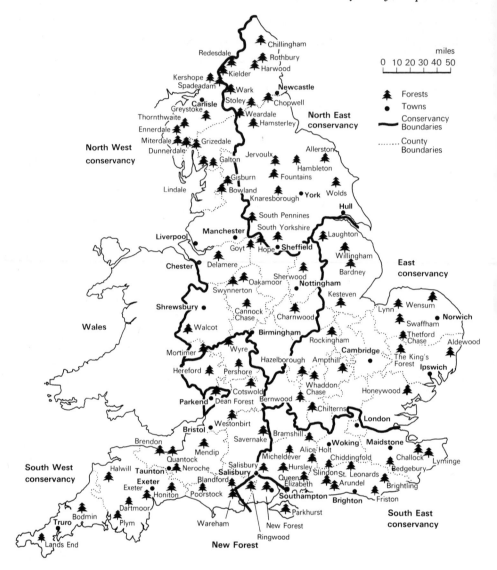

Fig. 5.11 The larger Commission Forests of England. (Source: Forestry Commission, 1963)

England, where the supply was limited. There were now three competing land uses, agriculture, forestry and recreation. The competition in England was such that in 1963 Forestry Commission policy favoured concentration of their activity on the uplands of Wales and Scotland in particular. However, the tax concessions and other benefits enjoyed by woodland owners attracted private finance syndicates which began to come into forestry in 1958 (Miles, 1967), and their activities were sometimes viewed with alarm by conservationists.

It would appear, therefore, that upland soils may be further affected by extensions of agriculture or forestry in the future, or they may be subjected to new stresses as recreation areas are increasingly opened up. Examples of the effects of recreation on soils will be discussed further in Chapter 16. In this chapter the general range of soils that will be affected has been outlined and references to selected detailed sources for the soils of upland areas are given in the recommended reading that follows.

FURTHER READING

Ball, D. F. (1960). 'The soils of the district around Rhyl and Denbigh', *Memoir of the Soil Survey of England and Wales.*

Ball, D. F. (1963). 'The Soils and Land Use of the district around Bangor and Beaumaris', *Memoir of the Soil Survey of England and Wales.*

Bridges, E. M. (1966). 'Upland environments', *Scottish Geographical Mag.*, 79(3), 173–5.

Clayden, B. and Manley, D. J. R. (1964). *The Soils of the Dartmoor Granite*, in *Dartmoor Essays* (I. G. Simmons, ed.), Devonshire Association, pp. 117–40.

Furness, R. R. and King, S. J. (1973). Soils in Westmorland I. Sheet SD 58 (Sedgwick) Soil Survey Record. Soil Survey of England and Wales.

Hall, B. R. and Folland, C. J. (1970). 'Soils of Lancashire', *Bulletin of Soil Survey of England and Wales.*

Mackney, D. and Burnham, C. P. (1966). 'The Soils of the Church Stretton District of Shropshire', *Memoir of the Soil Survey of England and Wales.*

Rudeforth, C. C. (1970). 'Soils of North Cardiganshire', *Memoir of the Soil Survey of England and Wales.*

Rudeforth, C. C. and Bradley, R. I. (1972). *Soils, land classification and land use of West and Central Pembrokeshire*, Special Survey of the Soil Survey of England and Wales.

6 THE KARST SOILS OF UPLAND BRITAIN

§6.1 SOIL FORMATION ON A BARE LIMESTONE SURFACE

When a bare limestone surface is exposed at the surface of the earth it is at once subjected to the action of rain water. Rain is slightly acid and can dissolve the limestone. Calcium carbonate ($CaCO_3$) constitutes over 90 per cent of most limestones and is easily dissolved and washed away by rain water. The insoluble residue (the impurities in the limestone) is usually washed away as well. Thus there is little chance of a soil forming from the insoluble residue unless it can be trapped in some way.

Vegetation plays a crucial role in the formation of soil on bare limestone because it helps to trap the insoluble residue. But a factor of greater importance is that the litter from the vegetation accumulates to form a layer of organic matter over the limestone. In this way soils that form on bare limestone are usually thin and organic and contain a proportion of insoluble minerals like quartz. Chips of limestone can be incorporated in the soil and these are usually produced by the splitting action of frost in cracks or the similar action of plant roots.

It is clear that vegetation plays an important part in soil formation on limestone. But how does the vegetation become established on a bare limestone surface?

With any fresh stone surface life quickly becomes established. Bacteria, algae and fungi are the first colonisers (even in desert conditions there are some algae that inhabit the pore spaces in rock, acquiring water from dew). Studies of new gravestones made out of limestone have shown that algae can colonise the stone in 6–18 months. The most rapid rate of colonisation is on the flat surfaces where moisture can collect. The stone quickly becomes discoloured as the algae flourish in the pore spaces between the individual grains that make up the rock.

Once the algae are established it is possible for lichens to grow. Lichens are composed both of algal and fungal cells. The fungal cells arrive as airborne spores and colonise the algal cells. The first lichens are extremely primitive and are embedded in the rock. They form a hard crust on the surface (about 0.25 mm deep). More advanced lichens can colonise the pioneers. These are leafy and are termed *foliose*, whereas the more primitive ones are termed *crustose* lichens.

If the conditions are suitable for plant growth then mosses can colonise the

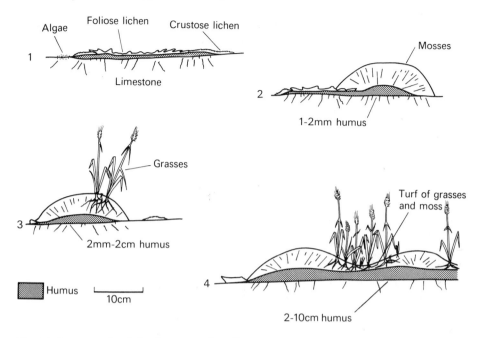

Fig. 6.1 Sequence of soil development on bare limestone.

lichens. The site must not be too exposed or too dry. Thus flatter sites where water may stand for a short while are more favourable to moss growth than steeply sloping exposed rocks. If moss becomes established then grasses can colonise the moss. The litter from the moss and grasses accumulates and an organic soil is formed. Furthermore the presence of a cushion of moss and grasses means that moisture is retained under the cushion. This aids the process of limestone solution because the rock is subject to the action of acid water longer than it would be on a completely bare surface.

The thin organic soil formed by the growth of vegetation and the solution of limestone is a *rendzina*.

Figure 6.1 demonstrates the sequence of vegetation colonisation and soil formation on bare limestone. This sequence may be arrested if the right conditions for plant growth are lacking. For example, the surface may remain covered with lichens if the site is too dry for mosses.

§6.2 SOIL DEVELOPMENT ON A DRIFT-COVERED SURFACE

Many limestone areas in Britain have been glaciated. This may mean that the limestone is covered by a layer of glacial drift. The crucial factors in soil development then become the depth of the drift and its mineral composition. If the drift is deep and acid the fact that limestone exists at some depth may be irrelevant to the progress of soil development: the soil type will be governed by the nature of the drift and the climate. Thus on deep drift over limestone soil types are found which are characteristic of non-limestone areas. In areas of shallow drift the influence of the limestone may be detected on the sequence of soil development.

In areas of Britain which escaped the last glaciation, notably the Mendip Hills (Somerset) and the Derbyshire Peak District, the limestone may also be covered with drift. In this case it is not glacial drift deposited under ice but is windblown drift. The material is thought to be loessial and of a similar nature to the much more extensive loessial deposits in Germany and China. The drift may be only as little as 4 in (10 cm) deep but may be thick mantling of up to 3—5 ft (1—1.5 m) thick. Once again the sequence of soil development depends upon the thickness of the drift cover.

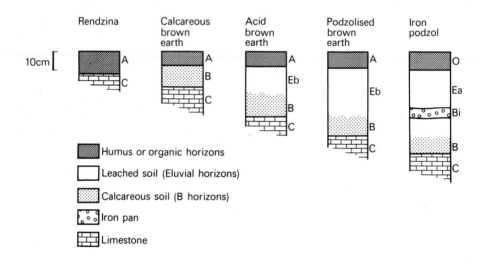

Fig. 6.2 Sequence of soils with increasing depth of drift cover. (After Bryan, 1967)

Figure 6.2 demonstrates the sequence of soils to be expected with increasing thickness of drift. The transition is one from soils rich in calcium carbonate (rendzinas and calcareous brown earths) where drift is thin or absent to deeper acid podzolised soils characteristic of non-limestone areas. This type of sequence can often be detected on a slope where varying thicknesses of drift have accumulated. Each soil type is associated with a particular vegetation type. *Sesleria* is characteristic of lime-rich soils while *Nardus* is adapted to grow only on acid soils. *Festuca* and *Agrostis* grasses are fairly wide ranging species but cannot compete successfully with other plants in extreme lime-rich or extreme acid conditions (Fig. 6.3).

In the succeeding sections the processes involved in soil formation and development will be described and the factors influencing the formation of different soil types will be discussed. To illustrate these general points four areas in Britain are given as

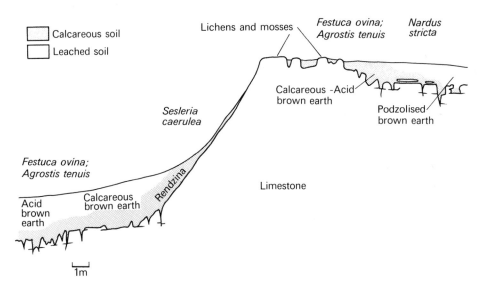

Fig. 6.3 **Sequence of soils and vegetation on slopes. (After Bullock, 1971)**

examples. Two glaciated areas, Co. Clare, Eire, and Yorkshire, are compared and these are contrasted with two areas which escaped the last glaciation: the Mendips and the Peak District.

§6.3 PROCESSES OF SOIL DEVELOPMENT ON LIMESTONE

Limestone is weathered by the dissolving action of acids. Although this is primarily a chemical reaction it is difficult in many cases to separate the chemical reaction from biological processes. On a bare limestone surface the limestone will be attacked by rain water and also by lichens. On a drift-covered surface the limestone will be attacked by soil water. Both the rain water and the soil water have been made acid by the incorporation of carbon dioxide. In the case of the former the carbon dioxide is simply derived from the atmosphere. In the latter the carbon dioxide is derived from the respiration of soil animals and plant roots as well as from the decay of litter. The carbon dioxide content of the air within the small pore spaces in the soil is higher than in the open atmosphere.

To understand the process of solution and to describe how carbon dioxide makes water acid it is necessary to describe the chemical reactions involved. Acids are compounds which split up or dissociate in water to yield hydrogen ions (H^+). Rain water becomes acidified as it falls through the atmosphere because it picks up atmospheric carbon dioxide. The chemical reaction can be written:

$$H_2O \quad + \quad CO_2 \quad \longrightarrow \quad H_2CO_3$$

rain water   carbon   carbonic
           dioxide   acid

Carbonic acid is unstable and dissociates into its constituent ions, H^+ and HCO_3;

$$H_2CO_3 \longrightarrow H^+ + HCO_3^-$$
carbonic
  acid

In this way the carbon dioxide in the atmosphere controls the acidity of rainfall and therefore the dissolving power of rainfall.

    Limestone dissociates very slowly in water; the reaction is simply:

$$CaCO_3 \longrightarrow Ca^{++} + CO_3^{--}$$
limestone   calcium   carbonate

However, this process is markedly speeded up by the presence of the acid hydrogen ions in water and limestone can be carried off in solution as calcium bicarbonate. We can combine all the above chemical equations to describe the limestone weathering process, in this way:

$$H_2O + CO_2 \longrightarrow H_2CO_3 \longrightarrow H^+ + HCO_3^-$$
$$CaCO_3 \longrightarrow Ca^{++} + CO_3^-$$
$$HCO_3$$

$$Ca(HCO_3)_2$$
calcium
bicarbonate

The rate at which limestone dissolves will depend upon the supply of hydrogen ions in water, i.e. the acidity of the water (see Fig. 1.3). Using a simple pH meter or coloured dyes that indicate pH it is often possible to measure pH changes in rain water once the

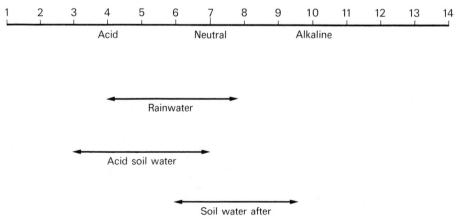

Fig. 6.4 **pH scale and limestone solution**

water has made contact with limestone. The loss of acidity can also be detected in water percolating down through lime-rich soils (Fig. 6.4).

It is clear that the rate of limestone solution will be lower on a bare pavement than under an acid soil. This is because the rate of solution depends upon the supply of hydrogen ions and that in turn depends upon the supply of carbon dioxide. Concentrations of soil carbon dioxide can be from ten to thirty times as high as that in the air. Soil animal and root respiration and also the decay of organic matter produces carbon dioxide which slowly diffuses to the surface through the soil pores. Because the soil pores are small and are not always connected to the surface this diffusion is slow. Soil water picks up large amounts of carbon dioxide from the soil air and therefore the dissolving power of soil water is much greater than that of rain water.

The amount of calcium carbonate that can be dissolved by water can be measured in mg $CaCO_3$ per litre of water. Table 6.1 demonstrates the relationship between carbon dioxide concentration and dissolving power of rain water and soil water.

Table 6.1 Calcium carbonate solubility and carbon dioxide concentrations in rain water and soil water

Water type	Carbon dioxide concentrations		Calcium carbonate solubility
Rain water	Atmosphere:	0.033%	75 mg/l
Soil water	Soil air:	0.3—1.0%	125—250 mg/l

The soil air concentrations are typical of measured ranges in Britain. The concentration drops towards the atmospheric figure in porous sandy soils where aeration is better, but in some wet clayey soils the concentration may rise to 2—10 per cent CO_2 at depths of 1—2 m from the surface.

Organic acids produced in the decay of leaves and wood can also acidify soil water by the direct production of hydrogen ions. The acids are usually very complex and thus the chemical formula for the reaction with calcium carbonate is usually written:

$$CaCO_3 + H(X) \longrightarrow Ca(X) + H_2O + CO_2$$

where (X) is the acid radicle. Note that more CO_2 is produced for further solution.

Another important biochemical process is that of *chelation*. Here the calcium from limestone can be directly incorporated into the molecular structure of an organic compound. Lichens, and the decay products of leaves and wood, can produce organic compounds capable of chelation. A mechanism of chelation for a lichen acid, lecanoric acid, is shown in Fig. 6.5.

It is difficult to separate biological, chemical and mechanical processes. Apart from the purely physical action of frost shattering a most significant bio-mechanical factor is the action of root growth. The action of tree roots in breaking up rocks is clearly visible in some areas, but less obvious is the action of lichens. The rootlets or *rhizoids* of lichens and mosses, attack bare limestone surfaces. They penetrate the limestone surface and increase the surface area over which chemical solution can occur (Fig. 6.6). The rhizoids penetrate the rock partly by chemical solution and partly by

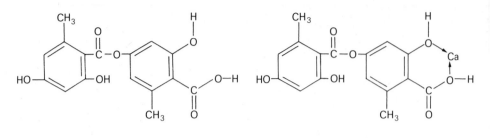

(a) Structure of lecanoric acid

(b) Suggested structure of lecanoric acid-calcium chelate

Fig. 6.5 Chelation by lichen acids (a) Structure of lecanoric acid. (b) Suggested structure of lecanoric acid—calcium chelate.

chelation, and their mechanical expansion, through growth, helps to break up the rock surface (Fig. 6.7).

§6.4 THE FACTORS INVOLVED IN SOIL DEVELOPMENT ON LIMESTONE

The processes involved in the development of soil can be influenced by factors concerned both with the past history of the soil site and the present day environmental conditions at the site.

6.4.1 Glacial action

In the glaciated areas the ice may have left a covering of glacial drift or it may have left the limestone clear of any deposit. In the latter case any pre-existing debris will have been removed by the ice. In other areas drift may have been present immediately after the ice retreated but has since been eroded away. Thus there are two ways in which the limestone may have become drift covered and two ways in which it may have become bare. These ways are summarised in Fig. 6.8.

On the bare limestone sites the soil development will depend upon those processes mentioned in §6.3 of the action of rain water and the action of lichens. On the drift-covered sites it is the nature of the drift that determines the type of soil. As mentioned in §6.2, if the drift is derived from limestone a lime-rich soil will develop, but if the drift is derived from rocks other than limestone, such as shale or sandstone, the soil development may bear no relationship to the limestone below the drift.

6.4.2 Slope

The slope factor is important because the angle of the slope and the position of a site on a slope determine (a) the stability of the soil, (b) the drainage of the soil and (c) the leaching of calcium carbonate from the soil.

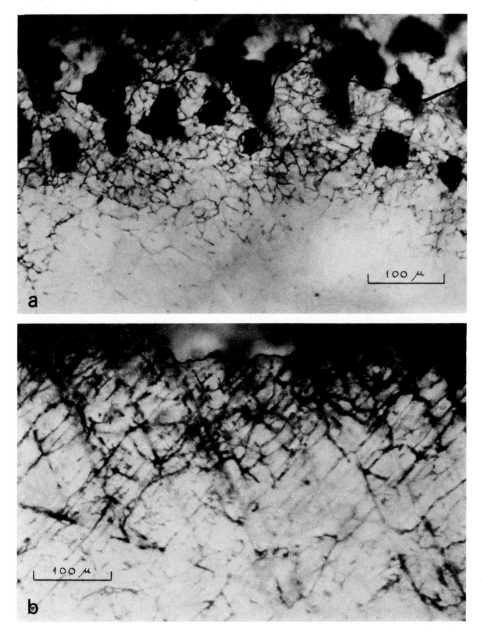

Fig. 6.6 Rhizoid penetration into limestone. (*a*) Photo-micrograph of endolithic lichen on Carboniferous Limestone. The black line marks the weathered edge beneath which the gonioidal layer appears as a row of black-stained masses. The lichen hyphae show as fine black lines penetrating the limestone. (For photographic reasons the weathered surface is not at right angles to the section and appears out of focus above the black line.) (*b*) The same section as above, showing a single calcite crystal and the marked preference of the lichen for the cleavage planes of the crystal. (Source: Jones, 1965)

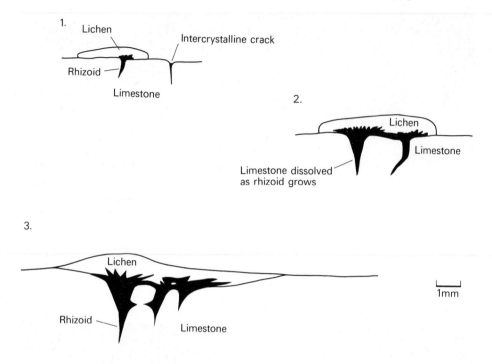

Fig. 6.7 The penetration of a limestone surface by lichen rhizoids.

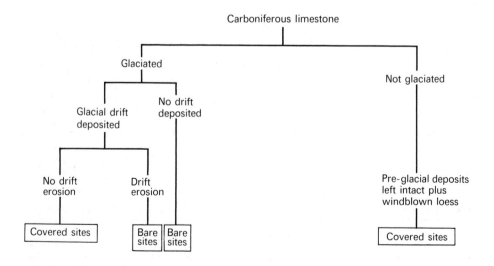

Fig. 6.8 History of sites of soil development on limestone.

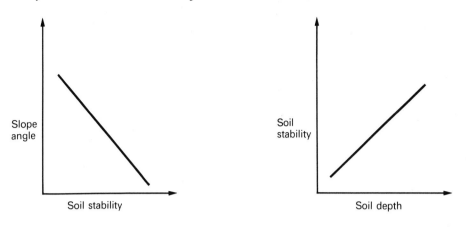

Fig. 6.9 Slope angle and soil stability.　　Fig. 6.10 Soil stability and soil depth.

The steeper the slope the less stable the soil will be and the less chance there will be for a soil of any depth to form. If the soil is constantly moving downslope it has little chance to form *in situ*: this is shown in Figs 6.9, 6.10 and 6.11.

Water running downslope will also carry calcium carbonate dissolved from the limestone and so calcareous soils will be present in the slope foot area. They may also occur on the steeper sections of the slope where the soils are thin and the bedrock is

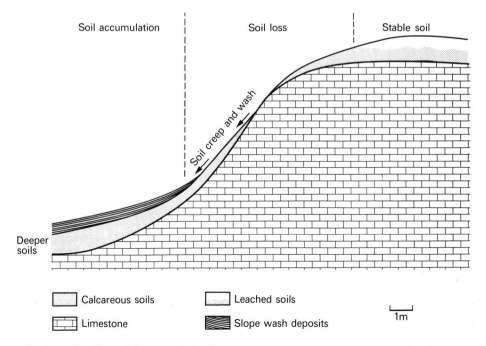

Fig. 6.11 The effect of slope on soil development.

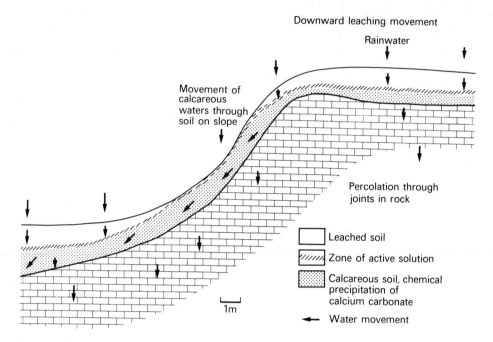

Fig. 6.12 Leaching and slope on limestone.

close to the surface. Flatter sites tend to become leached and more acid as the net water movement is downward through the soil rather than downslope (Fig. 6.12).

6.4.3 Limestone benches

The benches are found in areas of shallow limestone dip and each bench corresponds to a stratigraphic sequence of the rock. They are often associated with shale bands. The soil tends to accumulate at the back of the bench (Fig. 6.13).

6.4.4 Limestone pavement formation

The flat bench top in Fig. 6.13 is known as a limestone pavement. It owes its origin to the glacial action of scouring away of the previous rubble (see Williams, 1966). Joints in the limestone tend to be opened up by solution and cracks appear in the pavement and are termed 'grikes'. The intervening block is termed a 'clint'. These terms are from the Yorkshire dialect words for the features (Fig. 6.14).

Soil may be lost down the grikes and frequently finds its way into cave systems just under the surface. Soil washed down the walls of caves is visible to cavers entering caves which are just below the surface (about 6–33 ft (2–10 m) below ground). This means that when grikes are well developed soil formation is discouraged as soil is washed down the grikes.

Geological column

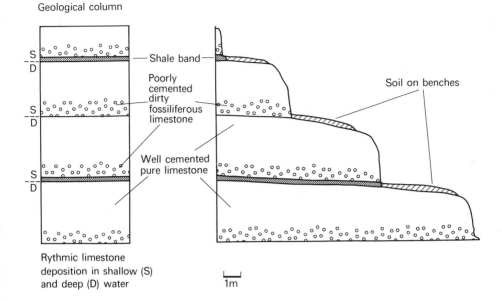

Rythmic limestone
deposition in shallow (S)
and deep (D) water

1m

Fig. 6.13 Limestone benches.

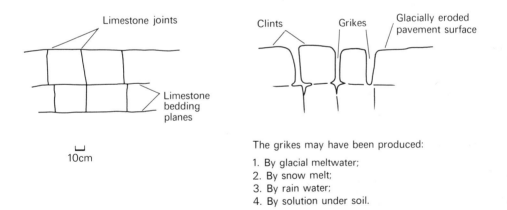

The grikes may have been produced:

1. By glacial meltwater;
2. By snow melt;
3. By rain water;
4. By solution under soil.

Fig. 6.14 Cross-section of limestone pavement with clints and grikes.

The type of soil can have an influence on grike formation. The solution of joints is greatest under the acid leached soils and therefore these soils tend to be unstable and can be washed down the grikes that are formed under the soil. In places on Ingleborough in Yorkshire and in Co. Clare, Eire, it is possible to see such areas where soils are retreating. Areas once soil covered are now bare.

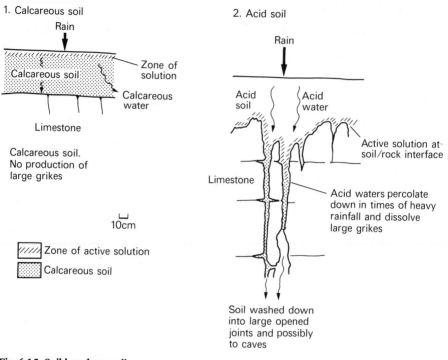

Fig. 6.15 Soil loss down grikes.

Grike formation is poor under calcareous soils. This is because the acid rain water uses up its dissolving power in the soil and the waters have no free hydrogen ions by the time they have percolated to the rock beneath the soil. These soils therefore tend to be more stable as the soils cannot be washed away down the grikes (Fig. 6.15).

6.4.5 The solubility and purity of the limestone

Limestones which are very soluble and have a high insoluble residue will give rise to the fastest rate of soil formation. Hard massive limestone generally has an insoluble residue of about 0.01 per cent and therefore it is necessary to dissolve a large amount of limestone to get a few centimetres of insoluble residue soil. This is why the formation of deep soil over limestone depends so much upon the presence of drift rather than only on the solution of limestone.

The solubility of limestone depends upon several factors, both chemical and physical. Chemically the most important factor appears to be the purity of the limestone, for instance trace elements like lead and zinc appear to inhibit solution. The effect of magnesium carbonate is less straightforward. The most soluble limestone will be that with 4—5 per cent magnesium carbonate in it but above this figure an increase in magnesium inhibits solution. But below this figure an increase in magnesium (say from 2—3%) encourages solution.

The key factor in the rate of solution is the surface area of limestone that water can attack. Clearly, a highly porous limestone, with a high reactive surface, will dissolve faster than a dense impervious limestone which water cannot penetrate. The percolation of water through the rock depends upon the degree of cementation and thus a poorly cemented limestone will allow the most solution. Dolomites (54% $CaCO_3$, 46% $MgCO_3$) are relatively insoluble, but may be eroded easily as they can be porous.

6.4.6 The mechanical strength of the limestone

Cementation often governs the strength of limestone. A poorly cemented limestone can be easily broken up by the action of frost and tree roots. Similarly a very jointed limestone or thinly bedded limestone will be prone to mechanical disintegration. Such weak limestones quickly break down into a rubble. This rubble will have a high reactive surface which water can attack. This has two results: (*a*) the limestone can dissolve faster, leaving more insoluble residue behind in the upper layers of the soil and (*b*) more calcium carbonate can move from the rock into the soil making the soil more calcareous, especially in the lower layers of the soil (Fig. 6.16).

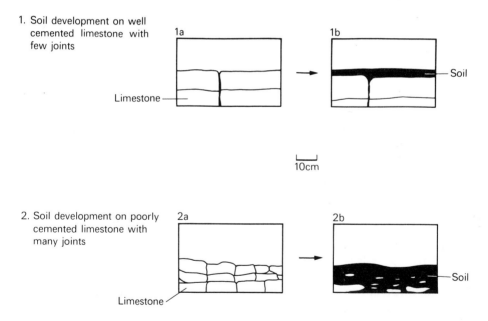

Fig. 6.16 Soil formation and the mechanical strength of limestone.

6.4.7 The amount of rainfall

The factors discussed so far will determine the progress of soil development at any one particular site. Regional variation in the development of limestone soils may be related to differing amounts of rainfall in various parts of Britain. As most limestones occur in regions of high rainfall it would be expected that most of the calcium carbonate should be quickly leached from the soils, at least from the upper horizons. Factors such as slope, the presence of rubble and thinness of soils, help to offset the effects of leaching by keeping the calcium carbonate near the surface.

6.4.8 Vegetation

Trees can offset the effects of leaching by 'pulling up' calcium and other nutrients from their root zone. The nutrients are then deposited back on the soil in leaf litter. Grasses and other lower plants also do this, but to a lesser extent, because their rooting depth is much shallower and they cannot bring nutrients up from any great depth. It can be argued that a nutrient-rich soil may be preserved under a well-developed vegetation cover as this cover will be efficient in recycling nutrients and preventing them from being leached away (Fig. 6.17).

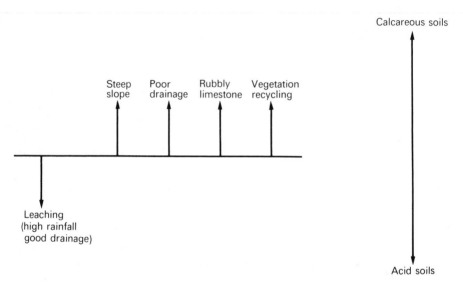

Fig. 6.17 Leaching and the opposing forces.

§6.5 LIMESTONE SOILS AND LAND-USE OF THE MENDIP DISTRICT OF SOMERSET

The Mendip Hills provide a good illustration of the influence of drift depth on soil formation. They lay just to the south of the southern edge of the last glaciation and are thought to have received windblown loess from the glacial areas further to the north. Confirmation for this idea comes from the fact that the drift has many minerals in it which are not found in the Carboniferous Limestone beneath the drift. Furthermore most of the drift is in the size range common in windblown loessial deposits (coarse silt to fine sand (0.02–0.06 mm size)). These observations support the idea

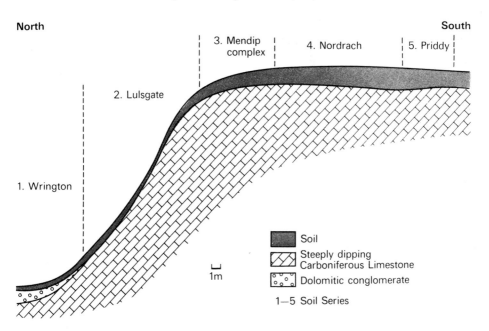

North   **South**

3. Mendip complex   4. Nordrach   5. Priddy

2. Lulsgate

1. Wrington

Soil

Steeply dipping Carboniferous Limestone

Dolomitic conglomerate

1m

1—5 Soil Series

Fig. 6.18 Soil series of the North Mendip limestone.

that the material is firstly, foreign and secondly, windblown as discussed in more detail in Chapter 9.

The drift is thickest on the Mendip Plateau surface. The influence of drift depth on soil type can be seen from Fig. 6.18. The *Lulsgate* series commonly has only a small proportion of foreign material in it, but the *Nordrach* series is developed almost exclusively in the drift. The latter's upper horizons are non-calcareous and it is only in the lower horizons that limestone fragments are seen. Quite frequently these are partly dissolved, leaving limestone 'ghosts' which appear as stones but are soft and crumbly to the touch.

The soil series can be listed:

1. *Wrington* series — calcareous brown earth on Dolomitic Conglomerate.

2. *Lulsgate* series — calcareous brown earth on Carboniferous Limestone.

3. *Mendip* complex — a mosaic of *Lulsgate* and *Nordrach*.

4. *Nordrach* series — deeper calcareous or leached brown earth.

5. *Priddy* series — peaty gleyed podzol.

Typical profiles of three of the soil series are shown in Fig. 6.19 (the *Lulsgate*, *Nordrach* and *Priddy* series).

Soil type, slope and exposure act to limit the uses of the land (Table 6.2). On the slopes occupied by the *Lulsgate* and *Wrington* series rough pasture, woodland and some

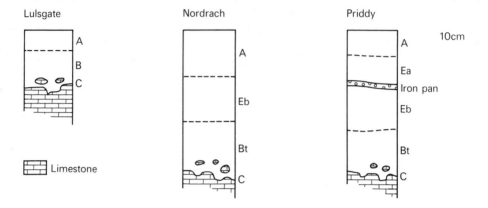

Fig. 6.19 Typical soil profiles of three soil series on Mendip. (After Findlay, 1965)

recreation were the dominant uses in 1960–70. Pasture, ley and some arable land-use is visible on the *Nordrach* and *Priddy* series. The better arable land is found on the eastern plateau, where the area is not so exposed.

Table 6.2 Land-use limitations, Mendip soil series

Soil series	Drainage	Limitations (after Findlay, 1965)
Wrington (clay loam)	Free	Steep slopes, stony soils
Lulsgate (silty clay loam)	Free — excessive	Steep slopes, shallow, often stony
Nordrach (silt loam)	Free	Occasionally stony, unstable structure, exposed plateau at 1 000 ft (305 m)
Priddy (Humose silt loam)	Poor above iron pan	Stony, wet surface. Most of this series has been ploughed out at the present day. The iron pan has been destroyed and the soil appears as a Nordrach

§6.6 SOILS AND LAND-USE OF THE LIMESTONE AREAS OF THE PEAK DISTRICT, DERBYSHIRE

In the Peak District limestone areas the soils are not always calcareous as the bedrock is covered by a non-calcareous silty drift. Like that of the Mendips, it is thought to be of loessial origin and can be up to 4 ft (1.5 m) deep. The area is thought to have escaped the last glaciation, though the surrounding low-lying areas were covered with ice. Acid brown earths and podzols are common and the soils usually associated with limestone, rendzinas and calcareous brown earths are less common. They usually occur on the steeper slopes where the drift is thin or absent.

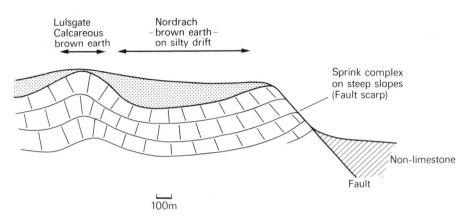

Fig. 6.20 Soils, drift and slope in Derbyshire limestone.

Both the *Nordrach* and the *Lulsgate* soil series described on the Mendips can be recognised in this area. An addition is the *Sprink* complex — a mixture of brown earths and calcareous soils on the steep scarp slopes (Fig. 6.20).

Much of the *Lulsgate* series is used as permanent pasture, as on the Mendips. These soils are, however, fairly shallow and tend to dry out fairly rapidly in summer and so the deeper *Nordrach* soils tend to give better grazing in dry periods. Ley pasture is an increasing use of the *Nordrach* series. On the *Sprink* complex steep slopes make the area suitable for rough grazing or woodland.

In areas where cultivation is absent certain types of vegetation are often associated with different soil types. Oakwoods are most frequently associated with brown earths. Under pastures the soil may be gleyed under conditions of impeded drainage or podzolised under freely draining conditions (Fig. 6.21).

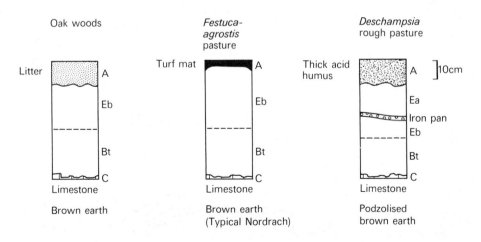

Fig. 6.21 Soil profiles and vegetation, Derbyshire limestone. (After Piggot, 1970)

§6.7 SOILS AND VEGETATION ON THE LIMESTONE AREAS OF YORKSHIRE – ADVANCING OR RETREATING?

In the limestone districts of Yorkshire it is possible to find areas where vegetation is retreating and soil is being eroded. In other areas vegetation is clearly colonising bare limestone areas and soil development is following (see §6.1).

In the long term the former may be explained by the disappearance of soil down grikes (Fig. 6.22) (see 6.4.4), but it is most probable that in the short-term grazing pressure is the most important factor. Overgrazing and hoof trampling can cause damage to vegetation and the thin humus rendzina soils (Fig. 6.23).

In areas deliberately enclosed to keep out grazing animals the proliferation of vegetation is visible. A stable humus soil accumulates, especially where tree seedlings can grow and are not nibbled off by grazing animals (Fig. 6.24).

In the Ingleborough district of Yorkshire the pattern of soils described in Fig. 6.3 (§6.2) is seen. A series of tabular blocks separated by bedding planes supports brown

Fig. 6.22 Scar Close: peat lying over a wide grike. (Photograph: S. T. Trudgill)

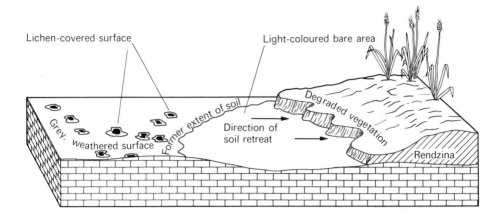

Fig. 6.23 Soil and vegetation retreat, Yorkshire.

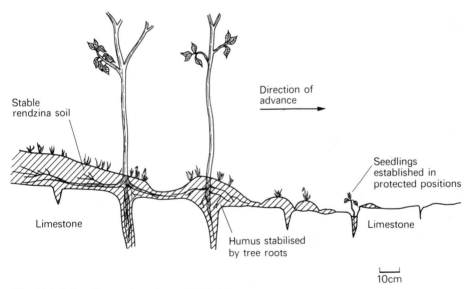

Fig. 6.24 Soil and vegetation advance, Yorkshire.

earths on the top and scree rendzinas at the edges. Areas of podzolisation can be picked out by the occurrence of moorland vegetation and heath.

§6.8 SOILS AND LAND-USE OF THE BURREN DISTRICT OF CO. CLARE, EIRE

Both Yorkshire (§6.7) and Co. Clare have been glaciated and therefore limestone pavements have developed (see 6.4.4). The occurrence of soil of any depth in the Burren

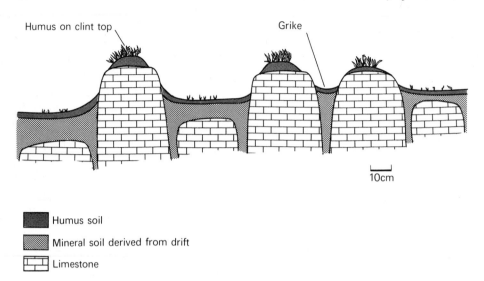

Fig. 6.25 Alternating mineral and humus rendzinas, *Burren* series, Co. Clare.

area is directly related to the presence of glacial drift, which has mostly been deposited in the valleys (Fig. 6.26). Away from the valleys the area is extremely bare and very little soil is present at all. Such soil as there is is found on the back of limestone benches (see 6.4.3 and Fig. 6.13) where drift has been deposited. Humus soils are found on clint tops. A thin peat (8–12 in; 20–30 cm) can accumulate due to the high rainfall of the region (about 100 in = 2 550 mm p.a.) which may waterlog the soil for

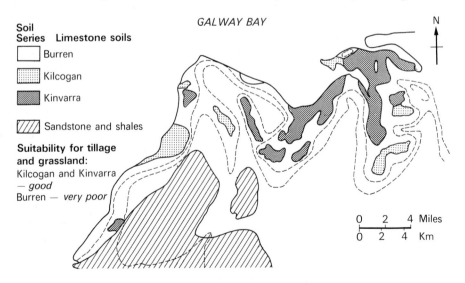

Fig. 6.26 Soils and land-use, N.W. Clare, Eire. (Source: Finch, 1971)

much of the year and prevent the aerobic decay of organic matter. Mineral soils may be found in the wider grikes (Fig. 6.25). Land-use is limited to pasture and hay cropping for winter fodder. Very little cultivation takes place. The land suitability is shown in Fig. 6.26. The soil series recognised on the map are the *Kilcogan* and *Kinvarra* series, both suitable for pasture and the *Burren* series, which although mostly rock may support some grazing on soil between the clints. Both the *Kilcogan* and *Burren* series are rendzinas but the *Kilcogan* is characteristically 9.8 in (25 cm) deep and the *Burren* series only around 10 cm, with a maximum of 7.8 in (20 cm). The *Kinvarra* series is a brown earth developed on glacial drift. It typically has a gravelly sandy loam texture up to 20 cm and a gravelly clayey loam from 7.8—15.7 in (20—40 cm). Its greater depth makes it more valuable for pasture, but its stoniness restricts its cultivation.

The Burren area is famous for many rare species of plants that have adapted to the harsh, dry environment with little soil. Many woodland species grow down the grikes where it is moist and shaded. Some arctic—alpine species flourish, notably *Dryas octopetala* (Mountain Avens). Orchids are frequently found; for example, the dark red Helleborine (*Epipactus atorubens*), and grass swards often contain a rich profusion of scented herbs. It is thought that the unusual assemblage of plants survives because of lack of competition from the commoner species which are unable to compete effectively in such an exacting environment.

FURTHER READING

Much of the material discussed in this chapter has been drawn from the sources listed below.

The Mendips

Findlay, D. C. (1965). 'The soils of the Mendip District of Somerset', *Memoir of the Soil Survey of Great Britain*, Harpenden.

Derbyshire

Bridges, E. M. (1966). 'The soils and land use of the district north of Derby', *Memoir of the Soil Survey of England and Wales*, Harpenden.

Bryan, R. (1967). 'Climosequences of soil development in the Peak District of Derbyshire', *East Midlands Geographer*, 4(4), 251—61.

Piggot, C. D. (1962). 'Soil formation and development on the Carboniferous Limestone of Derbyshire'. Part I, 'Pavement Materials', *J. Ecol.*, 50, 145—56.

Piggot, C. D. (1970). Part II, 'The relation of soil development to vegetation on the Plateau near Coombs Dale', *J. Ecol.*, 58, 529—41.

Yorkshire

Bullock, P. (1971). 'The soils of the Malham Tarn area', *Field Studies*, 3(3), 381—408.

Eyre, S. R. (1968). *Vegetation and Soils*, Edward Arnold, pp. 154—5, 187—9.

Jones, R. J. (1965). 'Aspects of the biological weathering of limestone pavements', *Proceedings of the Geologists' Association*, 76, 421–34.

Co. Clare

Finch, T. and Synge, F. M. (1966). 'The drifts and soils of West Clare and the adjoining parts of Counties Kerry and Limerick', *Irish Geographer*, 5, 161–72.
Finch, T. (1971). 'Soils of County Clare', *National Soil Survey of Ireland*, Dublin.
Williams, P. W. (1966). 'Limestone pavements with special reference to Western Ireland', *Transactions of the Institute of British Geographers*, 40, 155–72.

7 SOILS OF THE ENGLISH PLEISTOCENE DRIFTS

§7.1 THE NATURE AND EXTENT OF GLACIATION

This chapter examines the effects of the Pleistocene era in producing soil-forming materials in various parts of England. The detailed examples, which form most of the chapter, are used to try to evaluate how far origin and nature of the drifts affects the soils and how far this, in turn, is reflected in the land-use.

Full discussion of the geomorphological implications of glacial action are given by several writers, notably for Britain by Sparks and West (1972). In general terms the modes of origin of the various glacial features are not as important in an understanding of soil genesis and distribution as the final shape and configuration of the feature. So, for example, the detailed way in which a particular moraine was formed has less relevance to the pedologist than its final structure and the distribution of soil parent materials on its surface. (Nevertheless, it is of course clear that this distribution of soil parent materials will depend on the way in which the moraine was formed.)

The effects of glacial action on soil can be divided into three main areas:

1. Glaciers swept clean some parts of the landscape, and the soil in such areas has thus been formed since the retreat of the ice.

2. Depositional material was laid down in some areas, both by ice (morainic material), by melt-water from retreating and ablating ice (fluvio-glacial material) and by water under ice-dammed lakes (fluvio-lacustrine material).

3. Where ice or water did not cover the landscape, or in areas from which ice had retreated, cold periglacial conditions influenced the formation of soils.

The effect of periglacial climates will be discussed in Chapter 9, the present chapter being concerned mainly with soils resulting from the first two processes.

Most of the British Isles, apart from southern England and parts of southern Ireland, have been subjected to multiple glaciations. The approximate limits of glaciation in Britain are indicated in Fig. 7.1. Within these areas it is reasonable to suppose that much of the more exposed land was swept clean of soil by ice. In the lowland areas, while glaciers might initially have 'swept' the area, they would themselves have

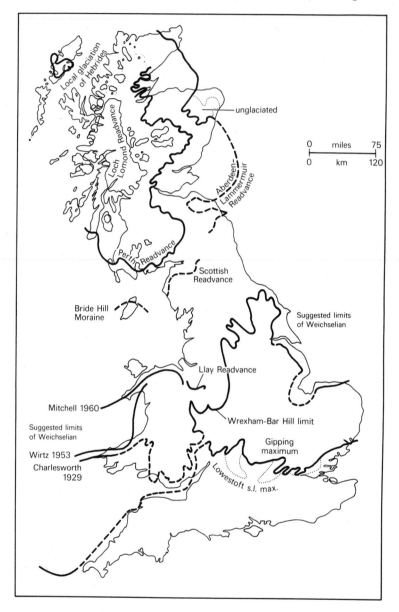

Fig. 7.1 Limits of the main glacial advances and of Weichselian retreat stages in Britain. (Source: West, 1968)

brought and deposited drift material. Perhaps more important, as they melted and retreated their melt-water would have deposited more drift material. Nevertheless there are probably some areas, even within the generally glaciated zones, where soil material may have escaped very much disturbance.

It is also important to stress that the glacial limits are tentative and by no means certain. In particular the maps predate recent discoveries in the Bristol area which suggest that the Wolstonian (Gippingian) advance extended from south Wales across the Severn estuary into north Somerset, south Gloucestershire and north Wiltshire (Hawkins and Kellaway, 1971; Kellaway, 1971).

§7.2 GLACIAL SOILS IN CENTRAL AND SOUTH YORKSHIRE

As with many areas of lowland Britain, this part of England has seen a general tendency for change in land-use from permanent pasture to arable cropping. In this section we will first consider the general origins of the soil parent materials and then examine the potential of the soils in two particular areas to cope with this transition in use. From the agricultural point of view there is a distinction — albeit rather imprecise — that can be made between farming on the glacial deposits north of the York— Escrick moraine and that possible on the lacustrine and other deposits further south. In general, the farming is mixed north of the moraine but dominantly arable to the south. The two detailed examples are therefore taken from within each of these areas.

7.2.1 Parent materials and glacial history

Most of the soils in the Vale of York are derived from Pleistocene drifts and Flandrian deposits (Crompton and Matthews, 1970). The underlying rocks are mostly Triassic and although very rarely exposed at the surface these rocks have contributed a great deal of material to the overlying drifts. It is thought that this accounts for the predominant reddish colours of the drifts.

As is the case in so many parts of Britain, the glacial history of the Yorkshire area is still far from clear. Only two of the five main glaciations are recognised, these being the equivalents of the Saale and Weichsel glaciations of continental Europe (see Table 7.1). During the Saale period (or possibly earlier) an ice sheet traversed the entire area but only limited amounts of its so-called *Older Drift* remain. These include thin cappings of eroded boulder clay[1] or till on the highest interfluves.

Table 7.1 Recommended correlations between British and Continental terminology (after Mitchell *et al.*, 1973)

British stages	Continental usage	Alps
Flandrian	Holocene	
Devensian	Weichselian	Wurm
Ipswichian	Eenian	
Wolstonian (= Gippingian)	? Saale	
Hoxian	Holstein	
Anglian	? Saale	
	? Elster	

The later glaciation (the 'Newer Dales') deposited the *Newer Drift* and can be sub-divided into two main stages — the Early Main Dales and the York—Escrick — separated by an interstadial (Raistrick, 1933). In some areas (for example, Wharfedale) the two stages are represented by distinct drifts. Work by Wills (1951) and Palmer (1966) suggests that ice of the second glaciation extended as far south as Doncaster, and south-west into lower Wharfedale. The main ice sheet penetrated south through the Vale of York and attendant glaciers probably pushed south-east down the Pennine valleys leaving remnants of morainic material.

Much of the present superficial material derives from the retreat of the Main Dales ice sheets. Lake Humber (with a shoreline at 100 ft; 31 m OD) was formed by blockage of melt-water by a North Sea ice sheet and Palmer (1966) has suggested that this was contemporaneous with the York—Escrick moraine. As the water level fell in the lake, current-bedded sands and gravels were deposited over the lacustrine clay by

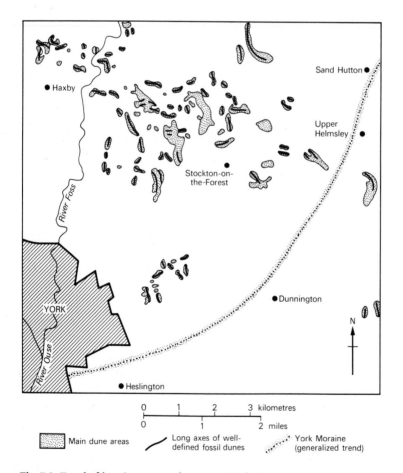

Fig. 7.2 **Trend of late Quaternary dunes near York. (Source: Matthews, 1971)**

rivers flowing east from the Pennines. As Crompton and Matthews (1970, p. 11) suggest:

> After the final disappearance of the lake these deposits would be entirely riverine. . . . Later under arid conditions, when permafrost still persisted . . . [the riverine deposits] appear to have been affected by aeolian action and a system of dunes may have been formed. Some are still recognisable in the . . . Selby district, but more so north-east of York where there is a system of barchan and seif-like dunes formed of fine sand. (See Fig. 7.2.)

After the final glacial retreat periglacial conditions continued for some time and were responsible for the production not only of sand but also of solifluction material. Ice wedges and other periglacial forms in the upper layers of drift material probably also date from this period. An indication of the complex nature of the soil parent material remaining after ice retreat is given in Fig. 7.3. This shows the stages in ablation of a glacier snout leaving a complex of moraine material.

Soil formation in the Pennine foothills probably postdates about 10,000 BP, after the return of more equable climatic conditions, but in parts of the Vale of York it probably started much later — perhaps about 2000 BP. This is suggested by the age of organic materials buried beneath sand at Stockbridge House (GR: SE561632) (Crompton and Matthews, 1970).

Because of the complex history of the glacial deposits, soils are in general distinguished by texture, base status, land-form and stratification rather than by the age of depositional material. However, differences in soil profile due to age can sometimes be distinguished. In particular this is true of the brown earths in this area where a chronosequence of more developed soils on the *Older Drift* and less developed soils on the *Newer Drift* can be distinguished.

The area of origin of the drift (provenance) is likely to be a determinant of the mineralogical characteristics of the soil and is therefore important. From the geologists point of view the area of origin and date of origin are closely linked and of equal importance, and it is only through knowing the date of deposition that one begins to get a full picture of the processes leading up to that deposition. In other words, the pedologist has to appreciate that the nature, mode of origin and date of deposition are inextricably linked.

The vast variety of drift materials can be summarised by eight categories of parent materials:

1. Till (Boulder Clay).   The till in the area is related not only to age but also to the nature of underlying rock. There are three main types of till in the area:

(a) *Newer Drift* on Carboniferous and Permain outcrops. This is usually a stiff grey clay which weathers yellow or brown and is very stony. It is closely related to the underlying formations.

(b) *Older Drift* is often stiff red clay of coarse gravel, containing gravel from the Lake District.

(c) *Vale of York Till* is much more variable and contains a very high number of

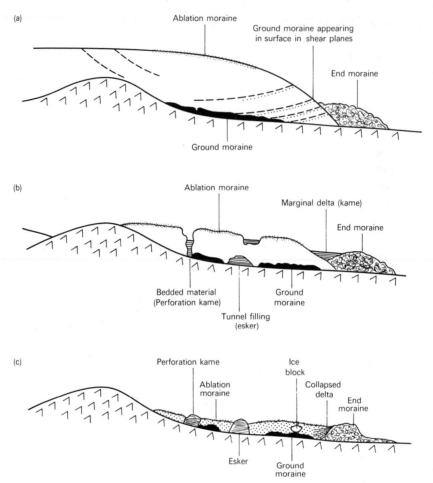

Fig 7.3 Formation of a moraine complex. (Source: Sparks and West, 1972)

erratics of many types and origins. In some areas it contains substantial contributions from the underlying Triassic rocks.

2. Morainic Drift. The upper layers of the drift forming the York, Escrick and Arthington moraines tends to be more sandy and gravelly than the surrounding till, but the arrangement is very complicated (see Fig. 7.3). The nature of the till also varies from east to west. In particular the York and Escrick moraines contain a great deal of Triassic material, whereas in Wharfedale — further west — Carboniferous material is present in the Arthington moraine.

3. Sands and gravels. The arrangement of sands and gravels is exceedingly complex. Many of the deposits are probably ice-marginal features and are thus highly variable. Patches of locally derived current-bedded gravel south of the Wharfe valley have, for example, been interpreted as beach deposits of Lake Humber (Crompton and

Matthews, 1970). Fluvio-glacial deposits cover much of the area and this material may vary from sand to sandy clay loam with occasional patches of clay. Some of the finer sand, derived from aeolian deposits, is still liable to blow.

4. Lacustrine deposits. The area of the glacial Lake Humber is now covered with laminated calcareous grey clay. North of the York moraine is slightly younger calcareous grey clay and silty clay, which was presumably formed in smaller ice-marginal lakes dammed by the moraine.

5. River terraces. In the Aire and Wharfe valleys are wide areas of river terraces. These may be up to 40 ft (12 m) above the present valley floor and are extremely variable. They range from sand and gravel to clay and loam.

6. Head and colluvium. The head in this area was mostly rearranged till which was presumably moved downslope under sub-arctic conditions.

7. Peat and **8. Alluvium** are both present in small restricted areas.

7.2.2 Soils and land-use of the Vale of York

In the glacial vale area east of Tadcaster the soils occur typically in distinct associations. North of the York moraine the flat lacustrine plain has mainly poorly drained soils (*Foggathorpe* series) with small patches of imperfectly to poorly drained gleys (*Biggin* series) where sandy layers occur in the clay. South of the River Wharfe a similar association occurs on the site of glacial Lake Humber. Areas of more loamy soils (*Ryther* series) are found around the edges, and in small patches within the former lake area.

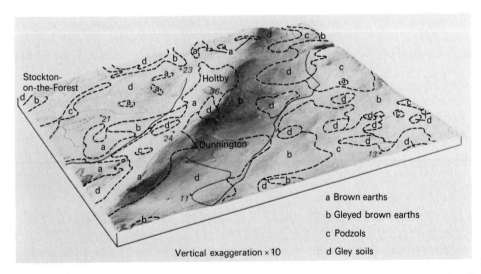

Fig. 7.4 Block diagram of the York moraine — soil groups. The area is 1.9 sq. mile (5 sq. km) and heights are in metres. (Source: Matthews, 1971)

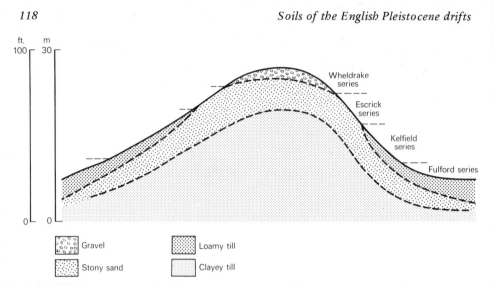

Fig. 7.5 Characteristic soil sequence on the York moraine. (Source: Matthews, 1971)

On the gravelly drift of the York and Escrick moraines the soils are better drained, and a more loamy argillic brown earth (*Wheldrake* series) covers parts of the crests and flanks. In less gravelly situations on these moraines loamy drift gives a brown earth (*Escrick* series) or a surface-water gley soil (*Wighill* series), while stony sands give rise to an acid brown earth (*Kelfield* series) or a ground-water gley (*Fulford* series). The complicated topographic relationships of these soils is illustrated in Fig. 7.4 while a generalised picture of soil—site relations is given in Fig. 7.5.

On the areas of stoneless fluvio-glacial drift and aeolian sand deposits, generally found north of the moraines, a hydrologic sequence of soils has been mapped (Matthews, 1971). In the area north of the York moraine, sand often overlies lacustrine clay or clayey till at a depth of less than 3 ft (1 m). This impedes water drainage and causes a perched water-table with soil gleying in areas of level topography. On fossil sand-dunes a sequence such as is shown in Fig. 7.6 is found. This consists of a well-drained acid brown earth (*Naburn* series) on summits. A humus—iron podzol (*Holme Moor* series) and a gleyed brown earth (*Kexby* series) occupy higher and lower sites, respectively, on the slopes while a poorly-drained ground-water gley (*Everingham* series) covers adjacent flat areas.

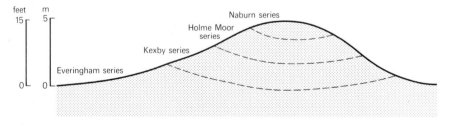

Fig. 7.6 Characteristic soil sequence on fossil dunes. (Source: Matthews, 1971)

As has already been noted, this is an area of mixed farming with both arable and permanent pasture land. The most popular arable crops are cash root crops, especially sugar beet and potatoes. These are generally farmed in a rotation with grain as follows: potatoes, wheat, barley, sugar beet, barley, oats. The oats is sometimes replaced by a 2-year ley (i.e. grassland). The presence of grazing livestock (chiefly beef cattle) is partly related to the nearness of 'ings' (river-meadows) for summer grazing, the stock being fattened in yards in winter. Sheep are far less common in the area than they were 20 years ago and there are now only a few large flocks.

Within individual farms, the areas of wetter soils (for example, *Foggathorpe* and *Biggin* series) are usually under permanent pasture, while brown earths (such as *Escrick* and *Wheldrake* series) are often subjected to continuous arable cropping. On the fossil dunes, if the soil is unprotected by vegetation, 'blow-outs' can occur. This wind erosion is particularly likely during spring gales but can usually be avoided if care is taken with cultivation.

Total rainfall in this area is relatively low — the annual mean rainfall at York is 24 in (627 mm). The summer (April–September) potential transpiration for the area is about 16 in (428 mm), while the average rainfall for the same period is only 12 in (321 mm). As Matthews (1971) points out, this indicates that there is a potential water deficit over these months and, in fact, this was the case in thirty-one out of thirty-five summers between 1921 and 1955. This fact helps to explain why arable farming is possible on some of the potentially wet soils, such as the *Everingham*, *Fulford* and *Ryther* series.

The *Everingham* series provides an interesting example of how the drift parent material can enhance the agricultural capability of a soil. Everingham soils are ground-water gleys with fine sandy textures throughout the profile and its reason for inclusion in the gley group is because of water held up by the underlying clayey or lacustrine till. This till lies below the soil profile (at depths of below 3 ft; 1 m) but water is held above it in the lower (Bg and Cg) horizons of the soil. This is an asset to the farmer because when there is a climatic water deficit irrigation of crops is rarely necessary. At these times plants can utilise this water held at depth. Only in the very driest years does the clay of the till crack and water shortage become very serious. However, on some soils, where the sand is deeper — for example, *Wheldrake* series — irrigation is required in the drier season for some crops.

7.2.3 The Doncaster Delta

About 30 miles (50 km) south of the York moraine lies an area of deltaic deposits, laid down marginal to a glacial lake or an extension of the North Sea. It is worth considering this area to see how far relationships noted in the Vale of York hold here, and how far local variations are causing soils and land-use to differ.

The deltaic drift in this area is 3–16 ft (1–5 m) thick and lies above Bunter sandstone, glacio-fluvial sand and gravel or clay (Jarvis, 1973). It is usually light reddish-brown sand containing rounded stones and gives rise to an acid brown earth (*Newport* series), which occasionally dries out sufficiently to 'blow', causing crop damage. The

Table 7.2 Particle size analysis for *Naburn* and *Newport* soils

Horizon depth (cm)	Naburn series			Newport series			
	Ap 0–33	B 33–75	C 75–85	A 0–22	B 22–47	B/C 47–65	C 65–90
Clay <2 μm, %	8.2	2.4	3.8	7.3	6.3	5.8	6.2
Silt 2–50 μm, %	4.7	1.7	2.0	19.1	11.7	9.8	5.6
Sand							
50–200 μm, %	76.6	87.1	80.1	37.7	41.3	40.2	44.2
200 μm–2 mm, %	10.5	8.8	14.1	35.9	40.7	44.2	44.0

Note: These figures are taken from Matthews (1971) (*Naburn* series) and Jarvis (1973) (*Newport* series) to whom reference should be made for detailed profile descriptions. Percentages refer to per cent total mineral particles in each horizon.

sand sometimes includes thin bands of gravel and clay knobs. The *Newport* series differs from the *Naburn* series, noted in the Vale of York, in that the former has a distinct reddish B horizon (*Naburn* is more yellowish) and is developed in lacustrine sand rather than the blown fine sand of *Naburn*. Table 7.2 shows particle size analyses for *Naburn* and *Newport* profiles and it is clear that the *Naburn* profile has a particularly high proportion of fine sand while *Newport* series has less fine sand but more coarse sand and also more silt.

Other soils occurring on the delta are generally subject to waterlogging and are mainly gleys, including ground-water gleys (*Stockbridge* and *Blackwood* series) which are also common in the undulating drift country of the Vale of York. Estuarine clayey and silty drift gives an association including non-calcareous gleys (*Foggathorpe* series) and where there are thin sandy layers another non-calcareous gley (*Biggin* series). In this respect parts of this area are very closely comparable to the lacustrine areas of the Vale of York. In the Doncaster area only the *Newport* series soil can be distinguished on the basis of topography: it generally lies at slightly higher sites than other series.

The land-use in this area contrasts with the Vale of York in having a far higher acreage of arable cropping. About 80 per cent of the area mapped by Jarvis (1973) is down to crops. Cereals, particularly spring barley, cover the largest acreage but root crops (especially potatoes and carrots) although covering a smaller area are generally more important in terms of farm income.

Lower yields for cereals are obtained on the coarse-textured (sandier) soils in the area, *Newport* series giving the lowest yield. This is also true for the root crops, sugar beet producing higher yields on the fine-textured (clayey) soils but posing the problem of difficult harvesting in the wetter years.

The annual mean rainfall for the area (22 in; 558 mm) is slightly lower than for York and so, as might be expected, here there is also a mean monthly rainfall deficit between April and August. The soils do not generally return to field capacity until mid-November. Crops yields on the various soil series show that there is some contrast between the water-retention properties of the various soils. Jarvis has shown that the coarse-textured *Newport* soil will usually dry out by July and a green crop will be subjected to water stress until September. On the other hand, the fine-textured *Foggathorpe* series never dries out in an average year.

However, these calculations are made by subtracting potential transpiration from rainfall and taking account of the available water capacity (calculated using the method of Salter and Williams, 1967). As we have already seen in the Vale of York, this does not give the whole picture when excess moisture is held in ground-water above an impervious layer at depth below the soil profile. As Jarvis (1973) points out, this is the case with many of the coarse-textured soils in the Doncaster area. For example, *Stockbridge* soils are sandy, yet there is almost always water at depth below the profile. In the past the height of the ground water-table demanded artificial drainage, but the general lowering of ground water-tables in the area has made much of the local drainage system redundant. There is no doubt that crop yields on some of the coarse-textured soils could be greatly improved by carefully controlled irrigation at certain seasons. However, this is not very widely practised at present simply because water is not available in sufficient quantities.

7.2.4 General conclusions

A full understanding of soils on the east side of the Pennines awaits both future work on the glacial history of the deposits and also further detailed soil mapping. So far a general pattern is beginning to emerge which is perhaps clearest in the Vale of York. The geological divisions into lacustrine clay lowland, moraines and stoneless fluvio-glacial drift areas are clearly recognisable in terms of soil distribution. Apparently there is some variation as one departs from this area and this is indicated by the work around Doncaster. Here, the presence of additional or alternative glacial activities (the formation of the delta) has added to the variation in soil-forming material. In particular, while a general pattern of lacustrine deposits can be related to the situation in the Vale of York, the presence of an extensive area of loamy sand causes a new soil-mapping division to be made (*Newport* series).

The main mapping problem in these areas is that soil—landscape relationships are exceedingly difficult to determine and call both for some generalisations on the part of the surveyor and also a recognisable degree of variation in the landscape. In these circumstances a great many observations by augering or digging pits are necessary for the delineation of soil units.

We will now proceed to examine the situation on the west side of the Pennines and try to establish whether any general conclusions may be applied to both flanks of the upland divide.

§7.3 GLACIAL SOILS OF THE LANCASHIRE AND CHESHIRE PLAIN AREAS

The distribution of drift and solid parent materials in Lancashire is shown in Fig. 7.7. The principal drift materials can be seen to be as follows:

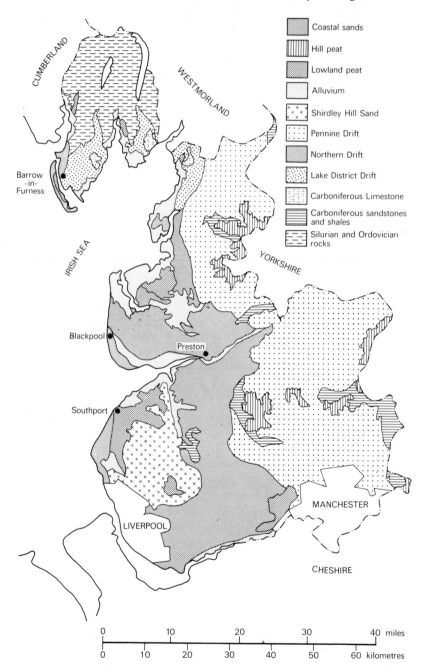

Fig. 7.7 Geology and parent materials of Lancashire. (Source: Hall and Folland, 1970)

1. Northern Drift. This forms the most extensive till in the area. It is a reddish-brown, slightly calcareous clay brought by Irish Sea ice. The reddish colour is thought to derive from the Triassic rocks flooring the Irish Sea. This drift covers much of the central lowland of Lancashire and extends into Cheshire, therefore forming an important parent material for soils of the area.

2. Lake District Drift. It is generally medium to coarse textured and is very stony. It is found both in the Lake District valleys and further south on the Furness and Cartmell peninsulas. It contains a variety of materials and usually takes its colour — often red or greyish-brown — from the softer rocks over which it has travelled.

3. Pennine Drift. Derived from Carboniferous rocks of the uplands, and includes sandstone and shale fragments in a medium to fine textured matrix. It also includes locally derived head material and is most frequent below the steep edges surrounding the Forests of Bowland, Pendle and Rossendale.

4. Glacial sands and gravels. These occur sporadically over the Plain and are particularly concentrated in certain areas. For example, there are fluvio-glacial sands and gravels in the Manchester embayment around Middleton and fluvio-glacial terrace gravels alongside the Mersey, especially between Warrington and the Glazebrook.

5. Head. This is common throughout the area and was probably mostly formed under periglacial conditions. It is derived both from the underlying solid geology and also from glacial till material and is most common at the bases of the steeper slopes.

6. Recent deposits. Other recent deposits also lie above the various glacial and fluvio-glacial materials in some areas. Examples are the coastal sands and dunes of the Southport area, and also the sandy veneer that covers older till in many parts of south Lancashire. In some areas these sand deposits are rather deeper, the Shirdley Hill Sands being one such deposit which, in this case, is thought to derive from blown coastal sand. The Downholland silt is thought to have a similar coastal origin, this material forming a bluish-grey silty clay on the southern shores of the Fylde and at a few other sites.

In Lancashire, Hall and Folland (1970) have distinguished soil associations mainly on the basis of till provenance, and to that extent the classification is slightly different from that for soil series in Yorkshire, which is based mainly on textural considerations. However, as is shown in the following chapter, this system is closely related to that employed for parts of Scotland.

The Northern Drift presents a hydrologic sequence of soils, the extent of gleying depending on the topographic position and site drainage of the soils. The most extensive association on this slightly calcareous reddish-brown till is the *Salop* association (poorly drained soils) which is combined in a complex fashion with the *Salwick* association (freely and imperfectly drained soils), as shown in Fig. 7.8. The former is generally found on flat or gentle slopes below about 250 ft (75 m) OD, and because of the fine-textured subsoil run-off from these soils is slow. The rainfall varies from about 30—40 in (760—1 015 mm) and *Salop* association soils are mostly under pasture, with

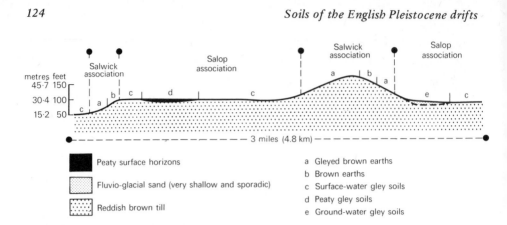

Fig. 7.8 Distribution of soils in the *Salwick* and *Salop* associations. (Source: Hall and Folland, 1970)

dairying and livestock rearing predominating in the Fylde. Further south, between Liverpool and Warrington, mixed farming is undertaken. The *Salwick* association is less extensive and run-off from these soils is much faster than from the Salop soils. Because of this they are generally more suited to arable cropping than the *Salop* association.

Over the Lake District drift two main soil associations are found. *Lindal* associations soils are developed over reddish medium to coarse textured very stony till (Fig. 7.9) in parts of the Furness district. The reddish colours of the till supporting this association result from staining by haematite which exists in the underlying Carboniferous limestone. (Haematite was originally mined and used as raw material for the local iron industry.) This till is permeable and so most of these soils are free draining with gley soils being restricted to topographic hollows. Much of this land is under permanent grassland, although some barley and root crops are grown for consumption

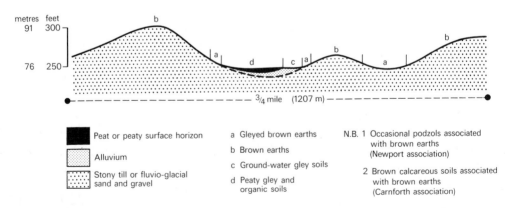

Fig. 7.9 Distribution of soils in the *Winmarleigh, Lowick, Lindal, Carnforth* and *Newport* associations. (Source: Hall and Folland, 1970)

by livestock on the farm (contrasting with the 'cash crop' production east of the Pennines).

The second association of soils on Lake District Drift are those of the *Lowick* association (also see Fig. 7.9). These are found on the browner coloured till, which is also medium to coarse textured and very stony. This till is more extensive than the reddish till supporting the *Lindal* association, and is common throughout Furness, around Lancaster and in valleys in the Lake District itself. Land-use is generally pastoral, but there are some areas of arable farming. The dominant controls on farming are rainfall, which rises to an average annual total of about 80 in (2 030 mm) near Ambleside, and also topography. Steep slopes and small enclosed wet hollows — chiefly north of Ulverstone — make cropping impossible.

The Pennine Drift gives rise to soils of the *Wilcocks* and *Brickfield* associations (see Fig. 7.10). The *Wilcocks* association contains peaty gley and organic soils which are developed in till derived from the underlying Carboniferous rocks. Generally the gleying results from impedance of soil water in the profile (i.e. surface-water gleying). These medium-textured soils are not sufficiently porous to allow the combination of high rainfall (up to 60 in; 1 520 mm) and excessive run-off from adjacent scarp slopes to permeate quickly through the profile. The peaty gley soils in this association have developed as a result of both the high rainfall and also the low temperatures experienced at their altitude (often above 700 ft; 213 m OD). Because of the low drainage capability of the soils in this association most of the area is under wet moorland or rough pasture.

The *Brickfield* association is dominated by surface-water gley soils and in general these are somewhat better for agriculture than are the soils of the *Wilcocks* association. Most of the *Brickfield* association is under permanent pasture used for dairy farming, but some areas suffer from rush infestation and this needs to be eradicated if pasture is to be improved. Tile drainage and subsoiling (see Chapter 12) can however have

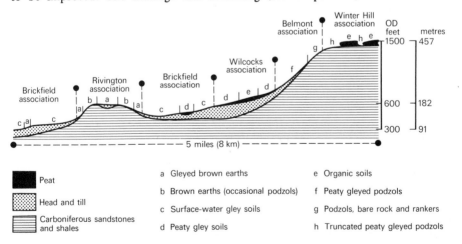

Fig. 7.10 Distribution of soils in the *Winter Hill, Belmont, Rivington, Wilcocks* **and** *Brickfield* **associations. (Source: Hall and Folland, 1970)**

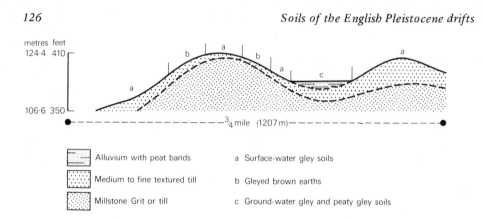

metres feet
124·4 410
106·6 350

¾ mile (1207m)

Alluvium with peat bands   a Surface-water gley soils

Medium to fine textured till   b Gleyed brown earths

Millstone Grit or till   c Ground-water gley and peaty gley soils

Fig. 7.11 Distribution of soils in the *Charnock* association. (Source: Hall and Folland, 1970)

considerable beneficial effects. Liming is also important because these soils are strongly leached.

Fluvio-glacial deposits give rise to a number of associations, of which the *Charnock* and *Newport* are the most important. The *Charnock* association (Fig. 7.11) is developed over medium to fine-textured till derived from Millstone Grit and occupies large areas north-east of Lancaster. This association contains various gley soils and the hydrologic sequence of soils on the sides of drumlins in this area contrasts with the system on fossil dunes found in the York area. In Lancashire the moderate rainfall (about 40 in; 1 016 mm in the area of the drumlins) and medium to fine-textured till parent material causes the development of gley soils, even on the short steep slopes. It was apparent (see Fig. 7.6) that on the Yorkshire fossil dunes downward percolation of water through the soil was sufficient to allow the development of podzol soils on some sites on slopes. Other parts of the slopes in that area are, however, covered with gleyed brown earth soils, as is the case here.

The *Charnock* association is mostly under permanent grassland, though at some of the higher elevations (400–500 ft; 122–152 m OD) this is rough unimproved pasture used for sheep grazing.

The *Newport* association occurs over a wide variety of glacial and fluvio-glacial sands and gravels throughout the Lancashire Plain. Although dominated by brown earths and gleyed brown earths in some areas it also contains podzols, ground-water gleys, peaty gleys and organic soils (see Fig. 7.9). These soils are also mostly under permanent grassland.

Further south, in Cheshire, Furness (1971) has mapped a fine-textured poorly drained surface-water gley soil (*Crewe* series) over the lacustrine clays that are thought to have resulted from glacial Lake Lapworth (see *inter alia* Taylor *et al.*, 1963). Where peat has developed over the clay a fine-textured peaty gley soil (*Warmington* series) is found. In this area, where rainfall is about 29 in (735 mm), the farming is dominantly dairying, over 87 per cent of the land being under grass, 70 per cent of which is permanent grass. One result of the dominantly fine-textured soil parent materials is that stock can easily cause poaching (see Chapter 12) if allowed on the land during wet conditions. This effectively limits the grazing season to about 6 months of the year, cattle being fed on hay and silage during the winter.

The differences in agricultural practice on either side of the Pennines is clearly partly a function of rainfall (Table 7.3). However, close examination of Table 7.3 shows that there is insufficient difference in rainfall to account for the almost total difference in emphasis of farming between the Doncaster area and the Cheshire Plain. The predominance of arable farming in the Doncaster area, as already noted, depends on the artificially lowered ground-water table and also on the medium-textured soils which allow relatively free drainage. In Cheshire the fine-textured soils and higher ground-water table do not allow such easy run-off and this, combined with the higher rainfall is sufficient to tip the scales.

Table 7.3 Rainfall and land-use in the example areas

Example area discussed in text	Approximate annual mean rainfall in inches (mm in parentheses)	Dominant land-use
York area	24 (627)	Mixed arable, some pasture
Doncaster area	22 (559)	Mostly arable
Lancashire	30–60 (760–1 520)	Mostly pasture, some arable
Cheshire Plain	29 (735)	Mostly pasture

The Vale of York and Lancashire have much in common, at least in the complexity of distribution of drift parent materials, but the very high rainfall of some of the upland parts of Lancashire dominates the control of land-use. In the Vale of York however the rainfall is low enough to permit economic cash-crop production in considerable areas. In Lancashire such crops as are produced are generally needed to supplement grazing for stock.

It might be reasonable to generalise that the further north, the less contrast there is in land-use between either flank of the Pennines. The following sections seek to clarify whether the converse of that statement holds true. In other words, the further south one goes does the distinction between land-use in west and east Britain on the drift soils become more marked?

§7.4 DRIFT SOILS OF THE WEST MIDLANDS

The extent of the so-called *Older* and *Newer Drifts* in the west Midlands is shown in Fig. 7.12. It will be recalled that the same terms were used for the Yorkshire drifts, but there are good reasons why Mitchell *et al.* (1973) suggest that the terms are confusing. the *Older Drift* in the west Midlands is related to the Northern Drift mentioned in §7.3 and are also derived from the Triassic rocks which form the floor of the Irish Sea. However, as Fig. 7.13 demonstrates, the distinction between *Older* and *Newer Drift* does not necessarily relate closely to till lithology.

Generally, however, there is a topographic relationship, with *Newer Drift* tending

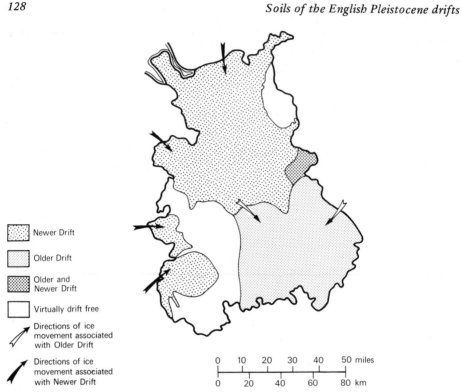

Newer Drift

Older Drift

Older and
Newer Drift

Virtually drift free

Directions of ice
movement associated
with Older Drift

Directions of ice
movement associated
with Newer Drift

0   10   20   30   40   50 miles

0   20   40   60   80 km

Fig. 7.12 Areas of *Older* and *Newer Drift* (the latter excluding terraces of the Rivers Severn and Avon) in the west Midlands, in part interpreted from Wills (1948, p. 112). (Source: Mackney and Burnham, 1964)

to blanket the lower ground, especially in Cheshire, Staffordshire and north Shrop-shire. In these areas the *Older Drift* is reserved for small isolated hilltops and plateaus. The *Older Drift* is somewhat more extensive further south but is still found primarily on hilltop sites.

As we have already seen, glacio-lacustrine deposits cover much of Cheshire. This cover extends south into Shropshire but the situation has been confused here by the movement of another ice sheet from Wales into this area. The moraine forming the edge of this sheet is today represented by a belt of hilly country around Ellesmere, where the landscape is pockmarked by kettle-holes, now containing meres and peat bogs. Loamy brown earths (*Baschurch* series) have been mapped on the moraines which contain a mixture of Triassic and Palaeozoic material (Crompton and Osmond, 1954). This soil is rather similar to the *Newport* series which was noted near Doncaster (§7.2) and which also occurs in the present area (north Shropshire). In both areas the *Newport* series is distinguished from the *Baschurch* by the browner and redder colours of the former. (The colour difference is, of course, a reflection of the origin of the drift parent materials and has a parallel in the distinction between *Lindal* and *Lowick* associations in Lancashire.)

Both Baschurch and Newport soils are chiefly under arable cropping. The farming

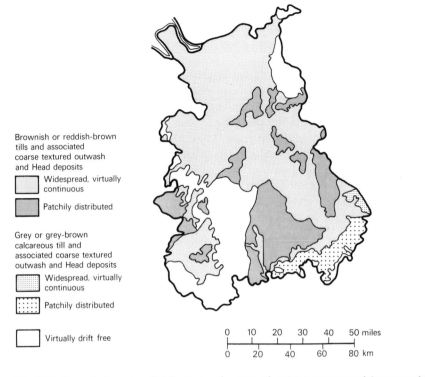

Brownish or reddish-brown
tills and associated
coarse textured outwash
and Head deposits

Widespread, virtually
continuous

Patchily distributed

Grey or grey-brown
calcareous till and
associated coarse textured
outwash and Head deposits

Widespread, virtually
continuous

Patchily distributed

Virtually drift free

```
0    10   20   30   40   50 miles
|----|----|----|----|----|
0      20      40      60     80 km
```

Fig. 7.13 General character, distribution and extent of major parent material groups within *Older* **and** *Newer Drift* **in the west Midlands. (Source: Mackney and Burnham, 1964)**

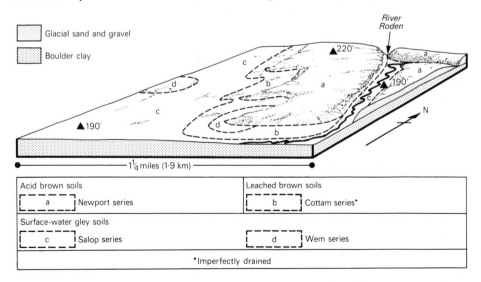

Glacial sand and gravel

Boulder clay

Acid brown soils		Leached brown soils	
a Newport series		b Cottam series*	
Surface-water gley soils			
c Salop series		d Wem series	
*Imperfectly drained			

Fig. 7.14 Soil–landscape relationships near High Ercall, Shropshire. (Source: Mackney and Burnham, 1964)

in north Shropshire is mixed arable and pasture and, as might be expected, farmers generally favour the coarser-textured soils for arable rotation (these being usually more freely draining) while the intervening poorly drained soils are under permanent pasture.

The soil—landscape relationships of coarse-textured drift soils are exemplified in Fig. 7.14. As we have already noted, detailed composition of the drift varies throughout the west Midlands, but there is a preponderance of Lake District material in north Shropshire and locally derived Bunter pebbles and flints in Staffordshire and Warwickshire. The Welsh glacial advance also brought Welsh grit and shale into Shropshire. In flat low-level sites where the parent material is more silty and slowly permeable, poorly drained surface-water gley soils of the *Crewe* series are mapped (Mackney and Burnham, 1964). As has already been explained, these soils occur in north Cheshire mainly in areas of lacustrine clays. Their presence in north Shropshire therefore might also suggest the sites of glacial lakes. As in the Cheshire Plain, *Crewe* soils are usually under permanent grass.

§7.5 GLACIAL EROSION AND DEPOSITION IN THE WELSH MARSHES

The Church Stretton and Ludlow areas provide very good examples of landscape at the edge of upland which has been affected by lowland ice sheets. The soils of both areas have been mapped in detail (Mackney and Burnham, 1966, and Hodgson, 1972, respectively) and the geology of the area has also been studied exhaustively over several decades.

The general distribution and direction of travel of the drift material is indicated in Fig. 7.15. It is now generally accepted that drift material in the area was brought in by at least two major ice sheets, and each deposition was later covered by fluvio-glacial sands and gravels as the ice withdrew. The first ice sheet laid down till brought from the Irish Sea via the Cheshire and Shropshire Plains and contains a contribution from Welsh sources (the 'Irish Sea—Welsh Drift'). The second came from the south-west and besides containing Welsh material also included a considerable amount of local material ('Local and Welsh Drift'). Throughout the period it is assumed that the summits of the upland areas remained largely free of ice ('nunataks') although they are now partially covered by quantities of locally derived head, probably produced under periglacial conditions. Further south similar events account for the distribution of drift in the lowland area west of Ludlow.

As might be expected, there are many similarities between the drift soils in the Church Stretton area and those further north in north Shropshire and on the Cheshire Plain. This is clearly a function of the origin of the parent material. *Baschurch* and *Newport* series soils are both present on outwash gravels and morainic deposits and a deep brown earth with gleying (*Cottam* series) is mapped over brown and reddish-brown till. This soil has also been mapped in west Lancashire, Cheshire and north Wales. It forms a hydrological sequence with a poorly-drained gley soil (*Salop* series) which has itself been mapped in Lancashire (see §7.3), in north Shropshire and also in

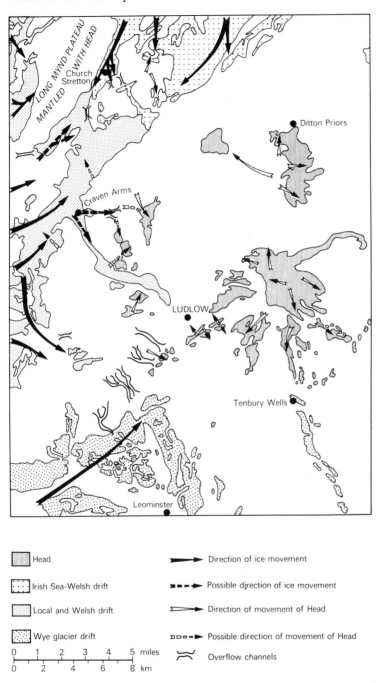

Head

Irish Sea-Welsh drift

Local and Welsh drift

Wye glacier drift

0   1   2   3   4   5 miles
0   2   4   6   8 km

Direction of ice movement

Possible direction of ice movement

Direction of movement of Head

Possible direction of movement of Head

Overflow channels

Fig. 7.15 Glacial drift deposits in the Church Stretton and Ludlow areas. (Source: Soil Survey of England and Wales, unpublished)

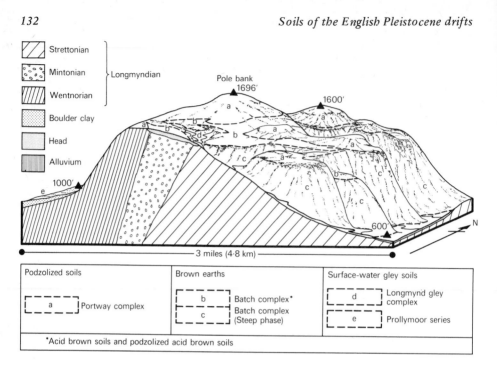

Fig. 7.16 Soils on the Long Mynd, Shropshire (solid geology after James, 1956). (Source: Mackney and Burnham, 1964)

Yorkshire on the reddish till. On fine-textured subsoils derived from fluvio-glacial material a poorly-drained clayey gley soil (*Prolleymoor* series) is mapped.

In the lowland area west of Ludlow relatively homogenous silty till and outwash material has been laid down (Local and Welsh Drift). This has given rise to brown earth soils which cover relatively large areas (*Wootten* series). In the limited wetter sites a surface-water gley soil (*Vernolds* series) is mapped.

In contrast to the lowland soils, the soils over the uplands in this area are closely related to topography. Figure 7.16 illustrates the soil—landscape relationships in the case of the Long Mynd and demonstrates the importance of the layer of head in determining the plateau-top pattern. It can be seen that the head completely obscures the underlying geology on the plateau. Conversely, soils on the lower hillsides were presumably subjected to intense solifluction during the glaciation and the present soil profile has developed since that date. Soils below about 1 000 ft (305 m) (which is thought to have been the maximum height of ice up the hillside) are generally extremely shallow (less than 20 cm deep) and many consist simply of AC horizons.

Annual mean rainfall in this area ranges from about 27 in (690 mm) in some of the lowland areas to over 40 in (980 mm) on the Long Mynd. Both the rainfall variation and that of the soil types are reflected in the land-use. Most of the land above 600 ft (183 m) is permanent pasture and the upland hills (Long Mynd and Brown Clee) are used either for rough grazing or are under forestry. Forestry also covers some of the steep slopes on the flanks of the hills.

The lower land is usually under mixed farming, which is typical of much of Shropshire. There are a number of dairy herds grazed in the summer months on the pastures over the wetter soils. There is also a considerable amount of beef stock raising and several flocks of sheep. Many farms are entirely under grass, but there are also limited amounts of arable acreage especially in the south-east of the area. In the past few years efforts have been made to produce winter wheat, although this requires very careful tillage, best undertaken when the soils are dry.

In this area it can be seen that glacial activity was considerably controlled by large-scale topographic influences. Nevertheless the general pattern of soil type in the lowland is closely related to elsewhere in the west Midlands and the north-west. The next section examines an area where topographic influences were, apparently, of much less importance in determining the movement of ice sheets and the consequent deposition of till.

§7.6 SOILS AND LAND-USE ON THE CHALKY BOULDER CLAY IN EAST ANGLIA

East Anglia is today one of the most important arable agricultural areas in Britain. Most of the soils that support this activity are developed on a variety of superficial drift materials which were laid down at various times during and subsequent to the Pleistocene glaciations. Although the stratigraphy of the various deposits has been studied in great detail by a number of workers over a relatively long period, the origin and genesis of the various deposits is only now becoming clear (see *inter alia* Sparks and West, 1972).

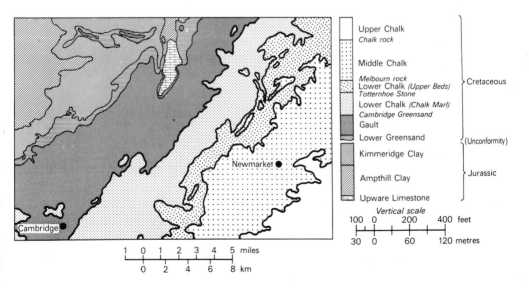

Fig. 7.17 Solid geology of the Cambridge—Newmarket area. (Source: Hodge and Seale, 1966)

This section considers the soils of one particular drift material — Chalky Boulder Clay. Soils over this material have been mapped in the Cambridge area (Hodge and Seale, 1966) and in south-west Essex (Sturdy, 1971) and these two areas will be used as detailed examples in this section.

The degraded chalk escarpment (see Chapter 10) runs along a south-west to north-east axis between Cambridge and Newmarket (Fig. 7.17). North of Cambridge the flat Fenland country extends to the Wash and into south Lincolnshire (see Chapter 13). East and south of Cambridge the land surface rises gently from about 25–250 ft (7–75 m) OD. Although there is no steep chalk escarpment here the change in landscape type from the Fenland north of Cambridge to the rolling Chalk country around Newmarket is nevertheless apparent. While the reclaimed Fenland provides excellent land for sugar beet and potatoes the Chalklands are mostly areas of cereal farming.

It is on the summits of the Upper Chalk that the most extensive areas of Chalky Boulder Clay occur. This is a grey chalky clay till which was probably laid down during the Anglian advance ('Lowestoft Till'). The fossils, colour and clay texture suggest that they are derived from Jurassic and Gault clays (Hodge and Seale, 1966). The diagram showing the direction of ice advance (Fig. 7.18) demonstrates that there were at least two stages in the advance of this ice and, in Norfolk at least, the movements appear to have been in slightly different directions. In addition to this so-called Lowestoft Till, there is a Chalky Boulder Clay till of later age ('Gipping Till') present in parts of East Anglia. However, Hodge and Seale think it more likely that the till in the Newmarket area is of the Lowestoft variety. The general suggested movement of ice during the Anglian (Lowestoft) and Wolstonian (Gipping) advances is indicated in Figs 7.19*a* and 7.19*b*. The distribution of the entire range of Chalky Boulder Clay

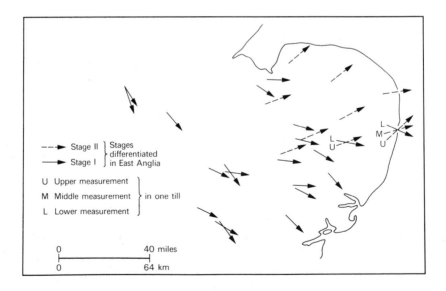

Fig. 7.18 Stone orientation of till of the Lowestoft ice advance. (Source: Sparks and West, 1972)

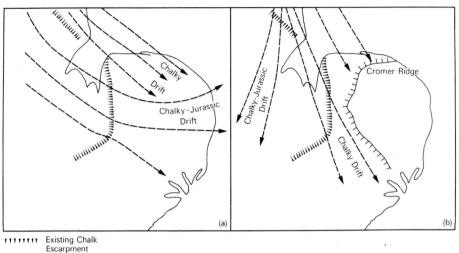

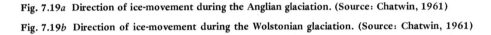

rrrrrrr   Existing Chalk
Escarpment

⊔⊔⊔⊔⊔   Ice-front

Fig. 7.19*a*   Direction of ice-movement during the Anglian glaciation. (Source: Chatwin, 1961)

Fig. 7.19*b*   Direction of ice-movement during the Wolstonian glaciation. (Source: Chatwin, 1961)

materials is shown in Fig. 7.19*c*, though in some places shown on the map this material does not form a soil parent material because it is covered with later deposits.

The soil—landscape relationships on the edge of the Chalk upland and Boulder Clay plateau in the Newmarket area are shown in Fig. 7.20. There are three soil series mapped over Chalky Boulder Clay. The *Ashley* series, a gleyed brown earth, covers a relatively small area where loam lies above the Chalky Boulder Clay. These soils are generally non-calcareous sandy loams. A much larger area is covered by the *Stretham* series, a gleyed brown calcareous soil. The presence of the underlying clay causes the soil to be imperfectly drained and artificial drainage has been used successfully. The calcareous nature of the *Stretham* series distinguishes it from the *Hanslope* series, a poorly drained gleyed calcareous soil. These latter soils have also been subjected to artificial drainage and both Stretham and Hanslope soils are now dominantly under cereal cropping. Most of the *Ashley* series mapped by Hodge and Seale is, however, used by stud-farms, nearby Newmarket being an important horse-rearing and racing centre. (As Hodge and Seale point out, however, the presence of this activity is probably more due to historical reasons than that the soils or climate is of special suitability. Much of the land now used for race-horse gallops would probably be equally suitable for arable cropping, as is the case in the Berkshire Downs, discussed in §10.8.)

Despite the use of Hanslope soils for arable cropping their wetness should not be underestimated and quite considerable areas remain under deciduous woodland for that reason. The annual mean rainfall for this area — 23 in (572 mm) — is much the same as in the areas of Yorkshire discussed in §7.2 (see also Table 7.3, p. 127). It is clear that the nature of agriculture in these areas over the drift soils is dependent on

Fig. 7.19c Distribution of Chalky Boulder Clay, eastern England. (Source: Chatwin, 1961)

both a relatively low annual rainfall and the presence of an impervious layer at depth to provide moisture during the period of soil water deficit during the summer.

As already noted, much of the Chalky Boulder Clay in East Anglia is covered by varying thicknesses of later deposits. At Saxmundham, Suffolk, the variable depth of the deposit, in this case a clay, is indicated by Fig. 7.21. The agricultural implications of such a variable soil depth above a calcareous parent material are important. Crop yields within individual fields vary considerably over small distances as a result of the variable moisture-retention properties and pH of the soil profile. Fertiliser application is clearly a difficult problem.

In south-west Essex, Sturdy (1971) has mapped Hanslope and Stretham soils over Chalky Boulder Clay. The general development of the landscape in this area has been described by Clayton (1957, 1960). An additional soil series – *Oak* – has been mapped by Sturdy. This is developed over leached Chalky Boulder Clay. *Oak* series is a non-calcareous surface-water gley with a marked texture profile, ranging from loam or clay loam in the surface to clay below 24 in (60 cm) depth. In this area of medium to

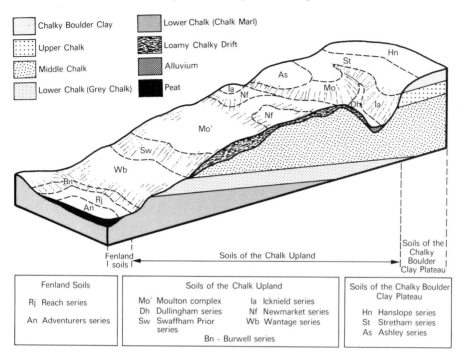

Fenland soils | Soils of the Chalk Upland | Soils of the Chalky Boulder Clay Plateau

Fenland Soils	Soils of the Chalk Upland		Soils of the Chalky Boulder Clay Plateau
Rj Reach series	Mo' Moulton complex	Ia Icknield series	Hn Hanslope series
An Adventurers series	Dh Dullingham series	Nf Newmarket series	St Stretham series
	Sw Swaffham Prior series	Wb Wantage series	As Ashley series
	Bn - Burwell series		

Fig. 7.20 Schematic section across the Chalk Upland from Fenland to Boulder Clay Plateau. (Source: Hodge and Seale, 1966)

low annual rainfall (25 in; 650 mm) there is a concentration on arable cropping, with all three noted soil series (*Hanslope, Stretham* and *Oak*) having large areas under arable rotation. Cereals, potatoes and beans are popular crops while about 20 per cent of the land is under permanent pasture.

Since south-west Essex is close to London, the area was, until the early part of this century, very important for supplying hay for the London horses, engaged in maintaining the city's transport. Today, of course, this trade has disappeared and the area has gradually changed over to a dominantly arable economy.

In concluding this chapter it is useful to consider the general relationship between land-use and the drift deposits. A dominating factor is clearly the availability or excess of water supply, and here there is an obvious contrast between the east and west of Britain which is largely accounted for by differences in rainfall. Most of eastern England from Yorkshire south to Essex suffers from a net water deficit in the summer months. The drift soils are particularly valuable at this stage because the lower subsoil layers, often being relatively impervious, retain water received from the upper layers during the winter. In western Britain, with higher rainfall, this facility is of little advantage and the frequent poorly drained characteristics of the drift soils only serve to make them more difficult to cultivate. As we have noted, in Cheshire they are often too wet even to allow grazing stock for up to 6 months of the year.

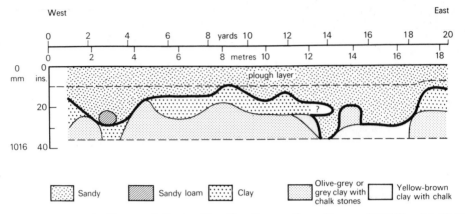

Fig. 7.21 Soil variability on Chalky Boulder Clay, Saxmundham. (Source: BSSS Norwich Field Meeting Notes, September 1970, unpublished)

Drift soils are probably most valuable to the farmer when there are large areas of relatively homogeneous material. The examples we have seen in north Lancashire, with wet topographic hollows, and in Suffolk, with a high degree of local variability, only serve to make the farmer's task more difficult.

FURTHER READING

Accounts of Pleistocene geology and stratigraphy are given by West (1968) and by Sparks and West (1972). More detailed treatment of individual areas is given in the various British Regional Geology Guides (HMSO), but some of the correlations of dating the various deposits must be made with care, since nomenclature has recently been modified. Details of the new system are given by Sparks and West (1972) and by Mitchell *et al.* (1973). (As far as possible the newer terms have been used throughout this chapter — see Table 7.1.)

Various soil surveys dealing with specific areas are detailed throughout the chapter, but some require special mention. Hall and Folland (1970) give a summary of the soils of Lancashire and this provides an indispensable introduction for students of that area. Crompton and Matthews (1970) in their Memoir for the Leeds area give a useful introduction to the soils of that part of Yorkshire. The Soil Survey Bulletin for the West Midlands (Mackney and Burnham, 1964), though rather less detailed, is a useful book with which to start a study of that area. The more detailed Soil Survey Records each tend to have a slightly different emphasis (although all include the basic soil formation), which can be helpful but sometimes makes information difficult to find. So, for example, the Record for the Doncaster area (Jarvis, 1973) has much useful detail on the irrigation requirements of the various soils, but, on the other hand, has very little discussion on either the geomorphology or the dating of the various soil parent materials.

At the time of writing a large number of Soil Survey Records were shortly due to be published on the soils of Norfolk. At present the only published Record is that for the Beccles area (Corbett and Tatler, 1970) and the Special Survey of Breckland (Corbett, 1973) referred to in Chapter 9.

NOTE

1. In referring to soil parent material the term 'boulder clay' is avoided as far as possible, and the word 'till' substituted. This is because the material is often not clay and rarely contains boulders.

8 THE DRIFT SOILS OF SCOTLAND AND IRELAND

§8.1 THE DRIFT SOILS OF SCOTLAND

Very few of the soils in Scotland are derived from the weathering of rocks *in situ*; most are developed on glacial drifts which have been transported by ice sheets from their original seat of weathering. Pleistocene glaciers have removed much of the weathered mantle which previously covered the area, leaving rugged bare rock scenery (Fig. 8.1). Ice sheets have also deposited the material on low-lying ground in thick

Fig. 8.1 Glacially-smoothed and striated rock, Coire Lair, Ross-shire. (Geological Survey photograph. Crown Copyright reserved)

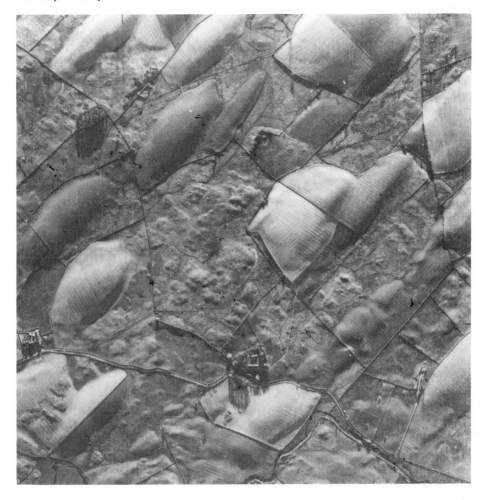

Fig. 8.2 Air photograph showing drumlins in an area of glacial deposition near New Galloway, Kirkcudbrightshire. (Ministry of Defence (Air Force Department) photograph. Crown Copyright reserved)

sheets of till, often with drumlins (Fig. 8.2). It is suggested by Ragg (1973) that a large proportion of the weathered material moved by glaciation, especially that of clay-sized particles, was produced by weathering either in the Tertiary era or possibly during warm interglacial periods. The tills therefore consist of clays produced before glaciation and rocks, sands and gravels, together with some clay, produced during the glaciations as a result of glacial and periglacial mechanical weathering. In order to understand the nature of till as a soil-forming parent material it is first necessary to study some of the geomorphological aspects of the ways in which till has been laid down in Scotland.

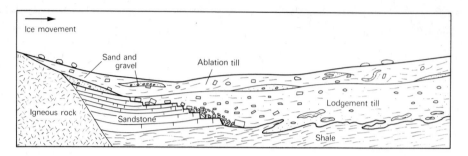

Fig. 8.3 Diagrammatic section showing relations of lodgement till to rock beneath and ablation till above. (Source: Sissons, 1967)

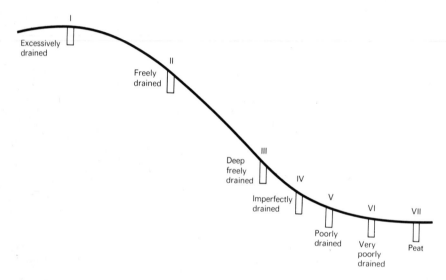

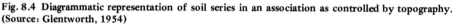

Fig. 8.4 Diagrammatic representation of soil series in an association as controlled by topography. (Source: Glentworth, 1954)

8.1.1 The origins of glacial tills as soil parent materials

Till may be classified according to its petrology, texture or origin. As far as soil formation is concerned, nutrient status and texture are important factors but place of origin can also be important since it affects both nutrient status and texture. Tills can be divided into two types, according to origin; these are lodgement till and ablation till (Sissons, 1967). Lodgement till is that deposited on the ground beneath moving glaciers. Ablation till is that laid down by decaying glaciers as they melt, either on the surfaces or edges of the glaciers. This distinction is significant to soil formation in that lodgement till is characteristically compacted, forming a tough, tenacious deposit of material of all sizes from clay to boulders. In contrast, ablation till is usually loosely packed and consists dominantly of coarser particles. Textures such as gritty sand with stones or sandy clay with stones are common. The generalised relationship between the two till types is shown in Fig. 8.3.

The second way that the origin of the till may affect soil formation is through the influence of the mode of deposition of the glacial material on the topography and drainage of the soil-formation site. Reference to Fig. 8.2 shows the hummocky topography characteristic of drumlin areas and this gives rise to differential drainage status across the slope as shown in the cross section in Fig. 8.4. The drainage status ranges from excessively drained on the hill crest to poorly drained at the foot of the slope. In this way soils on a similar till material may be differentiated according to drainage with a podzol on the crest slope, a brown earth on the backslope and gleys and peats at the base of the slope. This pattern forms the basis of the mapping of the soil associations in Scotland where soils on a given parent material are differentiated by drainage (Glentworth and Dion, 1950). Thus each soil association described below has a characteristic till as its parent material but may be sub-divided into soil series according to profile drainage or other factors.

8.1.2 Glacial tills and soil associations

The soils in Aberdeenshire have been mapped by Glentworth and Muir (1963) as falling into distinct associations according to the type of till forming the soil parent material. For example, the *Countesswells* association is developed on granite and granite gneiss tills and is therefore rich in quartz and feldspars. The coarse texture of the till imparts a sandy texture to the soils. As a contrast the *Foundland* association is developed on argillaceous rocks and weakly metamorphosed schists which have weathered to produce a high proportion of silt, giving the soils a loamy texture. Further examples of relationships between associations and parent material are given in Table 8.1.

Similarly in the southern uplands of Scotland (Fig. 8.5) Bown (1973) has been able to map the soils south of Girvan partly on the basis of the provenance and therefore composition of the till parent material. The *Ettrick* association is mapped on

Table 8.1 Glacial tills and soil associations

Soil association	Glacial till
Countesswells	Granite and granite gneiss till
Foundland	Argillaceous and schist till
Insch	Basic igneous and metamorphic rocks till
Leslie	Thin serpentine till
Tarves	Mixed basic and acid rocks (gabbro and granite) till
Strichen	Quartz—mica schist till
Ordley	Old Red Sandstone and schist till
Hatton	Conglomerate of Old Red Sandstone age till
Cuminestown	Old Red Sandstone till
Mormond	As above with schist, granite and sandstones
Peterhead	Sedimentary rocks (clays and sands) of Old Red Sandstone age till

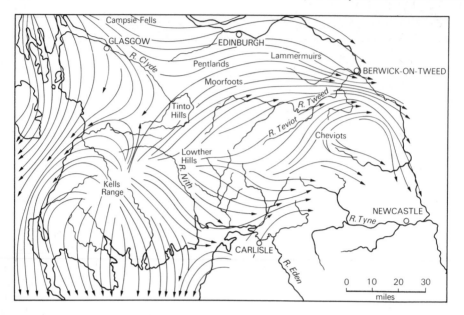

Fig. 8.5 Generalised direction of ice movement in the south of Scotland. (Source: Ragg, 1960)

deposits derived from Ordovician and Silurian greywackes (coarse sandstone) and shales. The *Darleith* association is mapped on tills derived from basic igneous rocks and the *Glenalmond* association on tills derived from Old Red Sandstone rocks. Some of these relationships are discussed in more detail by Ragg (1960) for the area around Kelso and Lauder. Figure 8.6 shows the solid geology for the Kelso and Lauder area and Fig.

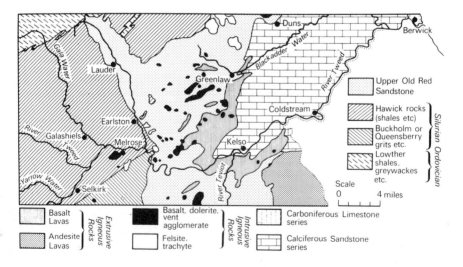

Fig. 8.6 Solid geology of the Kelso and Lauder area. (Source: Ragg, 1960)

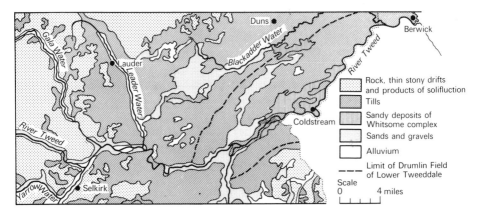

Fig. 8.7 **Distribution of drifts of the Kelso and Lauder area. (Source: Ragg, 1960)**

8.7 shows the drifts. Reference to Fig. 8.5 demonstrates that the ice in this area was moving largely eastwards and therefore the composition of the tills derived from the ice is mainly influenced by the geology to the west of any one particular location. Figure 8.8 shows the distribution of the major soil sub-groups in the Kelso and Lauder area and this can be compared with the maps of geology and drift. It can be seen that in many ways the pattern of soils is related to both drift and geology. Brown forest soils, for example, occur on tills overlying Old Red Sandstone whereas gleyed brown earths are associated with tills over calciferous sandstone.

In terms of agricultural potential the drift soils supply the major areas of culti-vated soils in Scotland (Bibby, 1973). In the north-east of Scotland the soils are dominantly stony, sandy loams or loams with small areas of clay. The conditions are suitable for a wide range of agricultural activities and good quality land occurs with

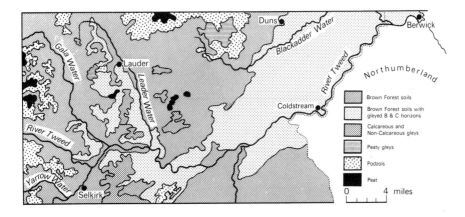

Fig. 8.8 **Distribution of major soil sub-groups of the Kelso and Lauder area. (Source: Ragg, 1960)**

land capability classes of 2 and below (Grade 1 is excluded because of climatic limitations). Much of the land in Aberdeenshire is ploughed, despite its inherent stoniness. Good drainage and a good thickness of soil combine to encourage agriculture in this area.

In Angus, Fife and the lowlands of East Lothian one of the most widespread soil types is derived from till of Carboniferous and Old Red Sandstone rocks. These soils tend to have a sandy topsoil over a sandy clay loam subsoil. Although there is a potential water deficit in this area of 1 in (25 mm) the fact that the soil drainage is slightly impeded because the texture becomes finer with depth makes the area favourable to agriculture as adequate moisture is retained at depth in the summer season.

In general the till soils often have a high clay content and, unless overlain by a coarser ablation deposit, can present problems to farmers in terms of drainage. Surface-water gley soils are common in the lowland areas of Scotland and such conditions may limit the choice a farmer has in terms of potential land-use. Nevertheless, in general, the drift soils of Scotland provide good grazing on the heavier or steeper land, and the coarser-textured areas provide good arable land.

Fig. 8.9 Glacial drift of Weichsel Age near Ringmoylan, Co. Limerick, showing thickness of drift cover and development of a soil profile in the top 1 m (3 ft). (Source: Finch and Ryan, 1966)

§8.2 THE DRIFT SOILS OF IRELAND

The majority of the soils of Ireland have developed from glacial drifts laid down by ice sheets or valley glaciers or from fluvio-glacial material deposited by waters derived from melting ice. From this, it follows that considerations of solid geology are relatively unimportant (Fig. 8.9), except in so far as that in each locality the drift will contain erratics characteristic of the provenance of the ice which gave rise to the drift.

Much of the glacial drift is thought to be of Weichsel age, but the earlier Saale glaciation has left several important deposits. The point concerning the provenance of the drift is illustrated in Fig. 8.10 from Gardiner and Ryan (1964) of the southern half of Ireland. Limestone from the Burren karst district of Co. Clare is deposited to the south-west of the area; granite from Galway over central-south Ireland and earlier easterly ice streams have brought some Scottish rocks to the south coast of Ireland. The extent of the cover of drift of Weichsel age can be deduced from Fig. 8.11. Comparison of Figs 8.10 and 8.11 illustrates that in the south of Ireland only drift of Saale age may occur, but in the rest of Ireland mixed drifts can occur. In west Ireland

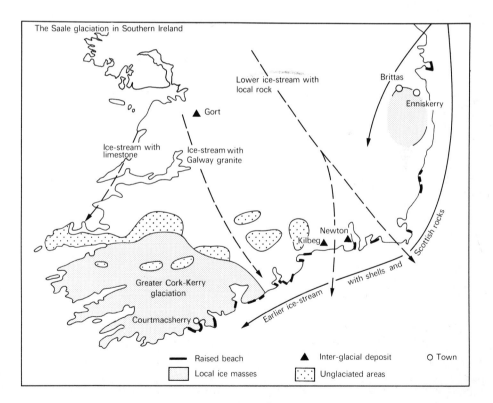

Fig. 8.10 Extent and distribution pattern of the Saale glaciation in Ireland. (Source: Gardiner and Ryan, 1964)

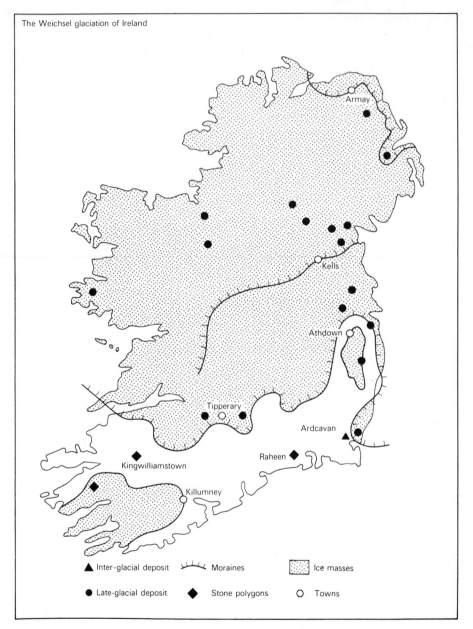

The Weichsel glaciation of Ireland

▲ Inter-glacial deposit   ⌇⌇⌇⌇ Moraines   ▦ Ice masses

● Late-glacial deposit   ◆ Stone polygons   ○ Towns

Fig. 8.11 Extent and distribution pattern of the Weichsel glaciation in Ireland. (Source: Gardiner and Ryan, 1964)

Finch and Ryan (1966) illustrate drift movement and composition (Figs 8.12 and 8.13).

Some relic, pre-glacial, soils may occur as shown by Gardiner and Ryan (1962). For example, some reddish soils are found in south Wexford on limestone parent

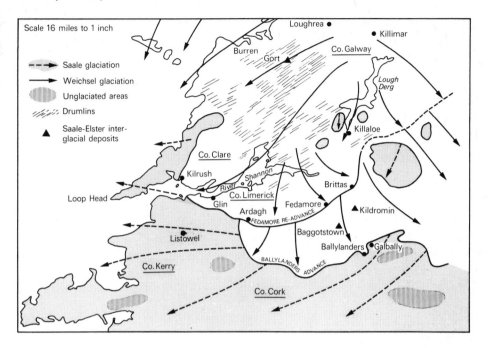

Fig. 8.12 Ice streams of the Saale and Weichsel glaciations in the Limerick–Clare region. (Source: Finch and Ryan, 1966)

material but the soils are covered with glacial drift. It is felt that the red soils may have been formed under the warmer climatic conditions of the late pre-glacial period because they bear a close resemblance to the Mediterranean soils or 'terra rossa' soils. Glaciation resulted in truncation of early soil and regolith cover and superimposition of Quaternary drift, leaving the red soil in protected pockets.

In Co. Clare, Finch and Synge (1966) have mapped a *Newer* and an *Older* (more weathered) *Drift* (Fig. 8.14) equivalent to the Saale and Weichsel glaciations. More recently, detailed work by Finch (1971) has grouped the soil series of Co. Clare according to the erratic composition and age of the drift on which they are found (Table 8.2).

It is interesting to note that the podzols occur in the older, probably more weathered, Saale drift. The soil series developed in the first group (Saale–Weischel drift of Silurian shale and Old Red Sandstone) are developed on similar parent material but are differentiated because of altitude and stoniness. The *Knockastanna* series is a podzol found only over 700 ft (213 m) in the Slieve Bernaghs and Slieve Aughty Mountains. The *Ballyanders* and *Ballynalackan* series are found below 700 ft in these localities in areas of mostly glacial kame, kettle and drumlin topography. They are both acid brown earths (pH values of the surface horizons ranging from 4.8 to 5.0 and 5.4 to 5.5. respectively) but the *Ballynalackan* series is differentiated because of its higher proportions of stones and boulders. Both the brown earths are classed as Grade

Table 8.2 Glacial deposits and soil series in Co. Clare. (Source: Finch, 1971)

	Drift composition	*Soil series mapped*
GLACIAL DRIFT DEPOSITS		
Saale–Weichsel age	Silurian shale and Devonian (Old Red) sandstone	Abbeyfeale, Kilrush (gleys); Kilfergus (brown earth); Knockanimpaha (podzol); Mountcollins (brown podzolic)
	Devonian (Old Red) sandstone	Gortaclareen, Puckane, Sellernaun (gleys)
	Upper Carboniferous sandstone and shale	Ballyanders, Ballynalackan (brown earth); Knockastanna (podzol)
Weichsel age	Carboniferous Limestone	Kilcolgan (rendzina); Kinvarra (brown earth)
	Carb. lst. + Old Red Sandstone	Knocknaskeha (brown earth)
	Carb. lst. + Carb. shales	Kilfenora (grey-brown podzolic)
	Carb. lst. + Carb. sst. and shale	Attyquin, Howardston (gley); Ballincurra (brown earth); Elton, Patrickswell (gleyed brown podzolic)
	Volcanic rocks, limestone, shale and sandstone	Derk (brown earth)
FLUVIO–GLACIAL DEPOSITS		
Weichsel age	Limestone, sandstone, shale and volcanic material	Baggotstown (brown earth)
	Old Red Sandstone, Silurian shale and limestone	Cooga (brown podzolic)

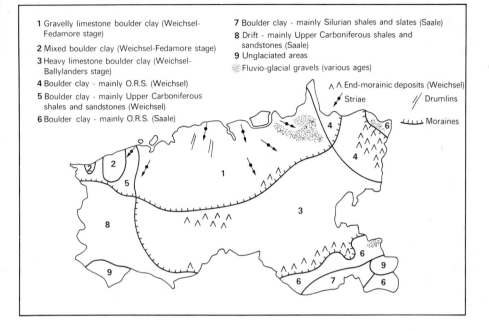

1 Gravelly limestone boulder clay (Weichsel-Fedamore stage)

2 Mixed boulder clay (Weichsel-Fedamore stage)

3 Heavy limestone boulder clay (Weichsel-Ballylanders stage)

4 Boulder clay - mainly O.R.S. (Weichsel)

5 Boulder clay - mainly Upper Carboniferous shales and sandstones (Weichsel)

6 Boulder clay - mainly O.R.S. (Saale)

7 Boulder clay - mainly Silurian shales and slates (Saale)

8 Drift - mainly Upper Carboniferous shales and sandstones (Saale)

9 Unglaciated areas

Fluvio-glacial gravels (various ages)

∧ ∧ End-morainic deposits (Weichsel)

Striae

Drumlins

Moraines

Fig. 8.13 Glacial pattern in Co. Limerick. (Source: Finch and Ryan, 1966)

1 in terms of suitability for both tillage and grassland, but they are mostly devoted to grassland which represents an under use of the agricultural potential. The reasons for this situation lie mainly in economics and the way of life of the inhabitants of the area, tied very much to a 'peasant' type of rural economy. Because of stoniness, exposure and infertility the *Knocknastanna* series is classed as only of moderate suitability for tillage.

Topography and drainage are the main factors differentiating the soil series in the Upper Carboniferous sandstone and shale group of drifts. The soils occur on areas of hummocky topography, kettles, kames, drumlins and terminal moraines together with more extensive, flatter, low-lying drift areas. The *Knockanimpaha* series is a podzol occurring on the higher areas, especially on the slightly more sandier drifts. The *Mountcollins* series is a podzolic soil and has developed on the better drained ridges, especially on the terminal moraines around the south and east of Liscannor Bay. On the steeper hillsides of the hummocky terrain brown earths are found, mapped as the *Kilfenora* series. In topographic hollows thin peats have accumulated over poorly drained parent materials giving rise to the *Abbeyfeale* peaty gley soils, while in the extensive level areas the combination of low-lying topography and high shale content of the drift leads to the development of the *Kilrush* gley series. This is the most widespread soil in the county, covering 19.5 per cent (61 385 ha) of the county but is of limited agricultural potential, the actual use being confined mostly to pasture.

In the Wexford area, Gardiner and Ryan (1964) demonstrate the significance of

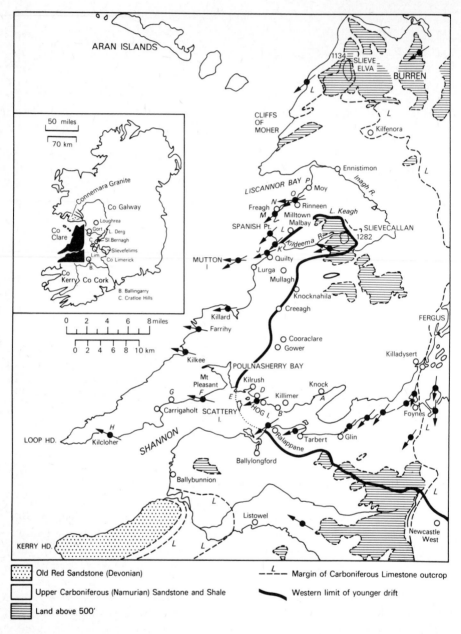

Fig. 8.14 Glacial pattern in west Clare and adjoining areas. (Source: Finch and Synge, 1966)

variations in drift type to soil development. Figure 8.15 shows the different drift types in Co. Wexford and Table 8.3 shows the major soil associations corresponding to the drift types.

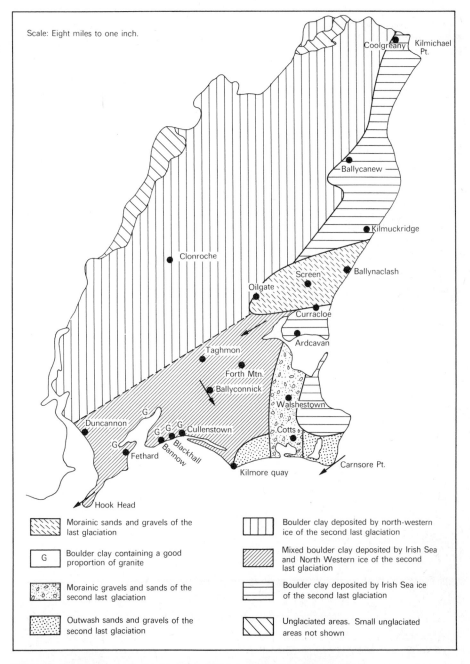

Scale: Eight miles to one inch.

Coolgreany
Kilmichael Pt.
Ballycanew
Kilmuckridge
Clonroche
Screen
Ballynaclash
Oilgate
Curracloe
Ardcavan
Taghmon
Forth Mtn.
Ballyconnick
Walshestown
G
Duncannon
G G Cullenstown
Cotts
G G
Blackhall
G Bannow
Fethard
Kilmore quay
Carnsore Pt.

Hook Head

Morainic sands and gravels of the last glaciation

G Boulder clay containing a good proportion of granite

Morainic gravels and sands of the second last glaciation

Outwash sands and gravels of the second last glaciation

Boulder clay deposited by north-western ice of the second last glaciation

Mixed boulder clay deposited by Irish Sea and North Western ice of the second last glaciation

Boulder clay deposited by Irish Sea ice of the second last glaciation

Unglaciated areas. Small unglaciated areas not shown

Fig. 8.15 Glacial pattern in Co. Wexford. (Source: Gardiner and Ryan, 1964)

Table 8.3 Glacial deposits and major soil associations of Co. Wexford. (Source: Gardiner and Ryan, 1964)

Glacial deposits	Major soil association
Boulder clay deposited by Irish Sea ice of the Saale glaciation, without other material deposited on or incorporated in it.	Macamore
Boulder clay containing material contributed both by the Irish Sea ice and by ice from the uplands of north-west Wexford (or from still further north-west) during the Saale glaciation.	Rathangan
Boulder clay deposited by ice of Saale Age from the uplands of north-west Wexford or from still further north-west. Boulder clay from the Wexford uplands becomes more granitic as the Leinster range is approached.	Clonroche
Morainic gravels and sands probably deposited at a late stage of the Saale glaciation.	Killinick
Outwash sands and gravels from the Killinick moraine of the Saale glaciation	Broadway
Morainic deposits (sands and gravels) of the Weichsel glaciation.	Screen

Fig. 8.16 Landscape of young (Weichsel) morainic drift in the Screen area, Co. Wexford (see Fig. 8.15). The poor nature of the soils is evidenced by the nature of the land use. (Source: Gardiner and Ryan, 1964)

Fig. 8.17 Soil profile from the *Macamore* series showing the poor permeability of the soil developed on the dense calcareous drift of Saale age. (Source: Gardiner and Ryan, 1964)

The glacial heritage has influenced the subsequent course of soil development. Podzols have developed in freely draining situations on coarse outwash sand and gravels (*Nethertown* series), on granitic glacial drift (*Blackstair* series) and on mica schist drift (*Black Rock Mountain* series), all parent material of Saale age. Brown earths of low base status (*Broomhill* series) have developed on Old Red Sandstone drift while brown earths of high base status (*Baldwinter* series) have developed on drift of mixed Carboniferous Limestone and Carboniferous shale composition Elsewhere gley soils, like the *Wexford Slob* series, have readily developed on the fine-textured estuarine alluvium in low-lying situations and on poorly drained drifts such as the drifts of Ordovician shale (*Knockroe* series). They have also developed on coarser grained drifts in low-lying situations where water-tables are high. Gleys cover some 30 per cent of the country, second to the brown earths of low-medium base status. The end moraines around Screen (Fig. 8.16; see map, Fig. 8.15) form sandy, excessively drained soils with a related low agricultural potential, nevertheless forming useful winter pastures. The main area of the country in the north-east is composed of brown earth and brown podzolic soils, with friable, easily worked soils developed on boulder clays deposited by the north-west ice of the penultimate (Saale) glaciation. Here pastures are suitable for sheep, milk or beef production, the soils forming some of the best agricultural potential in the county. However, in the north-east coastal region where the *Macamore* series (Fig. 8.17) is developed on dense calcareous drift of Irish Sea origin (Saale age), the poor stability of soil structures presents problems. The soils are heavy, retentive of moisture and the structures are weak, leading to dispersion and poor aeration. Rushes (*Juncus* spp.) are common and without reclamation a rather inferior pasture is the best use that this area can support. In conclusion, one may emphasise that in many ways the glacial heritage of Ireland has had a strong influence upon soil development and the agricultural potential of the land.

FURTHER READING

The book by Sissons (1967) gives a good general background, and the article by Ragg (1973) gives a general review of the soils developed on drifts in Scotland.

The Soil Memoirs by Finch (1971), Finch and Ryan (1966) and Gardiner and Ryan (1964) should be consulted for further information on soils developed on glacial drift in Ireland.

9 PERIGLACIAL CONDITIONS AND SOIL PATTERNS

§9.1 ENGLAND AND WALES

9.1.1 Periglacial processes

At the margins of the ice sheets the geomorphic processes moulding the landscape were very varied. Strong winds from the ice caps blew over largely unvegetated surfaces and were able to pick up sand and silt particles. Thus blown sand deposits and loessial silt are characteristic of many areas in periglacial regions. Melt-water from the ice- and snow-covered areas washed out debris from the higher ground and outwash fans and spreads of sand and gravel resulted from this redistribution of materials. Also the melt-water contributed to the saturation of soils which were often impermeable in the subsoil due to permafrost layers occurring at some depth below the surface. Saturated soils on sloping ground were therefore subject to mass movement and head deposits resulting from solifluction were widespread. In addition to the transport of material by wind and water the action of frost upon exposed rocks shattered them to form screes and clitters. In the finer sediments and soils the freezing of water contained in the pore spaces led to heaving and disturbance (cryoturbation). These freeze—thaw processes acting on rocks and soils led to the formation of a wide range of periglacial forms. The general outline of the geomorphic processes operating in periglacial regions has been described by Washburn (1973) and a report of observations of periglacial phenomena in the British Isles was given by Fitzpatrick (1956a). The range of periglacial features includes the following:

Asymmetric valleys — older valleys tend to be steeper on north-facing sides
Cryoturbation   — involutions and contortions in sediments and soils
Fossil stone polygons
Fossil stone stripes
Frost-shattered rocks
Frost wedges
Mass movements   — solifluction and head deposits, gravels spreads, coombe deposits
Mass movements   — block fields, cambering structures, rotational slips
Tors
Wind-blown sand and loess.

9.1.2 Loessial contributions to soils

The most widespread effect of the periglacial conditions acting on British soils has been the production of a veneer of wind-blown silt (loess) which has been recognised in many areas. Loess is typically stoneless, non-calcareous silt loam grading downwards to silty clay loam. It may be either of a uniform brown colour or greyish and rust mottled, depending on local water-table conditions. It is generally accepted that the widespread deposits of loess in western Europe, North America and Britain have a characteristic preponderance of particles in the coarse silt-fine sand size grade. Mechanical analyses generally show a dominant proportion in the range 0.002–0.06 mm. This loess cover imparts special characteristics which would not otherwise be expected of soils on particular parent materials. For example, Perrin (1956) found that loess explained the atypical acidity of soils occurring in heath areas on the chalk of southern England.

Wind-blown deposits often produce a uniformity in the soil pattern which would not be expected in areas of varied geology. Thus on the Mendip uplands, Findlay (1965) found that in spite of the geological variations the soil mantle was remarkably uniform and was characterised by high silt content. When the mineralogy of the soils (*Nordrach* series and *Lulsgate* series, see Chapter 6) was studied it was found that the surface layers of wind-blown silt contained heavy minerals (e.g. epidote and chlorite) and feldspars in substantial amounts. In contrast the weathering rock beneath the soil showed a low content of heavy minerals and feldspars. Likewise Crampton (1961) found that certain Glamorgan soils (*Lulsgate/Gower* series) were characterised by heavy minerals atypical of the underlying rock. Similar examples of wind-blown materials which have been incorporated in soils have been given for north Wales (Smithson, 1953), Yorkshire (Crampton, 1959) and Exmoor (Curtis, 1974).

9.1.3 Solifluction deposits and soil patterns

Solifluction resulting from mass movement of saturated soil downslope above frozen subsoils was often periodic rather than continuous in nature. As a result solifluction (head) deposits often contain layers of material of different textures and degrees of stoniness (Fig. 9.1), e.g. the head deposits over granite on Dartmoor. One also finds that head deposits vary somewhat according to the provenance of the material from which it is composed. This is well shown in the Mendip and Quantock uplands of Somerset where gravelly head of sandstone and limestone rubble extends as sloping fans and sheets from combes and gorges. The distribution suggests that there were melt-waters flowing through specific channels rather than sludge moving downhill, but contorted structures in the deposits reveal the effects of freeze–thaw processes. On the north side of the Mendips the gravels are mainly siliceous, being derived from Old Red Sandstone and silicified Lias rocks. To the south Carboniferous Limestone becomes the major constituent of the head deposits although minor quantities of sandstone and chert also occur. These deposits give rise to an important soil series (*Langford* series) which is used extensively for vegetable, salad and strawberry crops in sheltered sites around Cheddar and Axbridge. On the northern side of the Mendips the sandstone

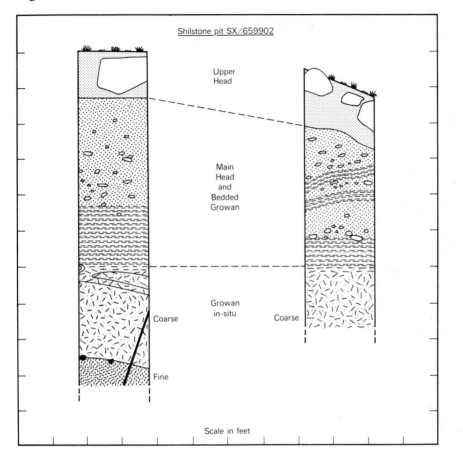

Fig. 9.1 Sections in superficial material in the Shilstone pit, Dartmoor. (Source: Waters, 1964)

gravels are less base rich and are exposed to northerly winds so that they tend to remain in mixed farming use.

Market gardening is also associated with another periglacial deposit in the Bristol region. This is the loamy head which can be found on the footslopes of many of the upland areas. It represents breccia and sand once banked against the limestone slopes but redistributed by solifluction processes. Soils based on this head are termed the *Tickenham* series. They are well suited to good quality orchards and intensive horti-culture although they also carry a good deal of pasture and arable crops (Figs 9.2 and 9.3).

Crampton and Taylor (1967) have described solifluction terraces in the upland valleys of south and central Wales carrying gleyed and podzolised soils (Fig. 9.4). These features reflect the considerable movement of material downslope in upland Wales. Such periglacial movement accompanied by freeze–thaw processes can often produce stone stripes consisting of alternating rows of stones and fine material with

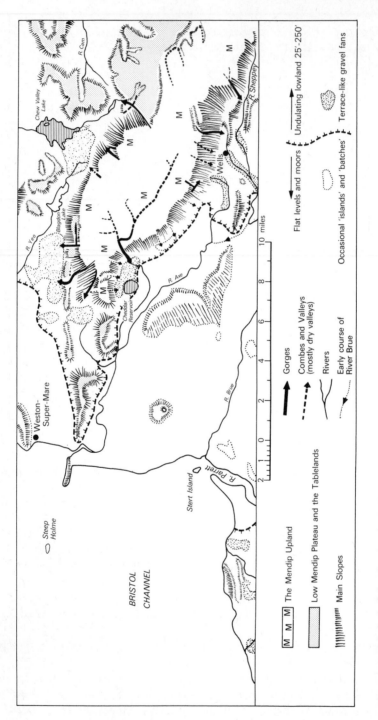

Fig. 9.2 Morphological sketch map of Mendip. (Source: Findlay, 1965)

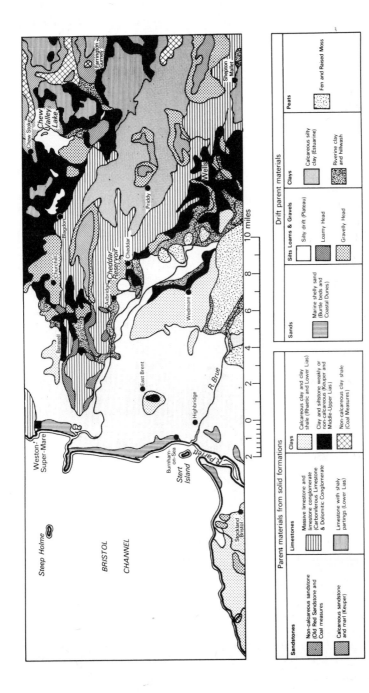

Fig. 9.3 Soil parent materials of the Mendip district. (Source: Findlay, 1965)

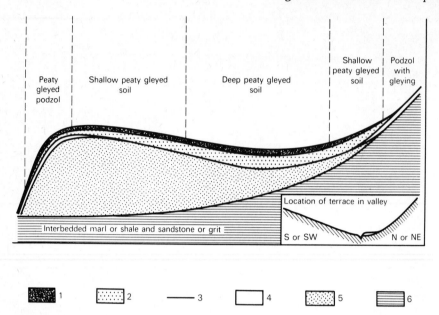

Fig. 9.4 Diagrammatic representation of transverse section across a solifluction terrace. Vertical scale considerably exaggerated. 1. horizon of organic matter; 2. horizon of gleying; 3. horizon of iron accumulation; 4. horizon of free drainage; 5. horizon of induration; 6. rock. (Source: Crampton and Taylor, 1967)

a downslope orientation (Washburn, 1956, 1973; Troll, 1958; Lundquist, 1962). Examples of fossil stone stripes in soils of the Rhinog Mountains, north Wales, show a transition from a well-developed peaty podzol in the fine earth zone through to a ranker soil in which peaty humus overlies boulders (Fig. 9.5).

Elsewhere in upland Wales the slopes are often covered by scree deposits. These are mostly relic formations resulting from periglacial conditions in the Late Glacial

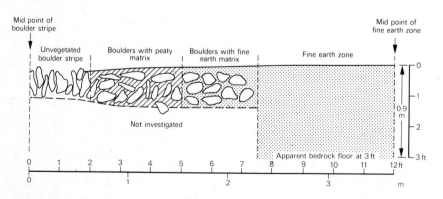

Fig. 9.5 Diagrammatic scale cross-section through a stripe pattern. (Source: Ball and Goodier, 1968)

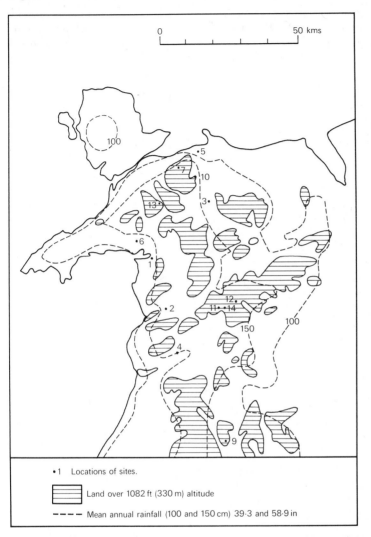

Fig. 9.6 Distribution of scree sites discussed from north and mid-Wales. (Source: Ball, 1966)

period. The locations of typical scree deposits and their widespread distribution have been noted by Ball (1966) (Fig. 9.6). An examination of the soil maps of north Wales (e.g. Bangor and Beaumaris sheets) will show that some 10–15 per cent of the land is classified as rock dominant and this provides witness to the effects of the Ice Age.

9.1.4 Soils on Brickearths and Coombe deposits

When one turns to consider the scarplands and lowlands of south and eastern England one finds that Coombe deposits are solifluction phenomena laid down in chalk areas (see Chapter 10). The main source of material forming the Coombe deposits was the

Chalk and overlying Clay-with-flints. The upper layers usually contain a high proportion of silt and very fine sand (0.002–0.06 mm) and even in unweathered, highly calcareous deposits this proportion is high. It is probable that this fraction is windborne loess which was added to the local material. Early geological memoirs (Reid, 1903) pointed out that Coombe deposits merged both 'horizontally and vertically by imperceptible gradations' into Brickearth deposits. These are typically stoneless, non-calcareous silt loams and Dines *et al.* (1954) recognised the resemblance of the Brickearth to continental loess, but they considered that much of it is essentially a head or solifluction deposit. The preponderance of fine sand and silt suggests that much of the material was initially loessial but the pellets of chalk in the unweathered layers show that it has been redistributed, probably by solifluction.

Brickearth forms a very valuable soil-forming material in the west Sussex coastal plain (Hodgson, 1967). The *Hamble, Hook* and *Park Gate* series based on the Brickearth deposits were first described by Kay (1939) in south Hampshire west of Portsmouth but they also occur extensively in Kent and Sussex (Fig. 9.7; Hodgson, 1967). The *Hamble* series consists of well-drained brown earths (*sols lessivés*) in almost stoneless silty drift. The *Hook* series comprises brown earth with gleying soils intermediate in character between the *Hamble* and *Park Gate* series, the latter being a range of gley soils developed in Brickearth. The deeper *Hamble* soils are probably the best horticultural soils in Sussex and are very suitable for fruit, glasshouse, intensive nursery or market garden crops. The *Hook* series is also used in market gardening and horticulture but it may show 'capping' problems when under arable management. The *Park Gate* series provides excellent arable land where field and regional drainage are good.

The Coombe deposits (flinty silty head) are widely represented in the chalk areas and give rise to a group of soils of which the *Charity* series is an important member. These latter are well-drained brown earths (*sols lessivés*) and typical profiles in natural or semi-natural woodland are decalcified and have an acid reaction. Normally about 18 in (45 cm) of flinty silt loam merges into reddish-brown flinty clay overlying chalky head, chalk or occasionally valley gravel. The *Charity* series provides good arable land and the less stony phases are suitable for vegetable and glasshouse crops as well as fruit. The extremely stony phases of the *Charity* series are much less valuable soils. Over 50 per cent of the volume of the soil is sometimes composed of flint and the capacity to hold water and nutrients is correspondingly reduced. In addition to its liability to 'burn' in dry periods the stony phase also causes excessive wear on agricultural implements.

As well as the Coombe deposits in the valleys one finds ancient drifts overlying much of the chalk containing the mixed and transported debris of earlier deposits. These ancient drifts were chiefly derived from the Reading beds which were formerly widespread on the chalk dip-slope. The materials were already exposed to weathering in late Tertiary or early Pleistocene times during which large quantities of nodular and shattered flint created by dissolution and erosion of the Upper Chalk were incorporated. There is evidence of much lateral movement of the drift by processes of solifluction and three distinct soil series are recognised, all conforming to the brown earth type. The *Batcombe* series occurs on plateau sites where the drift cover is thick. The

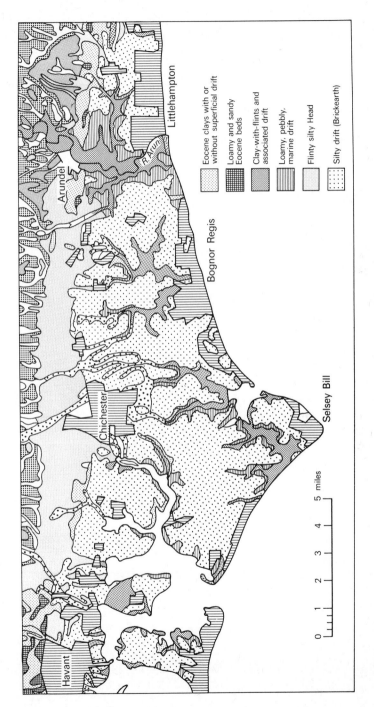

Fig. 9.7 Soil parent materials of the west Sussex coastal plain. (Source: Hodgson, 1967)

Legend:

- Eocene clays with or without superficial drift
- Loamy and sandy Eocene beds
- Clay-with-flints and associated drift
- Loamy, pebbly, marine drift
- Flinty silty Head
- Silty drift (Brickearth)

Littlehampton

Arundel

R. Arun

Bognor Regis

Chichester

Selsey Bill

Havant

0 1 2 3 4 5 miles

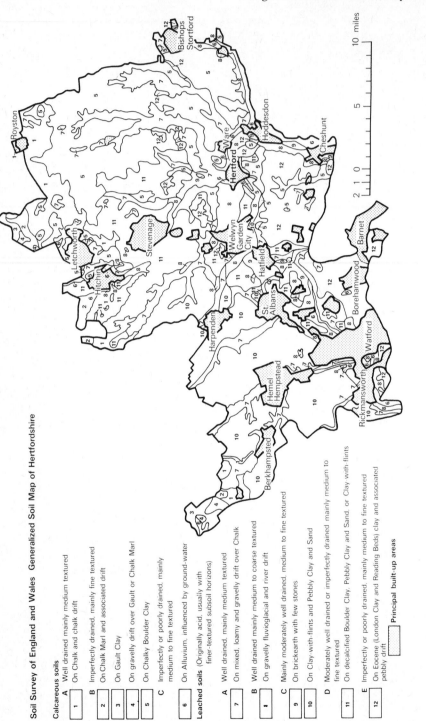

Soil Survey of England and Wales Generalized Soil Map of Hertfordshire

Calcareous soils

| 1 | A | Well drained mainly medium textured |
| | | On Chalk and chalk drift |

| 2 | B | Imperfectly drained, mainly fine textured |
| | | On Chalk Marl and associated drift |

| 3 | | On Gault Clay |

| 4 | | On gravelly drift over Gault or Chalk Marl |

| 5 | | On Chalky Boulder Clay |

| 6 | C | Imperfectly or poorly drained, mainly medium to fine textured |
| | | On Alluvium, influenced by ground-water |

Leached soils (Originally acid, usually with finer-textured subsoil horizons)

| 7 | A | Well drained, mainly medium textured |
| | | On mixed, loamy and gravelly drift over Chalk |

| 8 | B | Well drained mainly medium to coarse textured |
| | | On gravelly fluvioglacial and river drift |

| 9 | C | Mainly moderately well drained, medium to fine textured |
| | | On brickearth with few stones |

| 10 | | On Clay-with-flints and Pebbly Clay and Sand |

| 11 | D | Moderately well drained or imperfectly drained mainly medium to fine textured |
| | | On decalcified Boulder Clay, Pebbly Clay and Sand, or Clay with flints |

| 12 | E | Imperfectly or poorly drained, mainly medium to fine textured |
| | | On Eocene (London Clay and Reading Beds) clay and associated pebbly drift |

□ **Principal built-up areas**

Fig. 9.8 Generalised soil map of Hertfordshire. (Source: Thomasson and Avery, 1963)

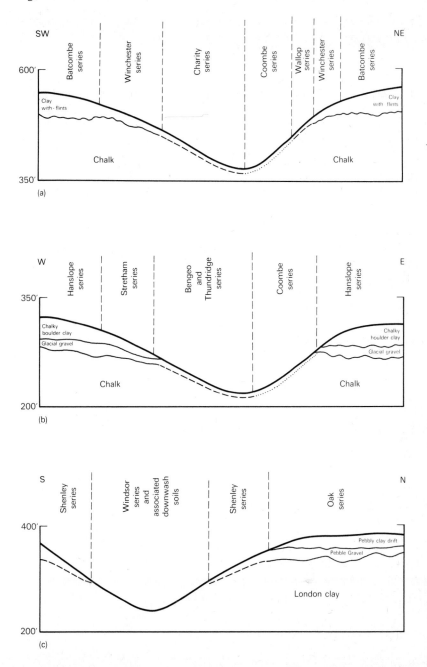

Fig. 9.9 Schematic sections of (*a*) a valley in the Clay-with-flints area of west Hertfordshire, (*b*) a valley in the Chalky Boulder Clay area of east Hertfordshire, (*c*) a valley in the London Clay area of south Hertfordshire. In all figures, sections are 1 mile long. The vertical scales are generally exaggerated. (Source: Thomasson and Avery, 1963)

Fig. 9.10 Soil patterns, 2½ miles west-south-west of Brettenham, Norfolk. (Photograph: Cambridge University Collection — copyright reserved)

Winchester series occurs on upper valley sides somewhat below the main plateau level, where the drift cover is rarely more than 5 ft (1.5 m) thick. The *Berkhamsted* series is typical of the Pebbly Clay and Sand shown on the New Series geological maps and is distinguished by the occurrence of numerous, well-rounded, bluish-grey or buff pebbles derived initially from the pebbly strata of the Reading beds (Avery, 1964; Thomasson and Avery, 1963). The extent of these soils around Hemel Hempstead and Harpenden can be seen in Fig. 9.8 and examples of the topographic relationships of the *Batcombe, Winchester* and *Charity* series are given in Fig. 9.9.

9.1.5 Patterning in soils due to freeze–thaw processes

British soils offer plentiful evidence of patterning due to freeze–thaw processes in the Pleistocene and Late Glacial periods. Some 500 localities in which fossil polygons and stripes occur in chalky materials have been described by Williams (1964) and maps of their distribution (R. B. G. Williams) are shown in Figs 9.11 and 9.12. Sites elsewhere in the south and west of Britain have been noted by Waters (1965). The main forms are polygons of the cellular type which are often replaced by stripe patterns on slopes greater than 1–2° (Fig. 9.10). Examples of these are the polygons and stripes in the Thetford Heath area which occur in the *Methwold* and *Worlington* series and are characterised by distinct vegetation patterns. *Calluna* is generally concentrated over the acid sandy infills, whereas varied plant covers are found on the areas where the calcareous subsoil comes close to the surface (Fig. 9.13). Even where there is no surface manifestation of patterning one may frequently find involutions in the subsoil horizons. For example, frost-heaved soils showing involutions have been described in soils developed upon the Lower Estuarine series of the Inferior Oolite in Rutland (Curtis and James, 1959). In these soils pockets of sandy loam and fine sandy clay are enclosed in grey clay which has been deformed by frost heaving (Fig. 9.14). These complex structures in the subsoils lead to rapid variation in textural and hydrological conditions over short distances. It is to be expected, therefore, that responses to fertilisers or to prolonged drought will vary considerably from place to place within a crop growing in such soils.

One of the most extensive areas in which periglacial deposition and processes have affected soil conditions can be found is Breckland. This comprises a distinctive low sandy plateau in north-west Suffolk and south-west Norfolk (Fig. 9.15). Today the area is largely under coniferous forest and little of the natural heathland remains. A major land-use change has taken place in the area in the last 50 years and it is this which makes a study of the local soil particularly interesting.

Most of the soils of Breckland are developed in chalk–sand drift which overlies solid bedded chalk at variable depths. Other parent materials include sand, gravel and small areas of peat. The underlying chalk contains large amounts of flints, many of which have worked their way upward into the overlying drift. (The flints were mined from the chalk during the Neolithic at the well-known site of Grimes Graves, near Brandon, and it is only recently that flint-knappers or masons have given up their very skilful work at Brandon.)

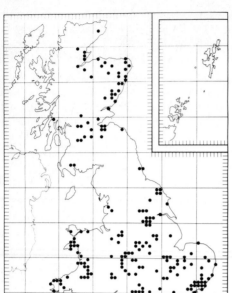

Fig. 9.11 Distribution of ice wedges of
Devensian age. (Source: R. B. G. Williams,
pers. comm.)

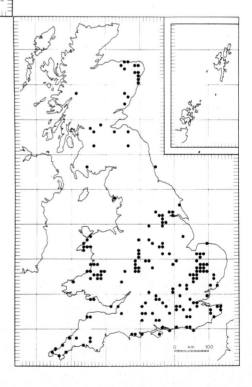

Fig. 9.12 Distribution of involutions of
Devensian age. (Source: R. B. G. Williams,
pers. comm.)

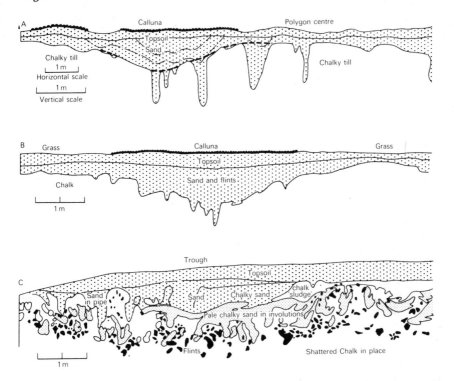

Fig. 9.13 Cross-sections through polygons and stripes illustrating subsoil features. A, polygon at Thetford Heath, Norfolk. B, stripes at Grimes Graves, Norfolk. C, stripes at Risby Poor's Heath, near Cavenham, Suffolk. Flints are shown in black. (Source: Williams, 1964)

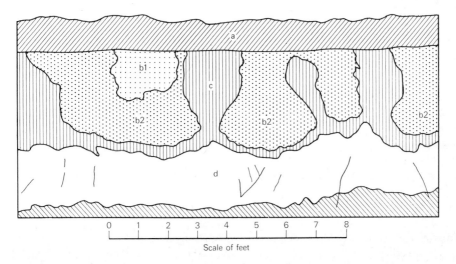

Fig. 9.14 Vertical section. a. Topsoil; b1. Reddish-brown sandy loam; b2. Yellow-brown fine sandy clay loam; c. Grey clay; d. Laminated clays and sands. (Source: Curtis and James, 1959)

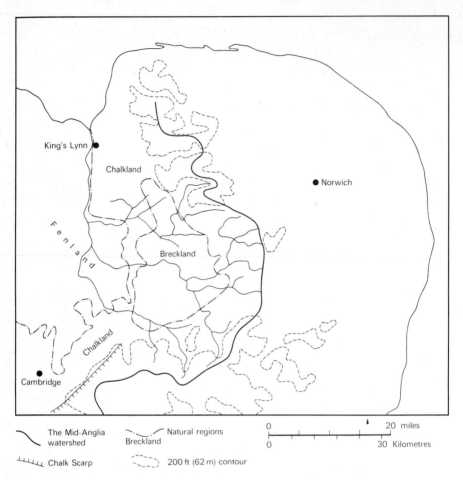

The Mid-Anglia watershed

Chalk Scarp

Natural regions
Breckland

200 ft (62 m) contour

0   20 miles

0   30 Kilometres

Fig. 9.15 Physiography of Breckland. (Source: Corbett, 1973)

Despite the fact that the superficial deposits have been studied in some detail (for example, see West and Donner, 1956; Perrin, 1955 and 1956) there are, at the time of writing, no firm conclusions regarding the age and origin of the chalk—sand drift. The stratigraphy of both this and the other deposits is highly complex (see Figs 9.16 and 9.17). This complexity is in great part due to the periglacial processes which acted in the area during the Wolstonian or a later glaciation (see Table 7.1 and *inter alia* Williams, 1964, and Sparks and West, 1972). The somewhat dramatic effect of some of these processes, when the results are viewed from the air, can be seen in Fig. 9.10. Much of the sand would appear to be aeolian in origin and, even now, wind blowing of soil material is not uncommon in dry seasons in East Anglia.

The soils of part of the Breckland area have been mapped by Corbett (1973) and he divides landscape in the area into five landscape facets, each having distinctive soil series (Table 9.1).

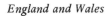
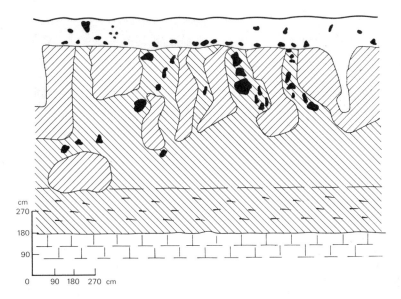

Fig. 9.16 Section across a slope at Brandon. (Source: Corbett, 1973)

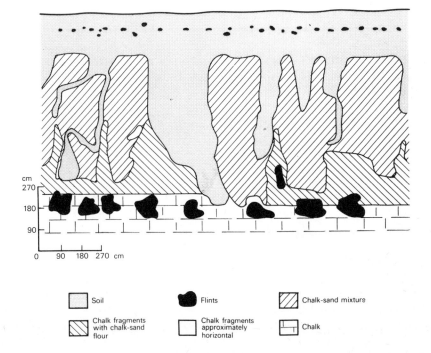

☐ Soil	⬤ Flints	⧄ Chalk-sand mixture
⧄ Chalk fragments with chalk-sand flour	☐ Chalk fragments approximately horizontal	⊟ Chalk

Fig. 9.17 Section across an upland site at Downham Highlodge Warren. (Source: Corbett, 1973)

Table 9.1 Landscape facets and soils

Facets	Soils
Upland	Worlington, Santon and Moulton series
Slopes	Newmarket, Methwold and shallow phase of Worlington series
Upland and terraces	Freckenham, Redlodge, Broomhouse and Croxton series
Dry valleys	Gravelly phase of the Worlington and Brandon series
Valley floor and low-lying sites	Row, Highlodge, Isleham, Adventurers', Wicken and Lakenheath series and peat soils

Although it might appear from the facet descriptions that a considerable variation in relief existed in Breckland, the relief is generally shallow with slopes of between 1° and 3° being common; particularly in the forest areas these variations are often difficult to detect in the field.

Corbett (1973) has divided the region into two main areas. These are the western and the eastern areas and each have distinct landscape types and combinations of land facets and soil types. The western area includes most of Breckland centred on Thetford and Brandon, while the eastern area is smaller and centred on Hockham. Figures 9.18*a* and 9.18*b* (Corbett, 1973) demonstrate the soil–landscape facet relationships in each area. Because of the complicated and convoluted nature of the parent materials many of the soils have been mapped in complexes. (In Fig. 9.18 this is indicated by a combination of soil names separated by oblique strokes, e.g. Wt/Mw.) An example of the detailed landscape arrangement of a soil complex is given in Fig. 9.19.

Table 9.2 classifies the series that have been mapped by Corbett (1973) in the Breckland area. The wide variety of soil groups represented is a further reflection of the complicating effect of the periglacial processes. The variation that exists even

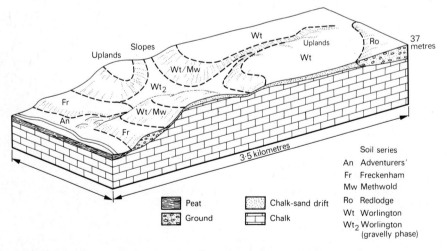

Fig. 9.18*a* Soil distribution in the western landscape. (Source: Corbett, 1973)

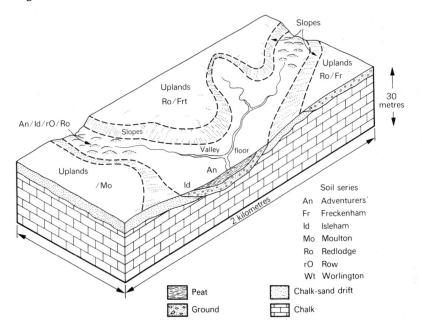

Fig. 9.18*b* **Soil distribution in the eastern landscape. (Source: Corbett, 1973)**

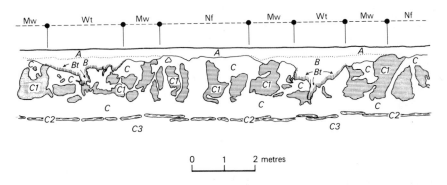

Fig. 9.19 Section across slope at Grimes Graves. A. topsoil; B. yellow-brown sand; Bt. clay-enriched layer; C. chalk-sludge with little sand; C1. chalk-sand mixture; C2. tabular flints almost *in situ***; C3. bedded chalk; Mw** *Methwold* **series; Nf** *Newmarket* **series; Wt** *Worlington* **series. (Source: Corbett, 1973)**

within some of the soil series is indicated in Fig. 9.20. The general agricultural poverty of the soil (in particular excessive drainage leading to leaching and podzolisation) and the complexity of the soil pattern have led to a very distinctive land-use in the Breckland region. An itinerary written in the 1890s describes Brandon as 'surrounded by miles of sandy heaths that form extensive rabbit warrens, from which many thousands of rabbits are sent annually to the London markets. The countryside . . . mostly

Table 9.2 Soil classification of the Breckland area. (Source: Corbett, 1973)

Major soil group	Soil group	Parent material	Drainage	Soil series and phases
Raw soils		Blown sand	Well drained (excessive)	Broomhouse
Calcareous soils	Rendzinas	Chalk—sand drift		Newmarket
	Brown calcareous soils			Methwold
Brown earths	Brown earths	Sand over gravel with finer layers	Well drained	Croxton
	Brown earths (*sols lessivés*)	Chalk—sand drift	Well drained (excessive)	Worlington
				Worlington (shallow phase)
		Thin stony sand over chalk—sand drift		Worlington (gravelly phase)
		Loamy chalky drift	Well drained	Moulton
		Sand over stony sand	Well drained (excessive)	Freckenham
				Santon
Podzols	Humus podzols	Chalk—sand drift		Redlodge
		Sand over gravel		Brandon
	Gley-podzols		Imperfect to moderately well drained	Lakenheath
				Row
Gley soils	Ground-water gley soils		Imperfect	Highlodge
	Clayey gley soils	Calcareous clay		Wicken
	Peaty and humose gley soils	Sandy and loamy drift	Poor	Isleham
Organic soils	Humified peat soils	Fen peat	Very poor to poor	Adventurers'
	Raw peat soils	Sedge-carr peat		
		Cotton-grass peat	Very poor	

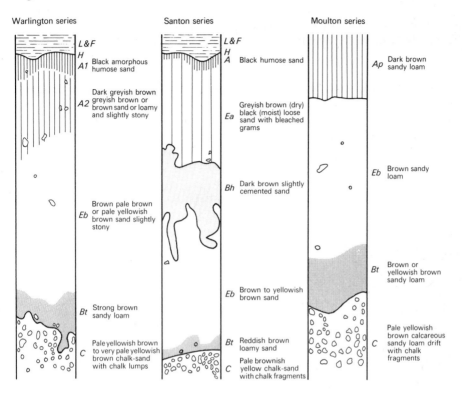

Fig. 9.20 Upland soils in the Breckland area. (Source: Corbett, 1973)

consists of sandy and flint-besprinkled commons dotted with clumps of trees and well known to sportsmen as the "rabbit and rye country".' (Great Eastern Railway, 1897.)

This state of affairs continued until the 1930s and the rabbits kept the grass-sward short thus preventing the invasion of shrubs and trees. As its description suggested, the vast population of rabbits in the area supported quite an industry. Regular trains of rabbit meat were conveyed from the small Breckland market towns (Thetford and Brandon), while rabbit fur was used in making garments at Brandon until very recently. However, there were also some areas of tillage on the better soils, rye being a popular crop. The only other major use of Breckland before the 1930s was as shooting country for the large estates — that of Lord Iveagh (a member of the Guinness family) being a particularly well-known example. To assist in the improvement of the shoots small coverts were planted.

For a number of strategic and economic reasons the late 1930s saw an increase throughout the country of planted coniferous woodland (see Chapter 5). At that time the Forestry Commission bought up large areas of the low-quality farmland and heath of the Breckland. Generally, in these depression years farmers were only too willing to sell. Another large area had already come under military usage and with the onset of the Second World War this was extended and the resident population were evacuated from a large area around Stanford (between Watton and Thetford). During and after the war most of those areas of heath that remained were improved considerably by modern agricultural practice. Today spring barley and sugar beet are the most important crops although, even now, yields from these crops are not high, when compared with more favourable parts of East Anglia. Soil moisture deficiency is still the main problem although it is now recognised that some soils also have trace-element deficiencies. Corbett (1973) suggests that both agriculturalists and foresters agree that more land could profitably be turned over to forestry. At the present time there is a considerable conservation problem in the area which, in many ways, is a microcosm of similar problems throughout the country as a whole. Thetford is a designated overspill town for London and the population has doubled in recent years. Thus there is obviously increased human pressure on the recreational facilities in the area, much of which is directed towards the forestry land. Large areas are now mature coniferous forest and the Forestry Commission has adopted a very enlightened attitude toward public access. It is also aware of the aesthetic and conservation effect of large-scale clearance of this mature woodland and is trying to clear and replant restricted areas at any given time.

The major problem in conservation is, however, the preservation of any of the areas of 'natural' heathland, particularly in view of the fact that they were at least partly maintained by the high rabbit population. The problem is mainly concerned with the difficulties of a relatively small number of species that used to be common on the open heath. One example is the stone-curlew, a bird with a very restricted breeding distribution elsewhere in the country. This bird makes its nest on the open flinty heathland and both its eggs and plumage are well adapted to blend in with the surrounding sandy soil with small flint fragments; so that a sitting bird is virtually indistinguishable from a few yards distance. Although once abundant the changes in the heathland have resulted in only a few birds regularly making their nests in Breckland.

Although a small number of relatively small heathland areas have been preserved as nature reserves the largest area of natural Breckland is 'preserved' by the Army. The presence of the Military has meant that most other human influence has been excluded. Even within the designated restricted area military activities have affected only a relatively small proportion of the natural habitat. This type of conservation is clearly only a relatively short-term solution to the conservation problem but had it not been for the influence of the Army all Breckland would have changed its character in the last 50 years. Now, at least, some of the heathland remains.

§9.2 SCOTLAND

9.2.1 Introduction

Many periglacial features can be seen in Scotland, some of which are fossil and some active today. They all have some degree of influence on the soils of the areas in which they are found. The features range from the larger-scale, distinctive ones such as

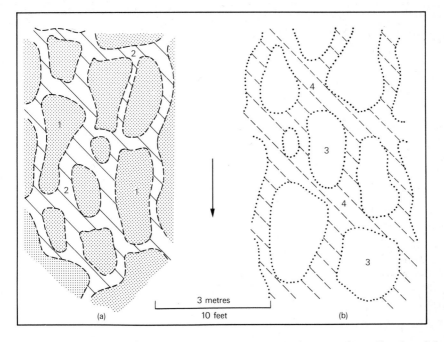

Fig. 9.21 Patterned ground (*a*) in Greenland, (*b*) on Ben Wyvis. Arrow shows direction of slope. 1. bare ground with few stones; 2. rim of large, platy stones; 3. low mounds of grey-brown *Rhacomitrium* moss growing in almost stone-free ground; 4. shallow hollows of green *Rhacomitrium* moss underlain by platy angular stones. (Source: Galloway, 1961*a*)

terraces, soil lobes, block fields, stone stripes and head deposits to the less readily recognisable, subtler features of indurated (compacted) layers occurring towards the base of soil profiles. The larger-scale effects of periglacial activity form noticeable geomorphological features and may also affect soils in respect of profile drainage, depth, texture and stoniness.

9.2.2 The larger-scale periglacial features

The distinctive solifluction features − terraces, lobes and block fields − have been studied in detail on Ben Wyvis (eastern Ross and Cromarty) by Galloway (1961*a*). Stone polygons are found here together with solifluction sheets on more or less flat or gently sloping sites at around 3 000 ft (1 000 m) (Fig. 9.21). On moderate slopes soil lobes with turf banks and terraces occur together with block fields whereas on steeper slopes (over 20°) large stone-banked lobes occur (Fig. 9.22).

Attention is drawn by Fitzpatrick (1958) to the widespread occurrence of head deposits on a rather more extensive scale than those features mentioned above. Head deposits have had a considerable effect on soil formation in Scotland because they constitute a large proportion of the parent materials from which soils have developed (glacial drift being the other major constituent). The distinction between head deposits and glacial drift may be difficult to make in some areas because the head deposits may

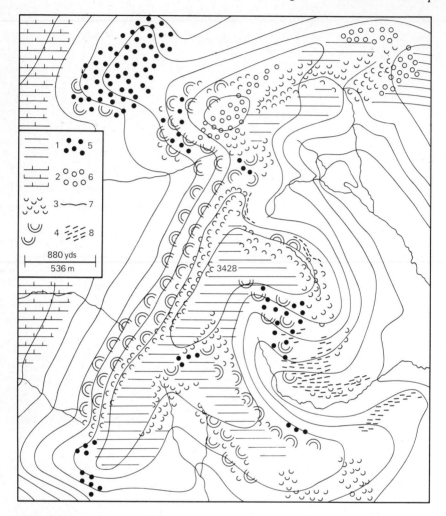

Fig. 9.22 Solifluction phenomena on Ben Wyvis. 1. uniform solifluction sheets on summit plateau (partially active); 2. uniform solifluction sheets at low-levels buried under peat (inactive); 3. turf-banked lobes (active). 4. stone-banked lobes (inactive); 5. block fields and stone streams (mostly active); 6. ring and stripe patterns in *Rhacomitrium*; 7. continuous turf-banked terrace (inactive); 8. discontinuous turf-banked terraces (active). (Source: Galloway, 1961a)

be those produced by periglacial process alone (frost shattering and solifluction) or they may be glacial drift which has been reworked by subsequent periglacial processes after the retreat of an ice cover.

Periglacial deposits are particularly common soil parent materials in the southern uplands and the eastern Highlands of Scotland. These areas lie outside the areas of the Perth readvance (Fig. 9.23) and were subject to intense periglacial activity while northern and western Scotland were still covered by ice. The products of frost shattering and solifluction mantle hillsides and footslopes, often up to depths of 65–100 ft

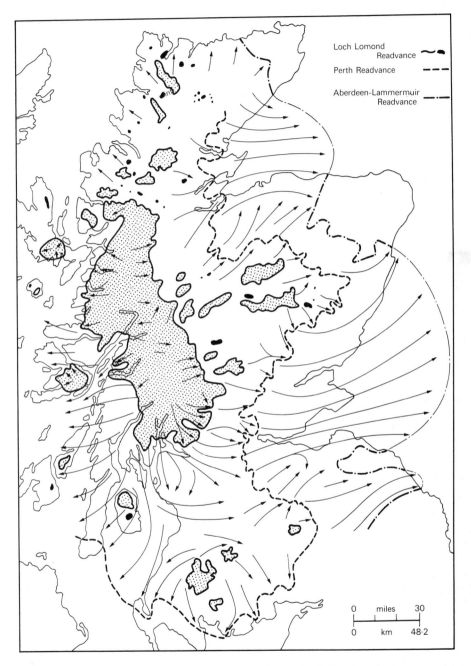

Fig. 9.23 Successive limits of the last ice sheet and associated directions of ice movement in Scotland. (Source: Sissons, 1967)

(20—30 m). The processes giving rise to the deposits often imparted properties to the deposit which affected subsequent soil development. Chief among these periglacial properties is a characteristic texture. Frost shattering is most effective in jointed rocks and continues to break the rocks up into granules. However, once the material has reached sand size (about 0.5—0.02 mm diameter) then the action is ineffective as further mechanical breakdown cannot occur. Thus these deposits are characteristically coarse textured, unless the rock material is easily weathered by subsequent chemical processes to form clays. Such coarse textures can be easily leached, encouraging the formation of acid and podzolised soils.

A very extensive soil series developed on soliflucted and frost shattered material is the *Linhope* series of the *Ettrick* association mapped west of Kelso by Ragg (1960). The parent material is frost shattered and soliflucted greywacke (sandstone composed of coarse, angular unsorted fragments) and shale. The material has a high stone content and is found on steep slopes and on hill tops. The soil textures are normally medium loams which are freely drained and strongly leached. They are, therefore, of low base status and typically show 3.5 per cent base saturation and pH 4.8 in the B horizon (Ragg, 1960). A profile description is given in Table 9.3 and the type of land-forms where the *Linhope* series occurs is seen where the smoothly contoured slopes characteristic of the southern uplands have been subject to considerable periglacial activity. On gentler slopes the *Linhope* series gives way to the *Minchmoor* series, an iron podzol in which figures for base saturation in the B horizon range from 1—2 per cent and pH values from 4.3 to 4.8 (Ragg, 1960). These soils may suffer from land-use limitations due to high stone content particularly on the steeper slopes, but on the gentler slopes the main limitation is that of fertility. The soils are typically deficient in lime and phosphate. When fertilised the soils provide good grazing for sheep and on the less stony parts of the *Linhope* series some arable rotation can be practised such as ley pasture alternating with oats (see also pp. 143—4).

There has been little extensive mapping of high mountain soils, such as those studied on Ben Wyvis by Galloway (1961*a*). This is chiefly because the Soil Survey of Scotland has confined its attention to the agriculturally important areas. However, Fitzpatrick (1964) has identified a group of soils which he terms the Mountain Tundra soils. These are high mountain soils composed of angular fragments of rock with little or no fine grained material, where no horizons can be distinguished except where a thin, poorly developed A horizon is present. Characteristic tundra structures such as stone circles, polygons and stripes together with small solifluction terraces with their step faces bound by vegetation are commonly found in these areas. At very high altitudes, over 5 000 ft (900 m), Ragg (1973) has observed thin, humus-ranker soils and mountain podzols as well as mountain tundra and rock-dominant soils. Such areas can also be found on the summits of the Cairngorm Mountains and the Grampian Mountains of the central Highlands of Scotland.

9.2.3 Small-scale periglacial features within the soil profile

While head deposits, terraces and polygons may be easily recognisable, and their effects upon soil formation identified with some certainty, closer study of soil profiles

Table 9.3 (Source: Ragg, 1960)

LINHOPE SERIES

		Profile description
Slope		Steep
Aspect		South
Altitude		1 000 ft (305 m)
Annual precipitation		35 in (890 mm)
Vegetation		*Pteridium aquilinum, Agrostis* spp.
Drainage class		Free
Horizon	*Depth* in (cm)	
L	½–0 (1–0)	Litter
A	0–7 (0–18)	Dark brown (7.5YR3/2) loam; medium crumb structure; very friable; low organic matter; stony; roots abundant; no mottling; gradual change into
B₂	7–18 (18–46)	Strong brown (7.5YR5/6) loam; fine crumb structure; very friable; low organic content; stony; roots frequent; no mottling; clear change into
B₃	18–27 (46–66)	Yellowish-brown (10YR5/6) loam; medium crumb structure; weakly indurated; friable; no organic matter; stony; roots occasional; no mottling; clear change into
C	27–40 (66–102)	Brown (10YR5/3) loam; very stony; no organic matter; roots rare; no mottling; sharp change into
D	40+ (102+)	Shattered greywacke with trace of C horizon material in interstices

SKELETAL SOILS

Slope		Very steep
Aspect		South
Altitude		1 000 ft (305 m)
Annual precipitation		30 in (760 mm)
Vegetation		*Agrostis tenuis, Festuca ovina* and *Deschampsia flexuosa*
Horizon	*Depth* in (cm)	
L	Trace	Litter
A₁	0–3 (0–8)	Brown (7.5YR4/3) gritty loam forming interstitial matter between stones
C	3+ (8+)	Shattered greywacke

may reveal features such as indurated horizons and fossil frost wedges. The indurated horizons normally occur towards the base of profiles (Fig. 9.24), and though Fitzpatrick (1956*b*) considers these to be due to permafrost only, it may also be postulated that some downward translocation of clay since the Pleistocene has contributed to their formation (Ragg, 1973). Fitzpatrick suggests that induration is produced by the freezing of a puddled mass of soil. Thus the lower surface of the indurated layer represents the lower limit of summer thawing with platy structures being produced by the formation and re-melting of small horizontal ice lenses. He supports his argument

Fig. 9.24 A soil profile showing the sharp line of demarcation between the surface soil and the indurated layer. The marked difference in colour is also indicated. (Source: Fitzpatrick, 1956*b*)

by reference to observations in present-day permafrost areas in Spitsbergen and photographs (Fig. 9.25) of the platy layers from Scottish indurated soils bear a close resemblance to photographs (Fig. 9.26) of platy structures in partially thawed permafrost blocks from Spitsbergen.

The occurrence of indurated horizons is widespread in Scotland and Glentworth (1954) has observed them in brown earth, podzol, peaty gley podzol, and brown podzolic profiles around Banff, Huntley and Turiff in north-east Scotland. They mostly occur in the area outside the Perth readvance shown on Fig. 9.23 (p. 181) and their effects upon soil conditions and plant growth may be marked. The platiness and compactness of the layers often leads to poor drainage and gleying above the indurated layers or to the deposition of iron and other solutes that have been washed down the profile until the indurated layer is reached. Root penetration of the indurated layer is

Fig. 9.25 Indurated layer showing marked development of the platy structure. (Source: Fitzpatrick, 1956*b*)

very uncommon. Thus, the shallow rooting of many conifers in Scotland is thought in many places to be due to the presence of indurated layers with the result that the trees are liable to 'wind throw' in violent gales.

Indurated horizons are found on a wide range of parent materials; for example, Mitchell and Jarvis (1956) found an indurated horizon at 9–21 in (23–53 cm) depth in the *Knockendon* soil series mapped west of Kilmarnock developed on sandy loam till derived from Old Red Sandstone with lavas. The coarse texture of these soils leads to free drainage and leaching in the upper parts of the profile but there is some slight mottling just above the indurated layer indicative of impeded drainage at that depth. Soils of the *Insch* association developed on till derived from base-rich igneous rocks also show strong induration at 16–24 in (40–61 cm) depth (Glentworth, 1954). Another example of an indurated soil in north-east Scotland is the *Strichen* association developed on till derived from schists. The principal agricultural limitations these

Fig. 9.26 A partially thawed block of permafrost showing a marked retention of the platy structure. (Source: Fitzpatrick, 1956*b*)

indurated horizons may impose are those of poor drainage and root restriction. If the compacted layers are near the surface they may, however, be broken up by subsoiling.

Locally other soil profile features which may be found consist of fossil frost wedges (Galloway, 1961*b*). These may appear as alignments of the soil particles extending in a vertical direction extending for up to 12–20 in (30–50 cm) in the soil profile. They are usually associated with tundra polygons (Fig. 9.27).

9.2.4 Present-day periglacial activity

While it is clear that periglacial conditions during the Pleistocene have left a legacy of features which have subsequently influenced the development of soils it is also true that periglacial activity occurs at the present day in high mountain areas. That this is so was confirmed in the fascinating, but simple, work of Miller *et al.* (1954). In October 1951 an area of about 3 ft (1 m) square of stone stripes present near the summit of Tinto Hill in Lanarkshire at 2 335 ft (710 m) were dug over, the stone stripes obliterated and the area trampled. The stripes were dug up to a depth of about 1 ft (30 cm). In March 1953 the area was revisited and it was found that perfect striping had re-established itself. This demonstrates that the processes leading to the formation of stone stripes on Tinto Hill are active at the present day (Fig. 9.28).

Similarly on the steep slopes of hills in the southern uplands observations may be made of large boulders with 'turf rolls' on their downslope side. These boulders are moved gradually down hill as a result of freezing and thawing of the soil beneath them.

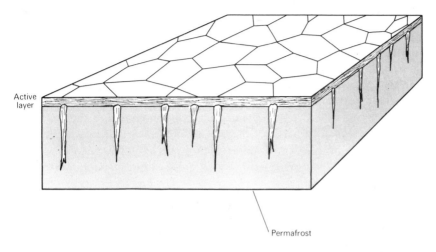

Active
layer

Permafrost

Fig. 9.27 Block diagram of tundra polygons and their associated frost wedges. (Source: Fitzpatrick, 1958)

Fig. 9.28 Stone stripes, Tinto Hill. (Source: Miller, Common and Galloway, 1954)

As they move the soil and vegetation in front of them is pushed up into a small roll and a scar of bare earth is left behind the boulders. This movement takes place because the stone is a better conductor of heat than the turf and so there are more freeze—thaw cycles under the stone than in the turf layers around. The stone slides under its own weight when the soil becomes semi-plastic upon melting.

Contemporary periglacial processes may well be taking place in stone polygons, especially at higher altitudes. It is clear that on the higher ground of the Grampians and the Cairngorms that disturbance of the ground by freezing and thawing is a very dominant physical process. Active formation of polygons and other features has been noted by Simpson (1932) and Watt and Jones (1950). Fitzpatrick (1958) asserts that present-day formation of head is taking place on Beinne Eighe in the western Highlands. Here the quartzite capping of Torridonian sandstone is being broken up by frost action with the resultant formation of a talus slope of comminuted rock debris and skeletal soils (Table 9.3, p. 183).

The significance of this contemporary periglacial action in respect of pedogenesis is that it militates against the formation of mature soil horizons. Frost action, freeze—thaw and solifluction all tend to mix the soil physically, especially by overturning layers when the soil is moving down slope. If drastic physical mixing occurs this tends to prevent chemical reorganisation from taking place. Thus leaching of soluble constituents from an A to a B horizon may be offset by the fact that under severe periglacial movement the B horizon may be inverted and rolled over the A horizon. Although this extreme example is not of frequent occurrence it serves to illustrate the point that in soils where periglacial activity is pronounced there is likely to be little or no horizonation. Such horizonation that does occur is mainly in the form of an Ao horizon, but even this is limited in a mountain environment. Even where climate and exposure do not hinder plant growth and organic accumulation the vegetation mats that form may be disturbed by frost heaving.

The relationship between soil formation and periglacial activity varies according to the severity of the periglacial action. Under severe conditions rock debris and angular stones are formed and active patterning is common. In slightly less severe situations away from mountain summits humus may accumulate from slow-growing arctic—alpine plants leading to the formation of a ranker soil. At lower altitudes, below about 3 000—2 000 ft (900—600 m), the aspect of the slope becomes important, there being more freeze—thaw cycles on south-facing than north-facing slopes which may be under permanent snow for much of the winter. Soil churning on the south-facing slopes leaves the soils loose, soft and structureless often to a depth of 24 in (60 cm) (Ragg, 1973). The vegetation surface is disturbed, forming terracettes on slopes of more than 50° although terracettes are most marked on slopes of 10—25° (Ragg, 1973).

FURTHER READING

Washburn, A. L. (1973). *Periglacial Processes and Environments*, Edward Arnold.

10 SOILS OF THE CHALK DOWNLAND

§10.1 THE EXTENT OF CHALK IN BRITAIN

Chalk outcrops at the surface over a large area of southern Britain, extending from the Dorset Downs through central-southern England to the Wash and into Lincolnshire (Fig. 10.1). North of Lincolnshire and in much of East Anglia it is largely obscured by glacial deposits. Chalk was laid down under marine conditions during the Cretaceous period and is composed of mainly porous sediments containing a very high proportion of calcium carbonate.

Large areas of the chalk outcrop in southern Britain are overlain by shallow superficial deposits of uncertain age, including Plateau Drift and Clay-with-flints. The soils occurring over these various deposits, together with those over the chalk itself, are described in this chapter. Although, as will be explained later in more detail (§10.5), the genesis of the superficial deposits is uncertain, they can generally be distinguished from the 'Chalky Boulder clay' common in East Anglia and the east Midlands. Soils over this latter material are therefore described separately in Chapter 7.

§10.2 CHALK AS A SOIL-FORMING MATERIAL

Although there are few detailed studies it seems likely that chalk weathers in much the same way as the harder limestones (see Chapter 6). In the case of *in situ* soil development the main determinants of the availability of weathered material are the amount of insoluble (non-calcareous) residue, the relative resistance of the rock and, finally, the topography at the site of formation.

Chalk in southern England may be up to 984 ft (300 m) in thickness and there is considerable vertical variation in hardness and purity. Because of the scale of this variation the main geological divisions of Lower, Middle and Upper Chalk can serve only as a general framework of reference for the pedologist. Lower Chalk usually forms 'Vale' country and has a distinct set of soils, whereas Middle and Upper Chalk form a distinct upland cuesta with its own very different association of soils.

Fig. 10.1 Distribution of solid Chalk outcrop in England. Note: This map merely shows the areas of chalk base material and takes no account of the various superficial deposits.

Lower Chalk. This overlies the Upper Greensand, and is, at its base, a soft glauconitic marl. Above this lies a highly calcareous material, known as Chalk Marl, which is the predominant formation. This often contains harder beds of compact chalk rock and, north and east of Berkshire, is overlain by another hard bed — the Totternhoe Stone. In Kent, the Lower Chalk forms the base of the Chalk scarp. In the Chilterns the Lower Chalk does not generally outcrop on the surface north and east of Hertfordshire. Further west it occurs along the main line of the Chalk scarp in Wiltshire, Hampshire and Dorset, and at the foot of the North and South Downs escarpments.

Middle Chalk. This is generally harder and more resistant that the Lower and Upper

Chalk and is often nodular. It is present throughout the chalk areas and may be up to 328 ft (100 m) thick, generally forming the Chalk scarp and adjacent upland.

Upper Chalk. This is usually softer than the Middle Chalk but the base consists of a hard band known as Chalk Rock, which is hard limestone containing green grains of glauconite. The Upper Chalk also commonly contains bands of flint — hard, silicic material usually found in nodular lumps. Flint is lacking from the Lower Chalk although found in parts of the Middle Chalk in some areas. The Upper Chalk may be up to 100 m thick.

Although there is considerable vertical variation in hardness, lateral variation in each of the stratigraphic levels is limited. This is presumably a result of the extensive nature of the depositional processes acting at the time of formation.

The majority of chalk (normal White Chalk) is permeable very pure limestone, usually containing more than 97 per cent calcium carbonate. On the other hand, Chalk Marl may contain over 40 per cent of non-calcareous matter. The non-carbonate residues in both White Chalk and Chalk Marl are mostly micaceous and montmorillonitic clay with small amounts of calcium phosphate (as collophane) and small rounded grains of quartz and other minerals. Flint, as found in the Upper Chalk, is a mesh-like aggregate of colloidal (opaline) silica and microcrystalline (chalcedonic) quartz. Flint is exceedingly durable and contributes angular fragments or water-worn pebbles to most of the succeeding formations. Flints are attacked very slowly by carbonated water, which results in preferential removal of the relatively soluble opaline component, leaving a porous whitish rind or patina (Hodgson, 1967).

The important features of chalk as a soil parent material can be summarised:

(*a*) The structure of chalk permits water to drain rapidly away through the small fissures in the rock. Chalk Marl is less permeable than White Chalk but, nevertheless, does not seriously impede drainage. Any indication of drainage impedance in the soil profile is therefore indicative of the inclusion of drift material, or, possibly, in scarp-foot sites, ground-water seepage.

(*b*) The high proportion of calcium carbonate means that soils derived solely from weathering of chalk are initially extremely calcareous. However, in humid climates, such as exist in Britain, carbonates are gradually leached out, leaving a non-calcareous clayey residue. The amount of residue present will vary according to the purity of the parent rock and the length of time it has undergone dissolution.

§10.3 THE CHALK LANDSCAPE

The general rolling topography of the Chalk Downlands is relatively well known. In particular the origin and morphology of the dry valleys has caused considerable speculation (see *inter alia* Morgan, 1971). Slopes tend to be steepest on the escarpment itself (usually over Middle Chalk), where they are often in excess of 10°. However, even on these slopes, it is unlikely that there is very much net movement of soil material downslope, except where affected by intense rabbit-burrowing or by sheep-tracks.

The chalk landscape as we know it today developed mainly during the Pleistocene period in which, firstly, overlying Tertiary sediments were stripped off to reveal the pre-Tertiary (sub-Eocene) surface. Secondly, the vast network of dry valleys was formed by solifluction and melt-water activity in glacial periods. In inter-glacial periods they were probably dry, as they are today. However, some may have contained springs during periods when the sea level was higher than at present.

In late Glacial times the Chalk Uplands were covered by a spread of loess which has now weathered to form the non-calcareous fraction of most soils over Upper and Middle Chalk (Perrin *et al.*, 1974).

§10.4 SOILS OVER CHALK ROCK

Some of the first detailed descriptions of chalk soils were those of Kay (1934). She recognised that in the Berkshire Downs cultivation appeared to be responsible for the greyish colours of the *Upton* series, a shallow, extremely calcareous, rendzina. (It will be recalled from Chapter 2 that the principal characteristics of the rendzina group of soils is that they usually consist of only an A/C horizon and that they are calcimorphic.) *Upton* series soils were considerably lighter in colour, more calcareous, but in other respects similar to the more extensive shallow rendzina called the *Icknield* series. This latter soil had a higher organic matter content (hence darker colours) and was most common under old permanent pasture. Later Kay (1940) recognised a brown, extremely calcareous cultivated soil called the *Andover* series.

The situation that exists on the Chalk Downs raises the problem of how to classify soils that have been altered by agricultural practice and whether or not to classify them in a different group from those remaining unaltered. Kay (1934) took the view that in the case of chalk soils there was a meaningful separation to be made, and the Soil Survey take the same view today. However, it is difficult to work out the best criteria on which to base such separations, especially as soil changes may have taken place over hundreds of years.

The attempted solution of this correlation problem has, as usual, been complicated by the fact that chalk soils have been examined in detailed profiles in widely-spaced areas and it is only recently that detailed comparisons between areas have been undertaken. Nevertheless, in this particular case confusion has been compounded by comparatively recent classification of similar profiles into separate soil series.

Hodgson (1967, p. 46) explains the classification existing in the mid-1930s:

The [*Icknield*] series as originally defined (Kay, 1934) comprised very shallow, very dark coloured, humose soils under old grassland, whereas shallow pale coloured cultivated soils on Middle and Upper Chalk were classed separately as the *Upton* series. Avery (1964) abandoned this separation as being dependent on land-use but distinguished a normal (rolling) phase of the series with average slopes of 11° or less and a steepland phase on steeper slopes. A similar separation was made in this area [west Sussex].

Surveys in the Cambridge (Hodge and Seale, 1966) and Reading (R. A. Jarvis, 1968) areas followed the same general arrangement, although in neither of these two cases are phases distinguished on the published map.

By 1972 the name Upton had been resurrected to describe soils with lighter colours which seem to result from prolonged arable activity (see Green and Fordham, 1973). However, the situation was further confused in 1973 where the Wantage memoir described Upton and Icknield soils as phases within the *Icknield* series (M. G. Jarvis, 1973). In the same year Green and Fordham (1973) had described *Upton* series as the main representative in the Ashford area, although *Icknield* series was locally present on some steep slopes.

The name *Andover* (used by Kay in 1940) was applied by Green and Fordham (1973) to soils that are somewhat deeper than the *Icknield* series and have appreciable amounts of drift giving the soil a slightly browner (10YR4/2) colour. This name has a similarly confusing background. It was first used by Kay in various unpublished pieces of work relating to Hampshire, but later Avery (1964) incorporated it within the *Icknield* map unit as a variant and was followed in this manner by Hodgson (1967). By 1972 this name also had been resurrected and detailed descriptions are given both by M. G. Jarvis (1973) and by Green and Fordham (1973). Table 10.1 summarises the use of the names *Andover, Icknield* and *Upton.*

There are two main causes of the confusion that exists with these particular soils. Firstly, soils over chalk seem to exhibit relatively minor variations within local areas, but when viewed on a national basis these apparently small variations can be seen as indications of large-scale trends. It would indeed have been helpful if it had been possible to reclassify all known chalk soil profiles after each new one had been examined! The second cause of confusion is that until 1970 the term 'soil series' was often being used synonymously with 'soil mapping unit', at least for the purpose of description. (Further discussion of soil-mapping problems and the use of these terms is given in Chapter 2.) For example, Hodgson (1967) in his description of west Sussex soils does not use the term 'mapping unit', although he notes that of a total of twenty-four soil series described in the area only nineteen are separately delineated and six are combined in various complexes. In 1973 *Icknield* and *Andover* series profiles are included within the *Icknield* mapping unit in the Wantage area (M. G. Jarvis, 1973), while in Kent, *Andover* and *Upton* series profiles are included within the '*Andover–Upton*' Map Unit. (The term 'map unit' is taken to be synonymous with 'mapping unit', although the use of this plurality of terms does not itself assist clarity!)

The classification of these chalk soils has been greatly aided by criteria introduced by Avery (1973). Using Avery's criterion that chalky material cannot give rise to B horizon soil, Cope (1976) describes the classification of chalk soils in the Wilton area of Wiltshire:

On gently sloping benches . . . there is . . . sufficient silty drift for it to form much of the parent material of shallow soils over Upper Chalk, giving areas of *Icknield,*

Table 10.1 Development of use of names for rendzina soils over chalk

Name:	Andover	Icknield	Upton
Variation in colour:	*Brown or dark greyish-brown over brown*	*Grey or black*	*Dark greyish-brown or paler*
First used:	Kay (1940)	Kay (1934)	Kay (1934)
Use by Avery (1964), subsequently followed by Hodge and Seale (1966), Hodgson (1967), R. A. Jarvis (1968):		Icknield, steepland phase → Icknield, brown variant → Icknield, normal phase	
Use by M. G. Jarvis (1973):	Brown variant called Andover	'Steep' and 'normal' remain Icknield	
Use by Green and Fordham (1973):	As M. G. Jarvis	Icknield present locally	'Normal' phase becomes Upton
Use by Fordham and Green (1973):	As M. G. Jarvis	Not present	Not present
Use by Cope (1976):	As M. G. Jarvis	Icknield present	Upton present
Classificatory subgroup, after Avery (1973):	Brown rendzina	Humic rendzina	Grey rendzina

flinty fine silty phase, and *Andover* soils. *Icknield* soils occupy rolling ground once under downland but now ploughed-up at various times since the 1940's, and also steep uncultivated dry valley sides. Rolling upland tilled for a somewhat longer period is under *Andover* soils which have lost sufficient organic matter for the dark brown colour of their mineral soil to be dominant. Moderately sloping middle valley sides, which have been cultivated for centuries, are characterised by *Upton, flinty phase*, soils with a pale (brown) coloured chalky fine earth.

To summarise, British soil surveyors have run into very difficult problems of correlation with soils over Chalk Rock. Because the drift overlying the chalk itself exhibits considerable variation, we might therefore expect that the classification of soils developed in this drift would prove even more complex. However, before examining these drift soils we will look at soils on softer chalk material — Chalk Marl.

§10.5 SOILS OVER CHALK MARL

Where the more argillaceous beds of the Lower Chalk are exposed at the surface a dark greyish-brown to grey rendzina — the *Wantage* series — has been mapped. These soils occur near the base of the Chalk escarpment over stiff marl and marly chalk in the Buckinghamshire Chilterns (Avery, 1964) and along a broad belt at the foot of the main escarpment in the Abingdon and Wantage areas (M. G. Jarvis, 1973). Depth of soil is very variable and in some of the valleys of the Chalk escarpment there is often rather deeper marly chalky drift on the valley floor. Soils over this deeper material have been separately mapped in the Wantage area as *Gore* series, which is generally similar to the *Wantage* series, but deeper than 15.7 in (40 cm).

In the North Downs near Ashford in Kent only the deeper soils have been found to be extensive. *Gore* series has been mapped, although shallower soils, very similar to the *Wantage* series, are noted as having a very restricted distribution at eroded sites (Green and Fordham, 1973).

§10.6 SUPERFICIAL DEPOSITS OVER CHALK IN SOUTHERN ENGLAND

In southern England shallow superficial deposits cover the chalk in many areas. At the present time their nature and distribution is not fully appreciated nor is there a full understanding of their origin, genesis and detailed structure. What is becoming clearer is that the deposits must have had a profound effect on the development of the landscape as we know it today (see *inter alia* Hodgson *et al.*, 1974). Quite clearly a full knowledge of the origin of the superficial deposits is crucial to an understanding of the soil profiles that have been developed in them and until we have this understanding classification and explanation of the soils is bound to be provisional.

Having made that proviso we can proceed to divide the deposits into four main soil parent material groups:

1. Clay-with-flints. This is characteristically a yellowish-red tenacious clay with large unworn and often unbroken flints. In south-central and south-east England it forms a soil parent material on upland slopes and plateaus. In the Chilterns and northwards it is often obscured on plateau sites and usually forms a soil parent material only on slopes. In the North Downs also Clay-with-flints occurs as a soil-forming material more commonly on slopes.

The origin of Clay-with-flints, as already noted, is not known with certainty (see *inter alia* Avery, 1958; Hodgson, 1967; Pepper, 1973), but it is thought to result from overlying Tertiary material, or drift, or a combination of the two, permitting free percolation of rain water. Most of the clay would, in this case, derive from the overlying material and not from the chalk itself. (The thickness of Clay-with-flints appears to deny the possibility that they are derived solely from the chalk. As early as 1906 Jukes-Brown pointed out that the dissolution of about 328 ft (100 m) of Middle and Upper Chalk would be needed to produce 3 ft (1 m) of Clay-with-flints, bearing in mind the purity of chalk.)

In the Salisbury area, Cope (1976) shows that the presence of two layers of *in situ* Clay-with-flints is probably a reflection of a two-stage origin. He suggests that the

Upper Clay-with-flints, which is a very flinty yellowish-red clay varying in thickness from 24—59 in (60—150 cm), has mainly formed from the weathered remains of Eocene deposits. Basal Clay-with-flints (which is thinner and contains no flints) was formed by translocation of clay into pores provided by chalk dissolution, and thus contains clay derived both from the upper horizons and from the chalk.

2. Plateau Drift.   This is the term applied by Loveday (1962) to the thicker hetero-geneous accumulations which occur over the Clay-with-flints. Plateau Drift is mostly yellow-red in colour and formed of more or less stony clay which often has greyish coatings around the stones. The deposits may be up to 50 ft (15 m) in thickness and may contain appreciable amounts of sand, either dispersed through the material or as a lens or wedge.

   While most of the Plateau Drift probably derives from remnants of Eocene beds — which are thought to have covered large areas of the chalk — deposits on the lower part of the dipslope of the Chilterns and more northerly chalk hills contain small Triassic pebbles and other glacial erratics, and are therefore thought to be glacial deposits. It is clear that the Plateau Drift is extremely heterogeneous and Gallois (1965) suggested that Plateau Drift should be regarded as the weathered remains of chalk, Eocene deposits and later sediments all more or less mixed or rearranged under the influence of local ice caps or snow-fields. The term 'Plateau Drift' is most com-monly applied to deposits on the Chilterns and southwards — north of this area drift deposits are altogether more extensive and various different terms are used (see Chapters 7 and 8).

3. Aeolian material.   These soils in blown sand over chalk in the Breckland area of East Anglia are described in Chapter 9. Over much of the North and South Downs the most extensive soil-forming material is known as Brickearth. This is a firm porous unbedded silty loam deposit of resorted loess originally laid down as part of the general European cover deposited during the Pleistocene era (Perrin *et al.*, 1974). Though in the past not often recorded as a soil parent material it would appear that nearly all the Chalk Upland is covered by a significant proportion of loessial material (D. W. Cope, pers. comm.). However, this latter material is not Brickearth in the commonly accepted sense.

4. Coombe Deposits.   These are flinty, solifluction or melt-water deposits laid down in the cold Pleistocene climate. Where the deposits have become cemented by secon-dary calcium concretions they are known by the name 'Coombe Rock'. Usually Coombe Deposits are formed of compact gravel containing flints and lumps of chalk in a matrix of finely divided chalk and silty material (called 'Chalky Head' by the Soil Survey). They may be over 50 ft (15 m) thick, being thickest in valley floors where they often overlie other head deposits or indeed buried soils. Evidence from the latter has suggested that most superficial Coombe Deposits are of late glacial age. Melting snow-fields and ice would cause weathered chalk and other material to move down-slope either to cover valley floors or, in the case of an escarpment, to form alluvial fans in the vale below. Such alluvial fans are, for example, found in the Wantage area of the northern escarpment of the Berkshire Downs (M. G. Jarvis, 1973). Coombe Deposits are found throughout the chalklands of southern England.

In addition to these four main groups of deposits over chalk there are various other materials which lie in more or less shallow layers on top of the chalk or on top of the superficial deposits. Here the chalk (or the material immediately above the chalk, such as Clay-with-flints) is still at a sufficiently shallow depth to exert an influence on the soil profile. For example, Eocene deposits overlie the chalk throughout the London Basin and at the edge of the Basin the deposits lens out to form a very shallow superficial covering. Soil profiles exist that penetrate through the Eocene deposits (predominantly sands and gravels) to the material immediately overlying the chalk — for example, Clay-with-flints — and sometimes to the chalk itself. Other examples are the various river-terrace deposits which also cover the chalk and drift material.

§10.7 SOILS ON SUPERFICIAL DEPOSITS OVER CHALK

The previous section has explained how variable the deposits over chalk are and it has already been suggested that classification of the soils derived from these deposits is very difficult. It is perhaps best to consider these soils in the context of areas where detailed surveys have been carried out and such description follows in §10.8 and §10.9. However, there are some soil types which are more or less widely distributed over the deposits and some comments can be made about these. It must be stressed that these remarks are highly generalised and give little recognition of the local variation.

Shallow drift over Clay-with-flints has given rise to *Winchester* soils which have been very widely mapped throughout the chalklands. These brown earths (*sols lessivés*) are about 24—27 in (60—70 cm) deep and are usually dark greyish-brown to brown flinty loam to silty clay in the surface horizons with a yellowish-red clayey Bt horizon at depths varying from 7—12 in (20—30 cm). Chalk lies beneath at a depth of less than 6 ft (2 m). In the Chilterns they occur on ridge crests and on upper slopes and spurs (Avery, 1964) while in the Berkshire Downs they are more common on plateau sites (M. G. Jarvis, 1973). The depth of these soils and the variation in texture through the profile suggests that drift material other than Clay-with-flints is partly involved in their genesis (although vertical texture variation is, of course, also related to clay translocation in the profile). The soils have been mapped in Sussex (Hodgson, 1967) and in Kent in shallow Brickearth drift over Clay-with-flints (Fordham and Green, 1973; Green and Fordham, 1973).

Winchester soils are principally distinguished from the *Charity* series (another brown earth (*sol lessivé*)) by the more silty texture of the latter. The non-calcareous silty texture of the *Charity* soils results from the weathering of loess in the Coombe Deposits from which they are derived. Broadly similar in other respects to the *Winchester* soils, *Charity* series has also been mapped in many parts of the Chalklands in southern Britain. It is found not only on plateau sites but more often in shallow valleys and also (in the west Sussex area at least) on the coastal plain (Hodgson, 1967).

In this particular area it occurs on the broad gently sloping Upper Coastal Plain and in the broad Downland valleys. In Kent it is found on gently sloping valley floors (Green and Fordham, 1973).

Where the A and Eb horizons of silt loam are more than 10 in (25 cm) thick *Rewell* series soils (brown earths otherwise similar to *Charity* series) are mapped. These soils have been shown to be extensive on the southern slopes of the South Downs (Hodgson, 1967) but of limited extent elsewhere. The thickness of the silt loam is presumed to be due to the loessial addition of Brickearth material.

Where the Coombe Deposits are shallow and contain enough chalk material to enable the soils to be placed in the brown calcareous group, *Coombe* series has been mapped. This is a silty clay loam seldom more than about 20 in (50 cm) deep, is highly calcareous throughout and may contain quantities of flint. In the Chilterns it has been mapped on the upper slopes of valley sides (Avery, 1964) while in Berkshire it occurs in restricted areas on the plateau (M. G. Jarvis, 1973). In Sussex and Kent it is generally found in valley floor sites (Hodgson, 1967 and Green and Fordham, 1973, respectively).

Whereas Clay-with-flints material is not normally sufficiently thick to cause significant drainage impedance, Plateau Drift, on the other hand, usually gives rise to gleyed brown earth soils. One such — the *Batcombe* series — forms the most extensive soil on much of the Chiltern dip-slope. *Batcombe* soils are yellow-brown, friable and normally flinty silty clay loam with, below 30 in (75 cm), a horizon of mottled reddish or yellow clay in Plateau Drift, more than 6 ft (2 m) thick. This soil is probably extensive throughout the Plateau Drift-covered chalklands of southern England, having first been mapped in Dorset (K. L. Robinson, 1948). Elsewhere it has been mapped over deposits occurring on Greensand in Devon (Harrod, 1971; Clayden, 1971) but is less prominent in the Berkshire Downs, where the more stony and clayey *Berkhamsted* series is more common (M. G. Jarvis, 1973).

Where the Plateau Drift presents no impedance to water movement, brown earth (*sol lessivé*) soils of the *St Albans* series are mapped. The Plateau Drift giving rise to these particular soils, although high level, is fluvio-glacial in origin and is usually of flint gravel with a sandy matrix; it may be up to 15 ft (5 m) thick. The A horizon of these soils is typically a greyish-brown sandy loam with a Bt horizon at about 24 in (60 cm). In Berkshire these soils occur on the Winter Hill Terraces (R. A. Jarvis, 1968) and also at scattered localities on the chalk dip-slope. Elsewhere they are not extensive but occur occasionally on the dip-slopes.

The foregoing account mentions only the major soil types so far surveyed over the chalklands. Combinations and variations of the main types have already been found during the course of the various surveys and it is certain that, as mapping proceeds, other soils will be discovered.

Important questions, the answers to which will help in understanding the soils over the superficial deposits can be summarised as follows:

(*a*) Which soils have entirely relic subsoils weathered before the last glaciation and what criteria are going to be used to define them?

(*b*) Which soil profiles show evidence of recent weathering?

(*c*) To what extent has clay translocation been a cyclic process?

These questions can perhaps best be answered by detailed micromorphological studies. When answered, our knowledge of the genesis of the superficial deposit soils will have been considerably advanced.

Two detailed examples of soil—landscape relationships over the chalklands are now given. These have been selected to show both the similarities and some of the variety that has so far been noted in soils over the chalkland.

§10.8 SOILS AND LANDSCAPE ON THE BERKSHIRE DOWNS

The Berkshire Downs near Wantage and Abingdon have been mapped in detail and given a written description (M. G. Jarvis, 1973) that incorporates recent ideas on the use of the terms 'series' and 'mapping unit'. Table 10.2 summarises the mapping units over the chalklands and shows their included series. Figures 10.2 and 10.3 show typical arrangements of soils in chalkland valleys. (Detailed notes regarding the nature of the soils themselves are not given here since they have been described in earlier sections of this chapter.)

The *Icknield* mapping unit is the most extensive on the Downland and is found mostly on the gently sloping broad ridge tops. Most of this land is under continuous arable cropping, growing cereals, though some areas are used for racehorse training gallops, where the vegetation is the typical chalkdown grassland.

Coombe mapping unit soils are less extensive and are generally found in valley bottoms and on adjacent concave slopes. These soils are also mostly under arable cropping, though some of the flinty patches are less successful.

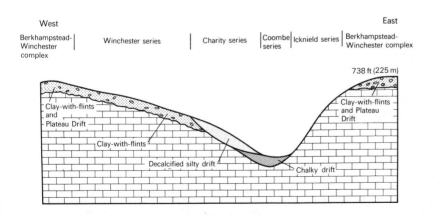

Fig. 10.2 Soils in a Chalk dry valley, with Plateau Drift capping the ridges. (After M. G. Jarvis, 1973)

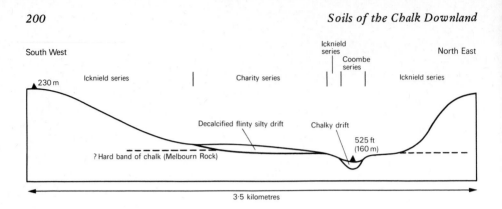

Fig 10.3 Soils in a broad Chalk dry valley. (After Jarvis, 1973)

Charity mapping unit soils cover limited areas and are either on valleyside slopes or on nearly level ground on terrace-like benches. Some of this land is used for cereal cropping or rotational grassland, though much of the area is covered by woodland. Large sarsen stones and occasional patches of acidity limit the potential of some areas. *Charity (Bouldery Phases)—Coombe* and *Charity—Coombe* mapping units cover only small areas. The former are found on valley bottoms or sides under mixed woodland and pasture: cultivation being precluded by the presence of large sarsen stones. The complexity of the *Charity—Coombe* mapping unit is illustrated in Fig 10.4. These latter soils are often under arable or rotational pasture but acute small-scale lateral pH variations make correct lime application a severe problem. These variations are caused

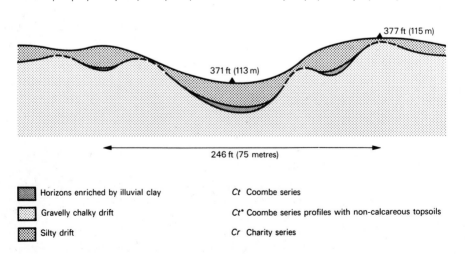

Fig. 10.4 Soils in the *Charity—Coombe* mapping unit. (After Jarvis, 1973)

Table 10.2 Mapping units on the Berkshire Downs in the Wantage area. (After M. G. Jarvis, 1973)

Mapping unit	*Included series* (dominant series is listed first)
Icknield	Icknield Icknield (Brown B variant) Andover Charity
Coombe	Coombe Andover Charity
Charity	Charity Andover Rewell
Charity (Bouldery Phases) Coombe	{ Charity (Bouldery Phases) { Coombe (Bouldery Phases) *Note*: Bouldery Phases mixed in a valley bottom association
Charity—Coombe	{ Charity { Coombe *Note*: Complexity arises from the undulating surface of the chalky drift covered with different thicknesses of non-calcareous drift (see Fig. 10.4)
Winchester	Winchester Charity Rewell
Berkhamsted—Winchester	Berkhamsted (*c.* 40%) Winchester (*c.* 15%) with small areas of eleven other series including: Batcombe Andover Charity Coombe *Note*: These soils are in Plateau Drift and include a variety of parent materials occurring in a very complex and unpredictable pattern, and enveloping small but distinct patches of Tertiary sands and gravels

by the varying depth of the underlying chalk and values may range from about 6.5 to about 8.2 in a distance of a few metres.

    Winchester mapping unit soils are found on the plateau but nearer the escarpment edge than *Berkhamsted—Winchester*. Because of their clayey profiles and natural acidity the *Winchester* unit soils have tended not to be used for arable cultivation but are more commonly under beech woodland or scrub. *Berkhamsted—Winchester* mapping unit soils are mostly on the higher dip-slope where their complexity makes them difficult to crop well. Again, much of the area is under woodland.

§10.9 SOILS AND LAND-USE ON THE SOUTH DOWNS IN WEST SUSSEX

The Chalk Downs near Worthing and Littlehampton in Sussex rise to a height of over 650 ft (200 m) OD and dip gently southwards. The main escarpment faces north and is outside the area mapped in detail by Hodgson (1967). A secondary escarpment exists within the area which divides the chalklands into two regions:

1. North of the secondary escarpment the chalk is dissected by valleys and is relatively free of drift, giving the typically rounded landscape of the South Downs with large farms, big arable fields and small valley settlements.

2. South of the secondary escarpment the valleys are wider and the broad interfluves are covered with drift deposits and are mostly wooded.

Throughout the area — in both regions — the flat valley floors have only intermittant streams in times when the water-table rises to the surface.

The block diagram (Fig. 10.5) shows the general pattern of soils over the dip-slope and valleys. *Winchester, Rewell* and *Wallop* soils are mapped over the Clay-with-flints, the first two of these soils having already been described (see p. 197). *Wallop* soils are shallow brown calcareous soils developed in a thin layer of Clay-with flints material,

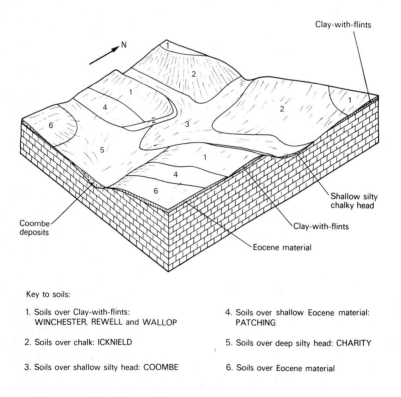

Key to soils:

1. Soils over Clay-with-flints:
   WINCHESTER, REWELL and WALLOP

2. Soils over chalk: ICKNIELD

3. Soils over shallow silty head: COOMBE

4. Soils over shallow Eocene material:
   PATCHING

5. Soils over deep silty head: CHARITY

6. Soils over Eocene material

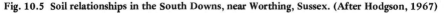

Fig. 10.5 Soil relationships in the South Downs, near Worthing, Sussex. (After Hodgson, 1967)

resting on chalk or chalk rubble at less than about 12 in (30 cm). Although most of the area covered by these three soils is under woodland of beech or conifers, cultivation of small areas takes place on *Winchester* and *Rewell* soils — but this is limited by extreme stoniness.

Shallow *Icknield* rendzina soils are present where the Chalk is not covered by the drift material. Most of these soils are today under arable cultivation or rotation pasture. However, on the valley-sides and escarpments, where the topography is too steep for cultivation, the soils support semi-natural grassland and scrubland, with some small areas of beechwoods and coniferous plantations. Under such semi-natural vegetation the soils are highly humose with an organic—carbon content of between 7 and 11 per cent.

Downslope the *Icknield* soils merge into *Coombe* series soils developed in silty chalky head deposits, which may be Coombe Deposits or, in some places, comparatively recent hill-wash. Some of the silt is calcareous and probably derived from finely divided chalk, but most is non-calcareous and loessial in origin. These soils are distinguished by their calcareous profiles and the presence of brown or yellowish-brown B horizons. Although more useful in arable cultivation than the *Icknield* soils, *Coombe* soils are occasionally limited by stoniness while horticultural potential is limited by high chalk content and high pH.

In the broad valleys and on the gentler footslopes of the degraded southern bluff of the chalk, *Charity* series soils are mapped. The parent materials of these soils contain varying proportions of:

(*a*) chalk and flints from the Upper Chalk;

(*b*) clay and flints from the Clay-with-flints;

(*c*) silty loessial material (Brickearth).

The soil is therefore rather variable, but generally the profile is flinty throughout, and almost all soils have a Bt horizon within 24 in (60 cm) of the surface. The parent material may be decalcified. The agricultural value of these soils depends largely on the degree of stoniness, with the least stony being suitable for horticulture and vegetables. The stonier soils are used for crops and grassland, though these latter have rather poor water-retention properties.

Further down the dip-slope *Patching* series soils are found. This soil (which is an imperfectly drained surface-water gley) is developed in thin flinty silty drift overlying disturbed remnants of basal Reading beds clay, which rests in turn on chalk. The composite parent material shows various evidence of disturbance under periglacial conditions, such as festooning. Silt loam textures predominate with a yellowish-brown gleyed Bt horizon being present at about 12 in (30 cm). Large areas of these soils are now under woodland although, in the past, areas of this soil together with areas of *Winchester* and *Rewell* soils on similar topography have been cultivated. Essentially the soils are too wet to provide good agricultural land and grazing stock easily cause poaching during the winter and spring. (For a further discussion of the poaching of grassland see Chapter 12.)

FURTHER READING

Until recently surprisingly little had been written about the nature and origin of the superficial deposits overlying chalk in southern England, though this probably reflects their heterogeneity and complexity. The classic work on geomorphological development in the south-east by Wooldridge and Linton (1955) is now somewhat outdated by more recent research such as that of Hodgson *et al.* (1974).

Reference to work on the soils has been given throughout the preceding text. The Memoirs of the Aylesbury and Hemel Hempstead area (Avery, 1964), the Cambridge area (Hodge and Seale, 1966) and the Reading area (R. A. Jarvis, 1968) are all moderately readable as is the more detailed Bulletin for West Sussex (Hodgson 1967). However, the Soil Records for Kent (Green and Fordham, 1973; Fordham and Green, 1973) are arranged on a more tabulated basis and are really only suitable for detailed reference. The Memoir for the Wantage and Abingdon District (M. G. Jarvis, 1973) combines both general reading and also a considerable amount of categorised information.

Very generalised surveys of Hertfordshire (Thomasson and Avery, 1970), the Luton and Bedford District (King, 1969) and the Saffron Walden District (Thomasson, 1969) cover areas of chalkland and provide a helpful introduction for those particular areas. Work is at the time of writing being presently carried out in various parts of Wiltshire. Detailed mapping has been completed for the Wilton area (Cope, 1976) and also for an area surrounding Marlow in Buckinghamshire. These and the County Survey (pre 1 April 1974 boundary) for Berkshire are due to be published in the near future.

11 THE WOLD SOILS

§11.1 JURASSIC LITHOLOGY

The previous chapter examined in detail some of the classificatory problems that have been caused partly by changes in land-use. The present chapter is concerned with broadly similar soils, which are today of equally high agricultural potential over the Jurassic limestones and clays.

The geological formations laid down in the Jurassic era outcrop at the surface in a wide band extending from north Somerset north-eastwards to Lincolnshire and Yorkshire (Fig. 11.1). As is the case with the Chalk, further east, the Jurassic limestone forms a broad cuesta (scarp and dip-slope), which dips gently south-eastwards. North of Gloucestershire, the cuesta becomes a much less prominent landscape feature and is often obscured by glacial deposits.

The Jurassic system has been the subject of several detailed geological studies (see *inter alia* Arkell, 1971), which have concentrated, to a large extent, on the correlation of the abundant marine fossils with the geological history of the system. From the point of view of soil-forming material, the Jurassic system is made up of calcareous and non-calcareous clays, and oolitic and sandy limestones, the limestones themselves containing discontinuous bands of sand and clay. Because of the nature of shallow marine deposition (with many beaches, bars and islets), many of the beds of the Jurassic era are thin and discontinuous. For example, submarine deltas were a fairly common feature of the Jurassic coasts and the nature of deltaic deposition is such that materials of different particle size will be deposited in different parts of the delta. When we come to examine the 'fossil delta' today, we find inextensive lenses of material of differing particle size. Hence, to sum up, the surface outcrop of the Jurassic formation presents a complex array of soil parent materials.

Topographically, the Jurassic system may be divided into the 'clay vale' areas and the 'limestone uplands'. However, the distinction between terrain type is by no means precise, even after making allowance for the fact that the upland will contain areas of clay covering the limestone, gradually increasing in thickness eastward, away from the scarp — forming the Forest Marble and Oxford Clay formations. The problem of trying to differentiate between the areas will be discussed further in §11.3 and §11.4, below.

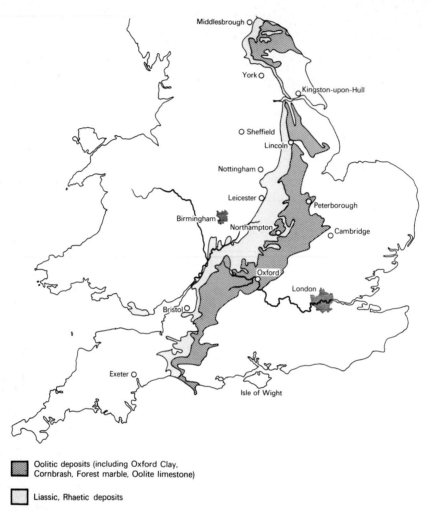

Oolitic deposits (including Oxford Clay,
Cornbrash, Forest marble, Oolite limestone)

Liassic, Rhaetic deposits

Fig. 11.1 Distribution of solid Jurassic outcrop in England. Note: This map merely shows Jurassic base materials and takes no account of the various superficial deposits.

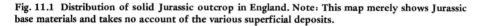

As already suggested, variation in lithology through the limestone is considerable and is perhaps best indicated by a geological section in a quarry in the north Cotswolds (Fig. 11.2). In the area of this section the general dip is about $2-3°$ south-east. Since the area is dissected by deep valleys, with slopes frequently in excess of $8°$, all the various formations shown in the diagram form soil parent materials within a small distance of the quarry. The primary parent materials therefore include hard nodular limestone, loose sandy limestone, sand, calcareous and non-calcareous clay and, on the dip-slope at least, the presence of these various materials is often unrelated to obvious breaks in slope.

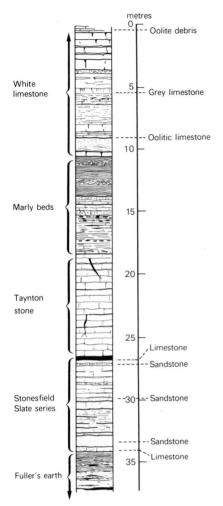

Fig. 11.2 Section of *Great Oolite* series, Hampen Railway Cutting, Glos. (After Richardson, 1929)

Looked at in a broader context there are certain variations in stratigraphic sequence between the south Cotswolds around Bath, Somerset and the Jurassic cuesta further north. Figure 11.3 is a diagrammatic representation of the Cotswold escarpment in north Somerset or south Gloucestershire. It can be seen from this that the scarp is a composite structure with harder bands of strata providing subsidiary scarp slopes within the general pattern. Although each slope and geological formation will often have a distinct soil type if left undisturbed, there are various movements which have confused this situation. In general terms these can be grouped as colluvial (downslope wash) and small-scale tectonic activity (cambering, etc.). Both these sets of processes serve to complicate the underlying geologic sequence of soil parent materials

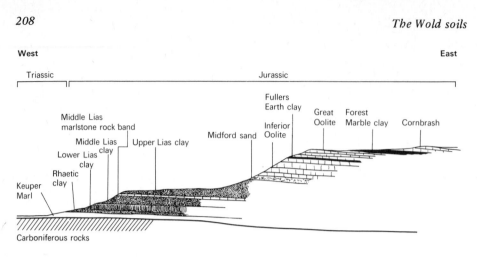

Fig. 11.3 Generalised cross-section of the Cotswold escarpment.

and are referred to in more detail in the next section. Some of the thinner formations are particularly difficult to identify as soil parent materials. Hence, while, for example, the Midford Sands form a relatively extensive soil parent material in the south Cotswolds (giving rise to *Atrim* soils), their north Cotswold counterpart — the Cotteswold Sands — is much thinner and harder to identify. In the Stow-on-the-Wold area this outcrop is less than 19 in (0.5 m) thick and is usually obscured on the scarp slope by cambered Inferior Oolite and superficial downwashed material. Nevertheless, downslope of the (usually inferred) stratigraphic position of the Cotteswold Sand, soil profiles often possess a more sandy texture. Usually, however, this difference is not sufficient to justify a separate mapping unit.

The dip-slope presents a rather different picture, although even here, with a shallow geologic dip, the various lithologies are not as extensive as might be expected. An example of the discontinuous nature of the surface outcrop of soil parent materials on the dip-slope is given in Fig. 11.4. This diagram shows the extent of the *Sherborne* mapping unit in the south Cotswolds. As is explained in §11.3 (below) *Sherborne* mapping unit contains over 70 per cent *Sherborne* series soil. This latter is a shallow calcareous brown earth and the map therefore generally indicates the distribution of limestone within 13 in (35 cm) of the surface in this area. It can be seen that apart from near the escarpment the soil is arranged in a rather complex pattern. Although this is partly a reflection of valley development it is also caused by the thin discontinuous spread of the next succeeding geologic formation above the limestone — Forest Marble Clay. Adding to the complexity is the fact that the Forest Marble itself (as its name suggests) contains a certain amount of limestone, although these are usually in thin discontinuous bands.

A further complicating factor is the extent of superficial drift covering the 'solid' geologic formation. In dry valley floors near Tetbury, Gloucestershire, on the Cotswold dip-slope, gravels of presumably fluvio-glacial origin have been noted (Findlay, pers. comm.). North of Stow-on-the-Wold glacial deposits become altogether

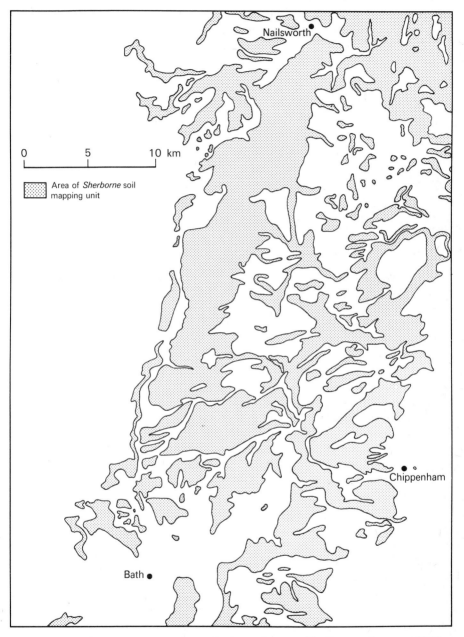

Fig. 11.4 The distribution of *Sherborne* soil in the south Cotswolds. (Source: Courtney, 1973)

more extensive on the dip-slope. It also seems likely that there is a considerable amount of loessial material included in some soil profiles, following similar conclusions for the Chalk areas, discussed in the previous chapter.

It is implicit in the foregoing discussion that there is an important relationship between soil type and parent material lithology. The detailed aspect of this has been noted by Findlay (1976, in press), who demonstrates the relationship between soil texture and type of limestone fabric:

> Oolites and oolitic-shelly rocks give rise to soils with clay contents of 35–45 per cent. Forest Marble shelly-oolitic limestones and some shelly or micritic Cornbrash is often associated with the finer textured soils with up to 55 per cent clay, but these two formations also include sandy limestones giving soils recognised . . . as . . . loamy.

These relationships will be referred to again in §11.3 and §11.5.

§11.2 GEOMORPHIC DEVELOPMENT OF THE LIMESTONE UPLAND

In the previous chapter we were rather hesitant about making generalisations regarding the effect of the Pleistocene climate on soil formation over the Chalk. As has already been suggested, it is difficult to make any conclusive statement about the amount of loessial material included in the soil profile, but there are on the other hand very clear geomorphic indications of the amount of periglacial activity that took place during the Pleistocene era. The extent and magnitude of that activity (as demonstrated by convolutions in the upper parts of the weathered rock) lead one to the tentative conclusion that most of the Jurassic cuesta was probably denuded of much superficial material during the course of the Pleistocene. Most of today's soil has therefore developed, in all probability, on a post-Pleistocene surface.

W. M. Davis wrote some of his classic geomorphological papers after visits to the Cotswolds (Davis, 1899; 1909) and the area has subsequently been studied by many other workers. They have been mainly concerned with the deeply incised river valleys (Beckinsale and Smith, 1953; Beckinsale, 1970). Dury (1960) has suggested that the major streams were formed under 'nival' (snow-melt) run-off conditions. Kellaway (1971), in common with some earlier writers, notably H. G. Dines (quoted by Richardson, 1929), has pointed to the possibility that ice overrode the Cotswolds during the Anglian glaciation. If this is so, the ice has left only very limited areas of superficial deposits on the upland plateau and dip-slope, although the deeply incised valleys and some of the vale areas contain extensive tracts of superficial material, though not necessarily of glacial origin.

There are extensive dry valley systems in the uplands, many of which are infilled with a deep locally-derived deposit, which seems to be a mixture of Pleistocene head and more recent colluvium. It has been suggested (Beckinsale, 1970) that some of the smaller vale areas, such as the Vale of Bourton, have been derived from very wide valleys which have become extended and coalesced through spring-sapping. The floors of these vales contain a variety of parent materials including substantial amounts of locally-derived limestone gravel and colluvial material. *In situ* Jurassic clay is only revealed on the floor of these vales in relatively small areas and one of the main problems of mapping these vales is in distinguishing transported clay from *in situ* clay. Soils derived from transported and *in situ* clay respectively often have differences which are not visually apparent in the profile but which give rise to subtle variations in agricultural potential.

The steep slopes in the upland valleys and on the escarpments have in the past been relatively unstable, there being a range of superficial tectonic activities which have altered the soil—landscape relationships in such sites. There were three main types of activity:

1. Valley bulges. Arkell (1947*a*) has explained the process of valley-bulging by which clay in the deep valley floors is forced above its proper stratigraphic position (Fig. 11.5). Flexure of the overlying Inferior Oolite under periglacial conditions is generally held to have been responsible.

2. Cambering, slumping and landslips. The processes causing such features have been

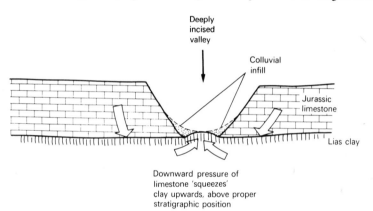

Fig. 11.5 **Diagrammatic cross-section of valley showing mechanism of valley bulging.**

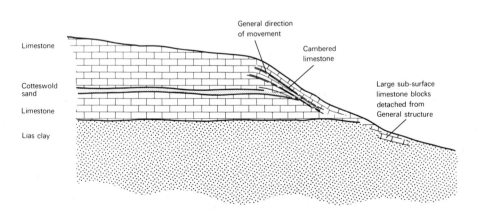

Fig. 11.6 **Diagrammatic section of Jurassic limestone slope showing cambering.**

discussed in detail by Ackermann and Cave (1967). Although such features are expressed dramatically at sites like Bredon Hill, near Cheltenham (Whitaker, 1973), confused sub-surface examples are very common, serving to complicate the soil—landscape pattern. So, for example, in the Stow-on-the-Wold area, cambering is responsible for downward flexure of the Inferior Oolite on steep slopes. Therefore, it not only outcrops apparently below its 'correct' stratigraphic position, but also thereby obscures the Cotteswold Sand outcrop (Fig. 11.6).

3. Ridge and trough features. These are described by Briggs and Courtney (1972) and consist of long ridges (160—320 ft; 50—100 m) and parallel troughs. It is suggested that they are the expression of the movement downslope of large longitudinally cohesive blocks of limestone due to large-scale cambering. Good examples are found in the Upper Windrush valley in north Gloucestershire.

§11.3 SOILS OF THE JURASSIC UPLAND

Although detailed soil—landscape relationships vary from area to area it is possible to make some general comments about the nature of the upland soils.

Over the limestone itself a shallow calcareous brown earth, the *Sherborne* series, is very extensive. These soils cover wide areas of the crest of the escarpment and the

Table 11.1 Inclusions in the *Sherborne* mapping unit in the south Cotswolds. (Source: Courtney, 1973)

Inclusion	Classificatory level	Soil features	Landscape features
1. Sherborne	series	Thin A and B horizons overlying bedded limestone rock or rubble within 40 cm	—
shallow	phase	<15 cm deep	—
deep	phase	>35 cm deep	—
loamy	phase	Soils derived from sandy limestones in the Forest Marble and Cornbrash characterised by very fine sand content	—
2. Yatton	series	Excessively drained shallow rendzina soils developed on Jurassic limestone head and scree	On steep slopes (>110)
3. Didmarton	series	Deep soils on colluvium and stony head	In valley bottoms
4. Haselor	series	Gleyed calcareous soils with slightly impeded drainage derived from inter-bedded limestone and clay	—
5. Shippon	series	Brown earth soils derived from similar materials to (4)	—

dip-slope plateau. They are less than 13 in (35 cm) deep and consist of A and thin B horizons overlying bedded limestone rock or rubble, which is usually *in situ* but may have arrived from a more distant source by solifluction. The soils are stony and have fine loamy to clayey textures.

The *Sherborne* mapping unit in the south Cotswolds was the subject of an intensive variability study (Courtney, 1972, 1973). The object of this was to investigate whether within the *Sherborne* mapping unit delineated by Findlay (1976, in press) there was any systematic variation of soil type. Table 11.1 shows the soils that Findlay included in his original mapping unit. Within the area of *Sherborne* mapping unit (Fig. 11.4) 184 sites were revisited according to a prearranged sampling plan. (Because areas of *Yatton* and *Didmarton* soils are readily distinguished by *site* — although they may be within the limits of *Sherborne* mapping unit — they were not included in the resampling programme. All other possibilities were.)

Table 11.2 Examples of profiles grouped in the modal group. (Source: Courtney, 1973)

Grid ref.	Land use	Depth of profile	Colour*	Texture†	Per cent stones	Base material
ST 862829	Ploughed	46 cm	7.5YR4/4	ZCL	1	Loose lst.
ST 822905	Pasture	22 cm	7.5YR4/4	ZCL	22	Loose lst.
ST 836958	Ploughed	15 cm	7.5YR4/4	CL	15	Loose lst.
ST 795698	Pasture	21 cm	7.5YR4/4	ZCL	16	Lst. with calc. matrix

* Colour — according to the Munsell system.
† Texture — ZCL, silty clay loam. CL, clay loam.

Multivariate statistics were used in the analysis of this data, and showed that about 70 per cent of soils corresponded to the original concept of the *Sherborne* series. Table 11.2 lists some of the characteristics of example profiles within the general definition of the series, while Fig. 11.7 is a diagrammatic representation of both a modal (*Sherborne* series) example and some of the non-modal (other series) profiles within the mapping unit.

One of the important conclusions of this study was that there were no apparent geographic trends in variation. The tentative conclusion to be drawn from this is that any general regional process (such as loessial deposition) has acted throughout the surveyed area, and not simply in part of the area. However, in some places there is sufficient depth of deposited material for a distinct mapping unit to be named. This is the *Waltham* series, which is a fine loamy brown earth, usually shallower than 15 in (40 cm). These soils which are fairly restricted in their distribution are far less extensive than the *Sherborne* soils.

If there is considerable thickness of calcareous clay (more than 35 in; 90 cm) a gleyed calcareous soil — *Evesham* series — is mapped. The clayey parent material for this soil may be either Forest Marble clay or Liassic clay — in other words, the strati-

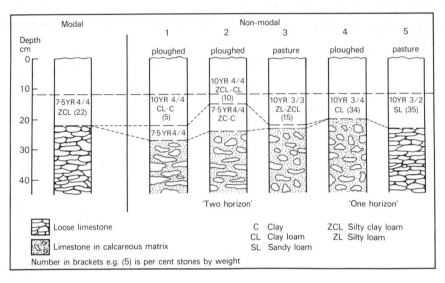

Fig. 11.7 Diagrammatic representation of selected profiles. (Source: Courtney, 1973)

graphic position of the clay is not important. Where the clay lies shallowly over limestone (within 24 in; 60 cm) another gleyed calcareous soil — the *Haselor* series — has
been noted. This soil has a dark brown or dark greyish-brown surface horizon with a
yellowish-brown subsoil. Although the nature of Jurassic lithology suggests that there
should be a continuum from *Evesham* (gleyed calcareous) to *Haselor* (gleyed brown
calcareous) soils, there is usually a quite distinct break within a space of a few metres.
Again, *Haselor* soils may be found in any stratigraphic position were shallow clay lies
over limestone, these situations being common in many parts of the Jurassic sequence.

Yatton series soils are found on very steep slopes over limestone. These shallow
soils are highly calcareous and are thus classed as rendzinas, while the deep stony soils
on the deeper valley floors (*Didmarton* series) are classed as brown calcareous soils.
These latter soils are mapped where the infill of colluvium and head is greater than
35 in (90 cm) deep.

§11.4 PARENT MATERIALS AND SOILS IN THE COTSWOLD VALES

As has already been mentioned, 'Vale' country in the Cotswold area can be divided
into firstly the Lias vales and secondly the Forest Marble and Oxford Clay country of
the dip-slope (§11.1). The Lias vales can again be sub-divided by their general position
in relation to the Jurassic cuesta.

The wider Lias vales are to the north and west of the main Jurassic escarpment
and, at the footslope of the Cotswolds, the Vales of Severn and Berkeley have large
areas where Lias clay is exposed at the surface. Further north, around Cheltenham and
Winchcombe, recent glacial and fluvio-glacial deposits largely obscure the underlying

Liassic strata. Moving still further north along the footslope of the Jurassic escarpment — into the east Midlands — the wide belt of Lias clay is generally obscured by glacial deposits.

The smaller vale areas are found in embayments in the escarpment. Here Beckinsale (1970) suggests the vales have been formed by rivers cutting down through the limestone and eventually causing the wide valleys to coalesce into vales. North of Cirencester these areas, such as the Vale of Bourton, have floors covered by varying amounts of drift material of different origins (see §11.2).

The areas of Forest Marble and Oxford Clay are not vales in the commonly accepted sense, but merely clay country on the general Cotswold dip-slope, lying stratigraphically above the limestone. They are considered here as part of the vales because their topography, soil and land-use characteristics are very similar to the Lias clay vales.

The Forest Marble clay is more extensive in the south Cotswolds and a large area of outcrop also exists near Frome in Somerset. Between the Forest Marble and the Oxford Clay lies an area of limestone — the Cornbrash — so-called because of the suitability of its soils for growing corn. The Cornbrash is in most respects similar to the other Jurassic limestone formations, and the soils over the rock are usually *Sherborne* series.

Despite the stratigraphic differences between the vale areas, the landscape presented by the clay substrates is generally similar. The relief is usually subdued and hedgerows replace the stone walls of the limestone areas. The amount of locally-derived colluvial and other drift material present is partly dependant on the distance from limestone upland. So, for example, the Vale of Bourton is closely surrounded by upland areas and parts of it are covered by a deep mantle of locally-derived drift. On the other hand, areas away from the limestone area, such as the western parts of the Severn Vale, are usually devoid of Jurassic limestone drift, although they may be (and often are) covered by fluvio or fluvio-glacial drift from elsewhere.

In the same way that the topography is broadly similar over the various Jurassic clay areas, so the soils also are very comparable. Until about 1969 it was thought that there were significant mineralogical variations between soils developed on Lias and Oxford clay, respectively. Imperfectly drained clayey soils on the Lias had been named *Charlton Bank* series in Somerset (Avery, 1955) while similar soils over Oxford Clay were called *Denchworth* series in Berkshire (Kay, 1934). However, with an extensive mapping programme over the Forest Marble clay it became clear that some clay soils over all three formations were so similar as to defeat division. Hence since 1970 these soils have all been called *Denchworth* (see *inter alia* M. G. Jarvis, 1973, and Findlay, 1976, in press).

In some areas of the footslope of escarpments, silty to clayey drift accumulates. This is often extremely difficult to distinguish from similar drift brought from a further distance by fluvio-glacial or fluvial processes. Surface-water gley soils in such drift material are named *Rowsham*, while soils over *in situ* clay or mudstone, which is more freely drained, are called *Martock* series. More obvious drift material may give rise to *Podimore* soils (fine loamy to clayey gleyed brown earth soils over clay or thin gravel). Over oolitic or shelly calcareous limestone gravel the brown calcareous *Badsey* soil is found.

§11.5 AGRICULTURAL POTENTIAL – CHANGES IN LAND-USE

The Cotswold Hills are an attractive and important tourist area and one can suggest three reasons for this. The first is concerned with the particular blend of land-form, with wide sweeping upland and deep valley, and woodland, with beechwoods clinging close to the valley side and breaking the monotony of the horizon. The second is concerned with the building stone: the characteristic soft yellow stone of the Cotswold farms and villages is well known. The stone is equally useful for forming field boundaries, stone walls being a common feature of the landscape where the stone is within a metre or so of the surface. (The nature of the subsoil is reliably indicated by the field boundary: where hedges are found instead of walls, one may conclude that a significant depth of clay overlies the limestone, making extraction of wall material too lengthy a process.)

The third reason is concerned with the apparent relative remoteness of the area and the fact that it seems 'undeveloped'. If one drives through the Cotswolds there appears to be a paradox here. The fields are about two-thirds planted to grain and the farms look prosperous while, on the other hand, the farms do not have the appearance of 'farming industry' given by the 'American Mid-west' buildings which are common in the rich agricultural lands of the Fens and parts of East Anglia. The answer to this apparent paradox is quite simple – it lies partly in the availability of building materials that we have already noted. But more important is the change in agricultural practice that has taken place in the area since the last war. Today the upland is dominantly under continuous arable cropping while in the 1930s it was the so-called 'sheep and barley country'. To understand the importance of this change in land-use it is necessary to trace the land-use history from early times.

Up to the Middle Ages much of the Cotswold dip-slope was probably forested. There is little to prove this apart from inferences one can draw from Domesday Book returns and place-name elements. In the case of the former, the returns were made for parish or manor holdings and most of these included a proportion of both upland and vale area. Thus a total return for woodland includes the amounts for both areas. However, if the total area of woodland was greater than the total area of vale in the manorial holding one may conclude that the upland included wooded areas. This is the case with many of the records for north Wiltshire and south Gloucestershire. In the case of place-name elements there are many upland village names including 'ley' or 'leigh' (clearing) such as Leighterton.

By the late eighteenth and early nineteenth centuries, Marshall (1809) and others were able to report that much of the upland consisted of sheep-runs. The sturdy Cotswold breed was developed in this area and it was with wool from the uplands that fortunes of the mill-owners in Wotton-under-Edge, Witney and other towns marginal to the Wolds were made. Enclosure came late to the upland areas, and enclosure acts as late as 1820 are recorded. Throughout this time and until the early years of the present century the Gloucestershire part of the upland at least has been an area of poverty and isolation. The meagre arable crops that were grown were produced on a rotation basis which included sheep pasture. It was thought that the sheep were vital to restore heart to the light shallow stony soil. However, as we have already seen (§11.3) most of the upland soils are not light textured: they are quite clayey. It is only the high proportion of calcareous material that makes them seem light, much of this material being of silt-sized fraction.

Experiments that took place in the last war showed that it was possible to grow excellent arable crops for long periods at a time, with no intermediate pasture phase in the rotation. Most arable farmers are now taking advantage of this fact and the land is now some of the most valuable in the country. Where land has changed hands in the last 10 or 15 years the farms have often been bought by businessmen from London or the Midlands who are using capital from industry to finance outlay on arable cropping machinery. Another result of this change is that farming in this area is now by no means labour-intensive and two or three men can now manage several hundred acres of arable land. This in turn has had the effect that a high proportion of cottages became available in the early and mid-1950s for purchase by non-agricultural workers. These houses were frequently bought by retiring city-dwellers, and this itself has caused a radical change in the sociological structure of the area. In this case, then, a deeper understanding of the soil resources, and an increase in exploitation of those resources, has had far-reaching social consequences.

§11.6 SOILS AND LAND-USE OF THE NORTH COTSWOLDS

Soils in the Stow-on-the-Wold area of Gloucestershire have been mapped in detail by Courtney and Findlay (1976, in press). West of Stow the Cotswold escarpment reaches a height of about 990 ft (300 m). It dips gently eastward for about 6 miles (10 km) but the dip-slope is then interrupted by the Vale of Bourton, which contains two south-east flowing streams, the Dikler and the Windrush. East of the Vale the dip-slope continues beyond a second escarpment (Fig. 11.8).

The geological parent materials are summarised in Table 11.3. In addition to the solid geology shown in the table the following superficial deposits are also present:

Plateau drift has been reported by Hull (1855) and Lucy (1872) but the drift seems to be very inextensive and Courtney and Findlay (1976) were unable to confirm its presence.

River gravels of Jurassic limestone material are present in the Vale of Bourton and are likely to be locally derived (Richardson and Sandford, 1960).

Alluvium of peaty and silty material is present in the lower reaches of the Dikler and Windrush valleys.

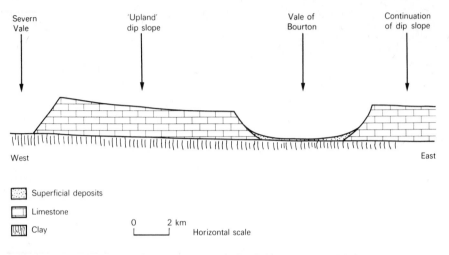

Fig. 11.8 Simplified geology section across the Vale of Bourton, Gloucestershire.

Figure 11.9 shows the general pattern of soil—landscape relationships in the upland area. Although *Sherborne* series soil cover much of the upland within this area, small patches of more loamy soil (*Waltham* series) are found in the general Sherborne area. Repeated cultivation of *Waltham* soils sometimes gives rise to a hard structure 'pan' at a depth of about 10 in (25 cm). This precludes downward root development and the crop in the particular small area (usually less than a hectare) may fail due to lack of moisture uptake. Because the sandy facies in the underlying limestone are not extensive the problem is not acute. Most farmers know which areas might give trouble and take care either to deep plough and thus break the pan or, alternatively, to leave the land to permanent pasture.

Soil distribution in the upland is closely related to lithological changes in the

Table 11.3 Summary of Jurassic succession in the north Cotswolds. (After Courtney and Findlay, 1976, in press)

Geologic period	Geologic formation	Lithology
Middle Jurassic	Forest Marble	Clays and limestones
	Great Oolite	Mostly limestone, with some marly intercalations
	Fullers' Earth	Intercalated limestone, clay and sand
	Chipping Norton limestone	Limestone
	Inferior Oolite	Mostly limestone, some intercalations of clay
Lower Jurassic	Cotteswold sand	Fine micaceous sands (thin or absent)
	Upper Lias	Clay with thin limestone
	Middle Lias	Marlstone, sandy ferruginous limestone and micaceous siltstone
	Lower Lias	Clay with thin nodular limestone

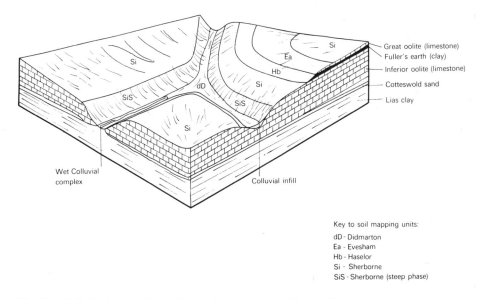

Key to soil mapping units:
dD - Didmarton
Ea - Evesham
Hb - Haselor
Si - Sherborne
SiS - Sherborne (steep phase)

Fig. 11.9 Soil—landscape relationships in the north Cotswold upland.

underlying rock and there is only a limited relationship with topography and natural vegetation. Occasional deep soils with texturally differentiated profiles (i.e. *Waltham*) appear to mark the places of maximum accumulation of insoluble residue of impure limestone in which clay has been translocated downwards through the profile under conditions of base unsaturation. In contrast, *Sherborne* soils contain calcareous material derived from purer limestone and are base saturated with little evidence of clay movement in the profile. In the soils over clay (*Evesham* and *Haselor*), clay leaching has produced a weakly or non-calcareous sub-surface horizon below which lie horizons containing redeposited carbonate concretions.

The arrangement of soils in the Vale accords with general discussion in §11.4. Where the *in situ* Lias clay is not covered by superficial deposits *Denchworth* soil (surface-water gley) is mapped. *Badsey* soils (stony brown calcareous soil) are found over deep limestone gravel, while *Thames* series soils (silty calcareous gleys) are mapped over the alluvium bordering the Dikler and Windrush rivers.

There is a clear distinction in land-use between the upland areas and the vales. The upland areas are mostly under arable cropping, for which the most important and useful soils are *Sherborne* series. Although the soils are shallow, the limestone substrate seems to offer little resistance to root development. In quarry sections roots have been observed penetrating fissures in the limestone down to a depth of 6—9 ft (2—3 m). Even in dry seasons crops rarely suffer from water deficiency. In the small areas of *Haselor* and *Evesham* soils (over clayey substrates) some drainage impedance occurs. With *Haselor* soils this is not often of great importance and most of the area covered by these soils is under arable cropping. *Evesham* soils are more often under permanent pasture. Because of discontinuous bands of limestone in these soils in the upland area proper planning and execution of drainage programmes is extremely difficult. So small patches of drainage-impeded soil have to be either drained on an *ad hoc* basis, or left under pasture. On the steeper valley sides the angle of slope sometimes precludes arable cropping, even over *Sherborne* soils, and in these sites the land remains either in permanent pasture or, in some places, under woodland.

In the Bourton Vale there is a higher proportion of permanent grassland. Much of this land was used for cultivation in the Middle Ages, evidence of which are the strips of 'ridge and furrow'. These longitudinal troughs and ridges have a vertical relief of about 3 ft (1 m). They are thought to have been formed by the peasant strip-farmer ploughing up and down his allotted strip, always turning the furrow toward the central axis of his holding and hence, over a long time, building up a ridge. Since that time soil development processes have continued and there is often a clear soil distribution pattern, with gley soils in the furrows and gleyed brown earth soils on the ridges. This situation is difficult to improve by conventional soil drainage (see Chapter 12) and even if the soil is drained satisfactorily there still remain the practical difficulties of using machinery over hummocky topography. Some farmers have attempted to 'plough out' the ridge and furrows, using the opposite technique from that employed by the original peasant farmers (i.e. turning the furrows into the troughs). However, this usually has the disadvantage that the soil at the base of the ridges (now exposed) is clayey and gleyed. For these reasons, much of this land with ridge and furrow has remained under permanent pasture.

In other areas of the Vale where ridge and furrow is absent, tile drainage has been used to some advantage. Some fields on *Denchworth* soils are now used for arable cropping, though usually in rotation with ley pasture.

FURTHER READING

Soil Survey publications dealing with soils over the Jurassic rocks have been noted throughout this chapter. One of the early surveys was that of the Glastonbury district of Somerset (Avery, 1955), while more recent work has been undertaken in the south Cotswolds by Findlay (1976, in press) and in the north Cotswolds by Courtney and Findlay (1976, in press). Cope (1972) has completed a survey of the soils in the Severn Vale area near Gloucester.

Further north, the Melton Mowbray area has been mapped (Thomasson, 1971) while Oxford Clay soils in the Wantage, Berkshire, area are discussed by M. G. Jarvis (1973).

12 SOILS OF THE CLAY AND MARL LOWLANDS

§12.1 THE EFFECTS OF EXCESSIVE SOIL WATER

One of the dominant features of areas of clay and marl parent materials is that the fine particle-size of the substrate often causes serious impedance of water movement. Many of the soils of these areas are therefore liable to waterlogging and this chapter considers some of the problems associated with the efficient farming of such soils.

The causes and effects of soil gleying were explained in Chapter 1. Briefly, if the soil is waterlogged then anaerobic conditions cause the reduction of certain chemicals in the soil profile. This is particularly noticeable in the case of iron, and 'rusty' mottles are characteristic of many topsoils under permanent pasture. More extensive grey colours reflect extensive reduction of iron to the ferrous state due to more prolonged waterlogging.

The chemical effect of waterlogging is complex and more detailed explanations are available elsewhere (see for example Russell, 1973). Under reduction conditions several gases are given off. These include nitrogen and nitrous oxide from the reduction of nitrates and also hydrogen and a range of hydrocarbons. Among the latter it is now thought that ethylene, in particular, has an effect on root development. Figure 12.1 demonstrates that ethylene concentration in some poorly drained soils over Oxford clay extends over a long period of time — certainly long enough to cause damage to crop growth.

Drainage improves the aeration of the soil in two possible ways. At best it allows the soil pores to drain free of water and allows free oxygen from the atmosphere to enter or, at worst, it allows the stagnant deoxygenated water to flow out and oxygenated rainfall water to enter. Either of these end results will assist crops to grow more extensive rooting systems and hence have available a larger area of soil for nutrient supply. Another effect of good drainage is that it allows the soil temperature to rise more quickly in the spring. This is because wet soil has a higher heat capacity (i.e. requires more heat to raise it through 1°) than a dry soil (Baver, 1956).

Prolonged waterlogging also has other effects which are, indeed, more obvious to the casual observer. Firstly, where the soil is totally waterlogged the addition of further surface moisture will cause surface flooding. Secondly, a high amount of water

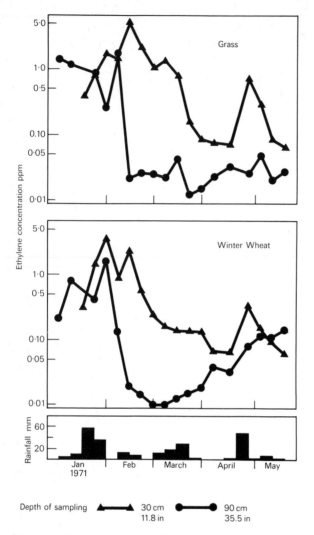

Fig. 12.1 Mean concentration of ethylene in the soil atmosphere of an Oxford clay soil under grass and under winter wheat. (Source: Russell, 1973)

in the topsoil will cause a reduction in the stability of soil structures and heavy compaction by animals or machinery will cause *poaching* or soiling of grassland. Thus, indirectly, an excess of soil water will have an adverse effect on both the variety and quantity of crops that can be produced or, alternatively, on the number of animals that it is possible to graze in a particular area (the *stocking rate*). The proneness of a soil to waterlogging may also cause a farmer to consider the *timeliness* of cultivation operations. So, for example, on some soils he may have to avoid the wet winter months when machinery may be bogged down or the soil structure may be damaged.

In Britain, generally speaking, an excess of soil water is more common than a deficiency and hence the most important hydro-agricultural issue is drainage. Nevertheless there are some areas where either rainfall is low or soil drainage is excessive and in these areas some crops require irrigation. The present chapter is, however, concerned with some of the areas in Britain where soil drainage is a dominant agricultural problem. In general these are areas of fine textured (clayey) parent materials. This is mainly for two reasons. Firstly, soil with small-sized particles generally have small diameter pore spaces which are likely to be less able to transmit excess water through the profile and hence will more easily become waterlogged. Secondly, because of lower resistance to denudational processes, clay usually forms lowland landscape where water collects more easily. However, this is a rather simplistic way of looking at the problem and more elaboration is needed to understand the reasons for different approaches to improving soil drainage (see, for example, Thomasson, 1975).

Russell (1973) suggests that there are two main conditions under which soil drainage is necessary. Firstly soils with a high water-table (ground-water gley soils — see §12.2) need to have the level of the water-table reduced to a depth which varies according to the nature of the crop they are to carry. The general rule is that the water-table should lie permanently below the depth of the crop roots. Hence pasture land can be allowed to have a higher water-table than arable land, arable crops usually having longer roots than grass. The second group of soils needing drainage are the surface-water gleys which do not allow surface water to drain away quickly enough. This may be cured by improved surface husbandry which allows larger pores to form in the surface layers and assists in the formation of aggregates. If a compacted layer in the soil is responsible for the impedance of water movement, then the problem may be solved by deep ploughing or otherwise destroying or breaking up the layer.

§12.2 THE ASSESSMENT OF SOIL WATER

Gleying in a particular profile may be caused either by a high ground-water table (ground-water gley soil) or by slow drainage within the soil profile which gives rise to a perched water-table (surface-water gley soil). Ground-water gley soils generally show mottling in the upper horizons but the subsoils are dominantly grey gleyed horizons and the underlying parent material is also gleyed (Cg horizon). Surface-water gley soils also normally display mottled surface horizons and subsoil horizons which merge into a dominantly grey gleyed horizon (Bg). However, beneath this horizon there is normally a gradual improvement in profile drainage, sometimes leading to brown coloration of the C horizon.

In winter, ground-water gley soils rapidly flood with water seeping in from the base and the lower (Cg) horizons. On the other hand surface-water gley soils generally fill with water seeping in from the upper (Ag) and subsoil (Bg) horizons.

The drainage capability of soil can be assessed by means of ring infiltrometers (Hills, 1968) or by measurements of the height of water in a tube inserted into the soil. Such peizometer tubes can be used to measure the potential pressure exerted by water in the soil and the direction of saturated flow can be calculated (Curtis and Trudgill, 1974).

§12.3 SOIL DRAINAGE IN THE MELTON MOWBRAY DISTRICT

Detailed assessment of the value of soil drainage in north Leicestershire and south Nottinghamshire has been undertaken by Miers and Thomasson (1971). The soil pattern in the area mapped by Thomasson (1971) is complicated, being caused by discontinuous layers of drift material overlying Jurassic and Triassic clays. One result of this is that many of the soil-mapping units are dominated by more than one series (complex mapping units). Because of this it is difficult to generalise about natural soil drainage characteristics and hence about artificial drainage requirements.

Early tile drainage in much of the area dates from the beginning of the nineteenth century. Even at that time the complicated soil pattern of the area had been noted by William Pitt (1809), quoted in Marshall (1809):

> The nature of the soil is very liable to vary much in short distances, respecting its stony or friable qualities . . . about Melton Mowbray, and to a great extent rich sound pasture land abounds, but being a heavy loam upon clay, mixed with small fragments of calcareous stone, it is very wet in winter, and liable to tread with heavy stock: [in the area from] Melton toward Grantham [is] strong clay loam, . . . Waltham [area has] a sound gray loam, Branston towards the Vale of Belvoir, a deep red or snuff coloured loam to some extent; Hathern, deep gray loam, roads heavy: this is the general characteristic of soils in the Vale of Belvoir.

Although many of the older drainage systems are by now ineffective, in several areas they continue to function efficiently. Their present-day efficiency depends on a number of factors including the quality of drainage outlets. But often more important is the nature of the soil itself and it is clear, for example, that old tile drains in soils of the *Ragdale* series (a surface-water gley soil over chalky boulder clay) are no longer effective. This is because the clayey surface horizons of this soil stop downward water movement.

As in all soil surveys the soil series mapped by Thomasson (1971 op. cit.) are classified by *profile drainage status*. This groups the soils on a basis of:

(*a*) annual duration of waterlogging and

(*b*) depth to waterlogged layers in the profile.

According to this scheme five main classes are used as shown in Table 12.1.

Usually a soil series falls within a single drainage class, although exceptionally it may include profiles of two or even three classes. Also, efficient drainage may serve to 'promote' a soil from one class to another. So, for example, the *Quorndon* series (a ground-water gley over terrace gravels) is naturally poorly drained but in most places has been improved by drainage to imperfectly drained.

Thomasson (1971) has undertaken detailed studies of the permeability and

Table 12.1 Soil drainage classes. (After Thomasson, 1971)

Depth (cm)	Well drained	Moderately well drained	Imperfectly drained	Poorly drained	Very poorly drained
<30	—	—	0–3	3–6	>6
<60	—	0–1	1–6	>6	—
<90	0–1	>1	—	—	—

(Figures are mean annual duration of waterlogging, months.)

drainage status of some of the soil series in this area, the *Ragdale* series (already referred to) being one such example. These clayey gley soils over chalky boulder clay cover much of the undulating plateau and valley sides in the east Midlands. In the natural condition they are poorly drained and much of the area covered by *Ragdale* series soils is used for pasture. However, with good drainage they can grow arable crops (particularly wheat and barley) though there is a high risk of soil structure damage from cultivation operations when wet. As Thomasson (1971, p. 71) points out:

> Ragdale soils are virtually impermeable below 60 cm and all excess rainfall in winter must be shed laterally via natural fissures between peds above this depth. Owing to the small hydraulic gradient (<60 cm) excess water tends to cause waterlogging in the plough layer.

Figure 12.2 shows the depth of water in experimental water-level holes at five sites in *Ragdale* series soils. These diagrams demonstrate that, in general, dry spells in winter had little effect on water levels, although where a drainage system existed it enabled the soil to 'dry out' and the water-level depth to fall. However, the drainage system did not prevent the soil becoming waterlogged later in the same spring (in Fig. 12.2 see profile Rq II for March 1964). As already noted, these soils have a low permeability and this somewhat offsets the effect of drainage.

When installing drains Meirs and Thomasson (1971) suggest that five main aspects of drainage design need consideration:

1. *Depth of drains.* Low permeability (as in the case of the *Ragdale* series soils) precludes placing drains below about 3 ft (90 cm).

2. *Spacing of drains.* This is also determined by permeability. In the case of the low-permeable clayey soils in the Melton Mowbray district artificial ways of increasing permeability are necessary.

3. *Permeable fill.* This consists of gravel, slag or clinker laid above the line of drains.

4. *Moling.* These are channels drawn by a mechanical 'moler' through the soil usually at a spacing of between 6–8 ft (2–2.75 m) at a depth of 20–25 in (50–65 cm) and roughly at right angles to the underlying drainage system. The moling operation both forms a system of temporary channels under the ground and promotes the development of fissures.

5. *Subsoiling.* This improves drainage by dragging a heavy cultivator through the subsoil (see Fig. 12.3) and hence intensively shattering the upper horizons.

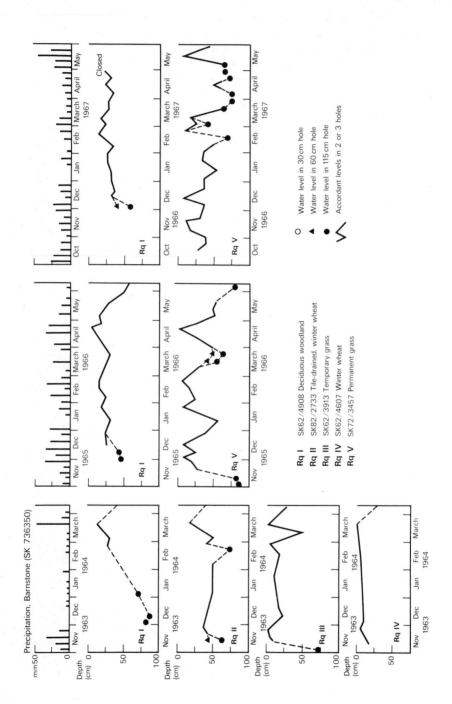

Fig. 12.2 Water levels in five Ragdale soils. (Source: Thomasson, 1971)

Fig. 12.3 The system of cracks produced by a subsoiler working in compacted soil at a suitable moisture content. (Photograph: Ransomes, Sims and Jefferies Ltd)

In the case of the *Ragdale* series soil permeable fill is necessary above the tile drains. The tile drains need to be laid fairly closely (again because of the impermeable nature of *Ragdale* soils) at a spacing of between 130—165 ft (20—40 m). The final decision on this is likely to be based on the closest possible spacing that is compatable with an adequate economic return. Moling is undertaken after laying the drains in order to get water flowing and if the soil is to be used for arable cropping then regular subsoiling will be necessary. However, subsoiling can only be carried out under relatively dry conditions and if there are a succession of wet seasons this will be impossible. In that case it is necessary to undertake more frequent renewal of the moling.

In some parts of the Melton Mowbray area Keuper Marl forms the parent material and gives rise to soils of the *Worcester* series. These are medium to heavy textured reddish-brown soils. Despite their heavy texture some soils of the *Worcester* series in the Midlands and also in the south-west are well drained where they occur on slopes (Avery, 1955). Elsewhere, however, these soils are imperfectly drained, showing mottled subsoil horizons above the unweathered marl. *Worcester* soils can be drained by deep subsoiling (at 24—29 in; 60—75 cm) because the lower horizons are permeable and subsoiling breaks up the upper impermeable horizons, thereby allowing water to drain away.

Generally, *Worcester* soils show well developed prismatic cracking in the subsoil but their high silt content makes the surface layers break down and cake after heavy rain. If dry weather then follows the surface can bake hard.

It is worth noting that the soils on Keuper Marl display certain similarities to the soils over Permian Marl. In the West Riding of Yorkshire these have been mapped in the Saxton Complex by Crompton and Matthews (1970). The main units in the complex are the well-drained *Saxton* series (brown calcareous soil), the gleyed and slightly acid to neutral *Watnall* series (surface-water gley soil) and the well-drained slightly to moderately acid *Micklefield* series (leached brown earth). The *Micklefield* series, in particular, resembles some of the well-drained soils occurring in the *Worcester* series.

§12.4 DRAINAGE PROBLEMS ON THE CULM SOILS IN DEVON

The shales of the Upper Culm Measures (Upper Carboniferous) cover a wide area in central and west Devon. A hydrologic sequence of soils over this parent material has been described by Clayden (1964, 1971) and is shown in diagrammatic form in Fig. 12.4. The sequence consists of a well-drained fine loamy brown earth (*Dunsford* series) on steep slopes, a more weathered clayey gleyed brown earth (*Halstow* series) and a surface-water gley (*Tedburn* series), both the latter series being found on more gentle slopes. Generally the *Halstow* series occurs more often on interfluve sites and the *Tedburn* series on lower 'receiving' sites.

The degree of weathering in the underlying shale is very variable, with both the *Halstow* and *Tedburn* soils occurring over more highly weathered clayey substrate than

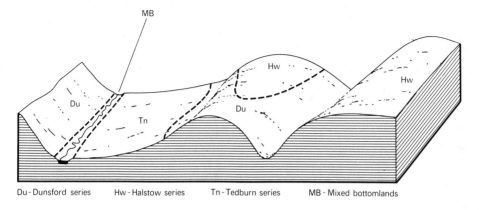

Du - Dunsford series   Hw - Halstow series   Tn - Tedburn series   MB - Mixed bottomlands

Clayey alluvium

Culm shales

Fig. 12.4 Soil pattern of the Exeter Shale Hills. (Source: Clayden, 1971)

the *Dunsford* series soil. The nature of the shale is also responsible for the chief drainage problem of these 'Dunland' areas. The general low permeability of the rock leads to water being held in the shale in several small discrete aquifers. These are in places where there are sufficient fissures to form such an aquifer. Where such an aquifer outcrops on a hillside then a spring may occur, though because it may only be apparent in wet seasons the discontinuous nature of the aquifer makes it difficult to predict.

Farming is based on dairy herds, though both farm and herd sizes tend to be small. Often each farm has in addition a store or fat-cattle herd and a flock of breeding ewes. Some arable crops (mostly barley) are grown on the part of the farm over *Dunsford* and *Halstow* soils but the *Tedburn* soils are usually left to pasture. Often these are long-term leys and it is on these soils that the most severe drainage problems exist.

The Ministry of Agriculture has conducted a detailed drainage experiment on *Tedburn* soils at Langabeare, near Hatherleigh (Trafford, 1970). A field has been laid with tile drains at a depth of 42 in (110 cm) and these were back-filled to within 17 in (45 cm) of the surface. Subsoiling was carried out later. The experiment compared this drained plot with an adjacent 'control' plot which was left in its original undrained condition. It was found that after wet weather the drained plot rapidly dried out while the control plot remained waterlogged for long periods. So far the experiment has not considered optimum spacing of drains and at present this depends very largely on rule-of-thumb.

Moling is not extensively carried out in the Dunland and this may be because of the frequent sandstone fragments found in the soil: these make production of an efficient continuous channel difficult.

Poaching (destruction of the upper structure of the soil — usually by animals — causing 'puddling') and soiling or damaging of grassland are also important problems on the *Tedburn* soils, particularly in the spring and autumn when the grass is growing but the soil structure is not sufficiently firm to carry the weight of stock. At these times the topsoil frequently breaks down into a slurry or unstructured particles which may spoil the grass and make it unpalatable. Experiments are now being undertaken to assess the extent of this problem (S. J. Staines, pers. comm.) but soil drainage, of course, effects very substantial improvements.

§12.5 DRAINAGE AND DROUGHT ON ROMNEY MARSH

The Romney Marsh area of Kent and east Sussex has been the subject of large-scale drainage enterprises since the thirteenth century and nearly all this low-lying area is now high quality farming land. Many parts of the area are below sea level, at least at high tide, and land drainage is therefore closely linked to sea defences (sea walls and sluices). Speed of internal drainage depends on tidal conditions, and during winter storms with consequent relatively higher sea-water levels, the effective time during which outflow is possible may be reduced. Figure 12.5 shows the extent of the major drainage channels at the present time.

The soil pattern in the area is extremely complex (R. D. Green, 1968), both because of the pattern of underlying soil parent materials and also because of the history of artificial drainage in the area. In some cases this latter factor has been sufficient to extensively modify the soils from their natural condition. The general pattern of soil parent materials is given in Fig. 13.12 while Figs 12.6 and 12.7 show sections through the deposits. These soils are further referred to in §13.8.

Table 12.2 summarises the way in which the soils in the area have been divided into series. Because of the problem of drainage improvement influencing soil development, R. D. Green (1968) has mapped soil *drainage phases* for each soil series. Hence, for example, the *Newchurch* series is sub-divided into a moderately well-drained phase and an imperfectly drained phase. The drainage characteristics of each series are summarised in Table 12.3.

Clearly not all the soils in the Romney Marsh area require drainage and, indeed, some — because of their excessive or well-drained status — require irrigation during dry seasons. The area is under a mixed arable and pasture economy, but since the last war the acreage of arable cropping has increased steadily. Since 1950 nearly 2 000 acres (808 ha) of permanent pasture has been turned to arable use (though this figure also includes land which has been turned into ley pasture). The sheep population has also increased, a fact which Green (1968) suggests is due to the high productivity of the new ley pasture.

There is some conflict between the drainage requirement of the grazier and that of the arable farmer. On the one hand the grazier has the problem that rainfall levels in

Table 12.2 The division into soil series of the soils in the Romney Marsh area. (After Green, 1968)

	Dominant texture of B horizon or between 12 and 30 in (30–75 cm)				
	Silty Clay or Clay	*Clay Loam over Silty Clay or Clay*	*Clay Loam*	*Sandy Loam or Loam*	*Loamy Sand or Sand*
Texture below 30 in (75 cm) similar to or finer than that of B horizon	Newchurch (C) Dymchurch (dC)	Walland (C) Brenzett (dC)	Agney (C) Finn (dC)	Romney (C) Snargate (dC)	Greatstone (C) Midley (dC)
Loam to sand between 24 and 42 in (60–106 cm)	Guldeford (C) Ivychurch (dC)				
Thick peat between 24 and 42 in (60–106 cm)			Fairfield (C) Dowels (dC)		
Thick peat above 24 in (106 cm)		Appledore (dC)			

Notes: (1) The soils are classified generally according to:

    (a) variations in the texture profile or lithology to a depth of 42 in (106 cm)

    (b) differences in intensity of gleying as shown by grey colours and/or ochreous mottling (in this case this is a direct reflection of dominant texture size), and

    (c) the content and distribution of native calcium carbonate.

  (2) C = calcareous soil,

    dC = decalcified soil.

Table 12.3 Texture and drainage of soils series. (Source: Green, 1968)

		Coarse		Medium		Fine	
Drainage	Sub-surface texture	Stony	Loamy sand or Sand	Loam or Sandy loam	Clay loam	Clay loam over Silty clay	Silty clay or Clay
Excessive or well drained	Similar to that between 12 and 30 in (30–76 cm)	Beach Bank* Dungeness	Lydd† Midley†‡				
Imperfectly drained	Similar to that between 12 and 30 in (30–76 cm)		Greatstone¶				
Imperfectly or moderately well drained	Similar or finer texture below about 30 in (76 cm)			Snargate†	Finn†	Brenzett† Walland¶	Dymchurch† Newchurch¶
	Loam to sand below about 30 in (76 cm) to at least 42 in (106 cm)			Romney¶	Agney¶	Ivychurch† Guldeford	
Poorly or very poorly drained	Thick peat occurs between 24 and 42 in (61–106 cm)					Dowels† Fairfield¶	
	Thick peat occurs at less than 24 in (61 cm)						Appledore†

* Beach Bank soils are locally calcareous.
† Decalcified soils.
‡ Midley soils are developed on loamy medium sand or sand.
¶ Calcareous soils.

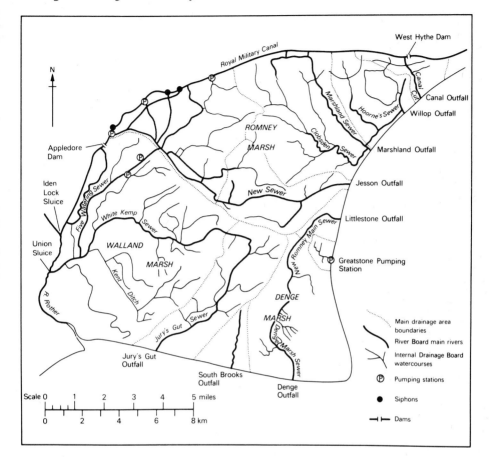

Fig. 12.5 Drainage of Romney Marsh. (Source: Green, 1968)

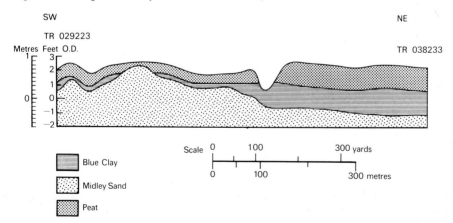

Fig. 12.6 Sketch-section showing variations in height and thickness of substrata near Midley. (Source: Green, 1968)

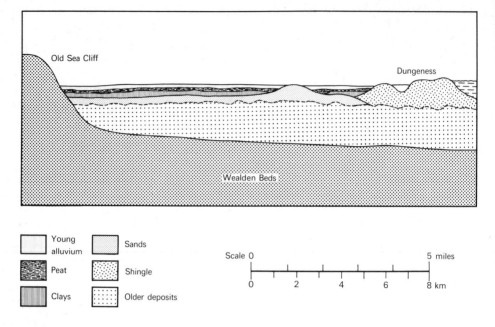

Fig. 12.7 **Sketch-section from the uplands to Dungeness. (Source: Green, 1968)**

summer are insufficient to keep the coarse- and medium-textured soils moist enough for optimum grass growth. The traditional method of maintaining moisture levels was to use stop-boards in the drains or 'sewers' to retain the last of the winter rain water. However, this meant that the water-table was too high for optimum grass growth in the finer textured (clayey) soils and certainly too high for effective arable cropping. The arable farmer would ideally like water levels to be as low as possible during winter and early spring. Many of the tile drain systems which have been installed are not efficient because they lie below the water-table level for long periods. If the sewers were to be increased in depth the permeable nature of the subsoil would in some cases render the tile drain systems obsolete.

Because of the complex nature of the soil pattern in this area it is not possible to produce an ideal soil drainage solution but selective summer irrigation combined with a lowered water-table would probably be generally beneficial.

§12.6 SOILS ON ESTUARINE CLAY AROUND THE BRISTOL CHANNEL

The soils of Romney Marsh have formed on marine sediments and on alluvial materials which accumulated in the old Romney Bay in the lee of offshore spits and beaches formed by the sea. There are, of course, other coastal areas where estuarine alluvium has been laid down to form the basis for soil development. For example, grey clays of the *Wentlloog* series are developed in recent estuarine alluvium bordering the Bristol

Channel. These soils are derived from estuarine mud accumulated on tidal flats, possibly during Romano—British times (Avery, 1955). Like most such deposits the material becomes lighter in texture at depth but typically the horizons are ill-defined. The surface soil of brownish-grey clay with a granular structure passes gradually into grey, silty clay which is mottled and displays a prismatic structure. The upper horizons are non-calcareous with pH values of about 6.0, but at depths of about 20 in (50 cm) the reaction becomes alkaline with the presence of finely divided calcium carbonate. The decalcification of the surface horizons is due to leaching over a period of some 2 000 years. Similar decalcification occurs on estuarine clays of the *Godney* series beneath peat accumulations (Curtis, 1968).

These estuarine clay lowlands often possess peat bands beneath the surface and examples of soil series based on such landscapes are the *Fladbury* and *Midelney* series (Avery, 1955). Clay on peat soils of this character are known as 'Rodoorn soils' in the northern Netherlands where they have developed as a result of deep drainage.

Although the flood hazards in Somerset and Monmouthshire have been greatly reduced in recent years these soils may still suffer from a great deal of surface water-logging in winter. It is desirable that the level of the water-table should be kept under control and for grassland on the deep clays of the *Wentlloog* and *Fladbury* series it should, if possible, be kept below 3 ft (1 m) throughout the year. On the *Midelney* series, however, there is some danger that the peat bands might dry out irreversibly in the summer if drainage were too extreme. These soils should ideally be drained so that the water level does not fall below 20 in (50 cm) in summer or rise to within 12 in (30 cm) of the surface in winter. It is a measure of the advances made in drainage works and technology that such control of landscape conditions is now within our reach.

FURTHER READING

As already noted, chemical and physical effects of soil waterlogging are detailed by Russell (1973). Several Soil Survey publications deal with areas where soil drainage is of particular importance and these have been detailed through this chapter. Green (1968) includes a great deal of historical information on the development of Romney Marsh. Soils on the Somerset Levels are discussed by Avery (1955) and Findlay (1965), while Crampton (1972) has mapped those on estuarine parent materials in south Wales. The recent Soil Survey Technical Monograph on 'Soils and Field Drainage' (Thomasson, 1975) contains useful detail regarding the drainage problems of various soils series.

13 THE SHRINKING SOILS AND LOWERING LANDSCAPES

§13.1 RECLAMATION OF THE FENS

When water is drained from soil shrinkage normally occurs and desiccation cracks are formed at the surface. Thus, following reclamation by drainage the low-lying peat and clay lands along our coasts and estuaries have assumed lower levels relative to the sea as shrinkage of the deposits has progressed. The principal areas affected in this way are the Fens, Somerset Levels and Moors, Romney Marsh and the Lancashire lowlands. Other smaller areas occur at other coastal and inland sites where organic and sedimentary materials are drained.

The deposits found in such areas consist mainly of layers of marine clays, estuarine sands, silts and shelly facies together with interbedded peat bands. Since the last glaciation there have been variations in sea levels which have led to phases of marsh development during still-stand periods followed by periodic inundations which have resulted in the deposition of sands, silts and clays. Thus, complex histories of stratigraphic development can be traced in these areas (Godwin, 1955; M. Williams, 1970).

Over the centuries a long battle has been waged to reclaim these ill-drained areas for use under agriculture. The reclamation of the Fens did not begin until the mid-seventeenth century, prior to which they were of little agricultural value. At this time they were only of interest as fisheries and for wild fowling and turf cutting (Darby, 1969). It was in 1630 that the fourth Earl of Bedford, together with co-adventurers (of capital) engaged the Dutchman, Sir Cornelius Vermuyden, to drain these areas of sedge and reed interspersed with occasional meres (Fig. 13.1). Following the Civil War, the fifth Earl resumed the work and new cuts were made by Vermuyden to take out water more quickly and prevent flooding (Fig. 13.2).

After the initial drying of the peat surface had rendered it suitable for agriculture, the peat began to shrink. As far as can be judged the general level of the peat was originally some 15–20 ft (5 m) higher than at present and, as the fields lowered, the drainage channels became increasingly ineffective. Initial drainage by gravity gave way at the end of the seventeenth century to pumping in order to lift the water from the lowered fields in to the drainage dykes. At first pumping was carried out by horse

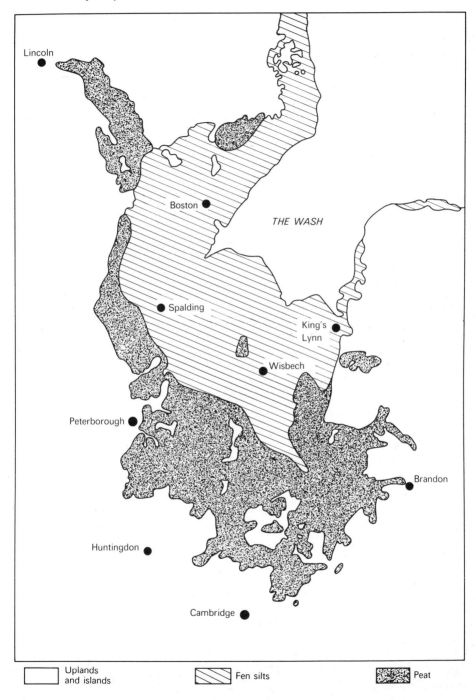

Fig. 13.1 Generalised map of fenland peats and silts: the Bedford levels. (After Darby, 1969)

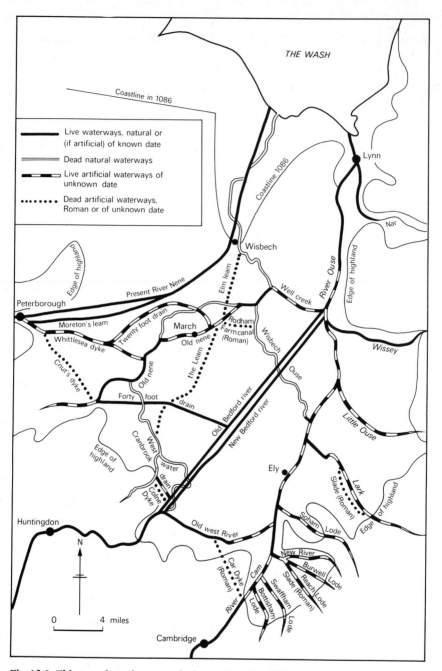

THE WASH

Coastline in 1086

Live waterways, natural or
(if artificial) of known date

Dead natural waterways

Live artificial waterways of
unknown date

Dead artificial waterways,
Roman or of unknown date

Lynn

Coastline 1086

Nar

Edge of highland

Wisbech

Elm leam

River Ouse

Edge of highland

Present River Nene

Well creek

Peterborough

Moreton's leam

Twenty foot drain

March

Rodham
Farm canal
(Roman)

Wisbech

Wissey

Whittlesea dyke

Old nene

The Leam

Ouse

Cnut's dyke

Old nene

Forty foot

drain

Old Bedford river

Little Ouse

Edge of
highland

Cranbrook

West water

New Bedford river

Lark

Colne
Dyke

drain

Ely

Slade (Roman)

Edge of highland

Huntingdon

Old west River

Soham Lode

N

Car Dyke
(Roman)

New River

Burwell Lode

0   4 miles

Cam

Swaffham

Slade (Roman)

Reach Lode

River

Bottisham Lode

Lode

Cambridge

Fig. 13.2 This map shows how great is the number of artificial waterways in the fens whose origins are unknown. Even so it included less than half the artificial channels known to be in existence before the seventeenth-century period of drainage, and an even smaller proportion of those which became extinct before that period. Some of the waterways marked — the Twenty Foot Drain, for example — although recommissioned by the seventeenth-century drainers, were first dug before their time. (Source: Astbury, 1958)

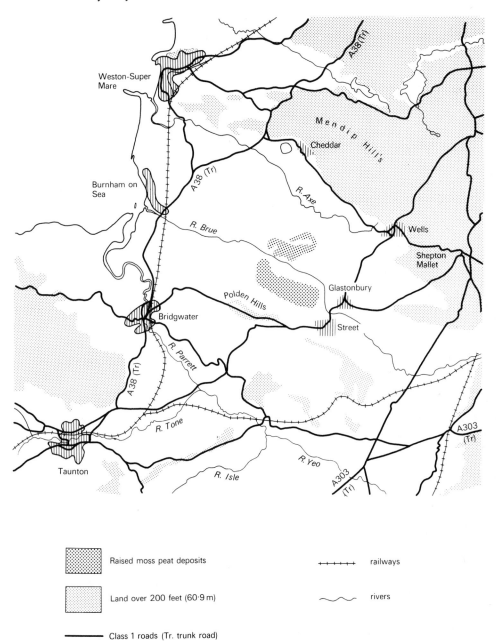

Fig. 13.3 Raised peat deposits of north Somerset. (Source: Somerset County Planning Office, 1967)

power but this was quickly replaced by windmills. As the landscape lowered these too became less effective and some areas reverted to their marshy state. It was not until 1820, when the first Watt engine was installed at Bottisham, that steam engines provided greater lifting capacity. As more and more water was removed from the peat, it shrank still further. This continual shrinkage resulted in a progressive increase in the diameters of the steam-driven scoop wheels used from 36—50 ft (11—15 m). Some of these large wheels weighed as much as 75 tons. However, as the introduction of the centrifugal pumps and diesel engines provided further efficiency, the lift of water increased from an average of 7—15 ft (2—5 m) in 1913 to 21 ft (7 m) in 1938. Only small areas of original surface fen remain within the main Fenland basin (e.g. Wicken Sedge Fen and St Edmunds Fen) and these, though doubtless generally lowered over the centuries, still stand 6—8 ft (2 m) above the surrounding drained land.

§13.2 RECLAMATION OF SOMERSET LEVELS AND MOORS

Reclamation schemes elsewhere in Britain generally began later. For example, in Somerset the major drainage schemes were started in the late eighteenth century, some 150 years later than in the Fens. The Axe and Brue valleys (Fig. 13.3) were mostly reclaimed and drained in the period 1770—1800, but drainage of the whole area of the Somerset Levels did not become fully effective until the last 50 years.

The peat deposits mostly began as reed swamp occupying the fresh water areas which flooded the marine and estuarine deposits. In course of time, and possibly following a climatic change towards dryness, the reed swamps were invaded by fenwoods of alder, birch and willow (Fig. 13.4), resulting in the development of wood peat. At about 2000 B.C. different vegetation developed in the form of *Sphagnum* moss, cotton grass (*Eriophorum* sp.), ling (*Calluna* sp.), and deer grass (*Trichophorum* sp.). These quickly built up a complex of gently domed raised deposits rising several feet above the level of the water-table. Raised bogs of this type are dependent on rainfall for water supply, drawing none from ground-water. Thus, little mineral matter is taken up by the plants except from the atmosphere and the deposits become strongly acidic and base deficient. The raised bog vegetation cover being sensitive to changes in climate, gradually changed in periods of dryness towards vegetation types with abundant ling, cotton grass and trees. This vegetation was subject to humification under aerobic conditions, so that a well-developed, decomposed, chocolate brown, colloidal peat was formed which was cheese-like and dense. This type of peat has been recognised for centuries as that which is best for peat fuel.

In periods of wetness, however, there was growth of *Sphagnum* moss on the wetter areas of the raised deposits. These gave rise to pale, humified, soft peat of low density. Thus, in peat deposits one may recognise a number of soil horizons which vary according to (*a*) vegetation types, (*b*) degree of decomposition. For example, if one considers the Somerset area it is possible to describe a generalised profile for peats (Fig. 13.5) as recognised by those working in peat-cutting industries. Local terms, e.g. 'upper light' and 'top black', are used to indicate various layers of commercial

Fig. 13.4 Fenwoods of alder, birch and willow on peat deposits in central Somerset. (Source: Somerset County Planning Office, 1967)

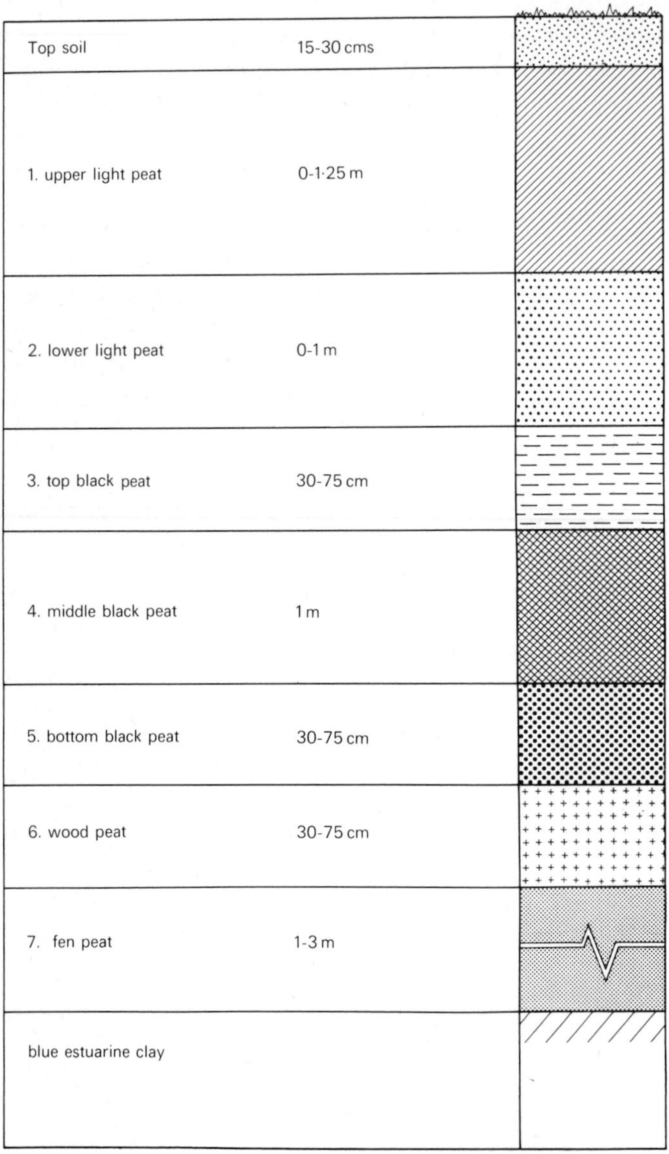

Top soil	15-30 cms	
1. upper light peat	0-1·25 m	
2. lower light peat	0-1 m	
3. top black peat	30-75 cm	
4. middle black peat	1 m	
5. bottom black peat	30-75 cm	
6. wood peat	30-75 cm	
7. fen peat	1-3 m	
blue estuarine clay		

Fig. 13.5 Section of Somerset peat deposits. The section is essentially diagrammatic. Individual layers are not always easily identified. All layers vary in depths and individual layers are not always present. (Source: Somerset County Planning Office, 1967)

significance. These peats have been surveyed and in areas not affected by late marine transgressions organic soils of three distinct types have been mapped. The *Turbary Moor* complex consists of raised moss peat with acid reaction (pH 4.9 at surface). Much of this has been cut for fuel or, recently, for horticultural peat. Therefore, the area has a somewhat uneven surface and a tendency for fields to be elongated, reflecting old turbary or peat-cutting rights. Peat depth is up to 15 ft (5 m), and is strongly fibrous with *Sphagnum*, *Calluna*, *Eriophorum* remains in the upper layers. Although it is moisture retentive, this soil may suffer from drought in dry summers and grassland upon it is often of poor quality.

The most extensive organic soil mapped in the area is the *Sedgemoor* series. This series includes organic soils with high base status, which are often referred to locally as 'black earth' and cover about 14 000 acres (5 656 ha). All this flat land is drained by rhines and ditches and, in most places, the water level in the rhines remains about 1 ft below the land surface for the greater part of the year. Base-rich waters flood this peat where it is adjacent to the Lias or limestone uplands and it often includes remains of *Phragmites communis* at depth and *Cladium mariscus* and *Menyanthes trifoliata* nearer the surface. The surface layer normally contains more mineral matter than the *Turbary Moor* complex peats, and where this layer exceeds 9 in in thickness and is predominantly clayey, the soils are classified as *Midelney* series.

§13.3 ORGANO–MINERAL DEPOSITS OF THE SOMERSET LEVELS AND MOORS

On Godney Moor and in other areas, peat or organo–mineral deposits occur above recent marine transgression clays. The peat varies in thickness (two phases have been mapped (2—5 ft; 1—1.5 m and 12—24 ft; 4—8 m)) (Avery, 1955). These soils are termed the *Godney* series and they are mainly strongly acid. An interesting feature of this series is that it carries distinctive surface patterning due to channels (Curtis, 1968). These channels appear to represent adjustment to shrinkage combined with surface erosion of the peat (Fig. 13.6).

The mineral soils of the Somerset Levels (coastal clay areas) and Moors (inlying peat areas) are all imperfectly and poorly-drained ground-water gley soils. Alluvium derived from the Triassic Marl has been mapped as *Compton* series and these areas give reddish-brown soils of neutral reaction. Streams draining from the Liassic deposits give rise to small areas of imperfectly and poorly drained soils: *Butleigh* and *Fladbury* series respectively. The coastal Levels are dominated by imperfectly drained soils of the *Wentlloog* series, which was so named because it was first described on the other side of the Severn Estuary in south Wales. This clay is very uniform and was laid down when estuarine waters flooded the coastal lands in recent times. Godwin (1941) dated this flooding episode to the Romano—British period but the dating of this clay is somewhat uncertain.

§13.4 VARIATIONS IN PEAT SOILS

It is important to recognise the marked differences existing between fen peat formed under ground-water conditions and raised bog peat. The former is often formed in water rich in bases and it is, therefore, eutrophic in character, whereas raised peats are

Fig. 13.6 Air photograph, Godney Moor, Somerset. Channelling in peat (A, B, C). Peat cuttings (D). Note rectangular fields characteristic of drained areas. (Photograph: J. K. St Joseph. Cambridge University Collection)

oligotrophic. The distinction between the 'low moor' stage of development under the influence of ground-water and the 'raised bog' phase dependent on rainfall can often be made in peat profiles. The *Adventurers'* series (so named after the Adventurers' Fen in Cambridgeshire) is typical of the eutrophic, low-moor, peat and organic mud in the Fenland basin (see Table 13.1). It covers about 15 320 acres (6 189 ha), of which 8 180 acres (3 304 ha) are less than 3 ft (1 m) deep, whereas about 7 140 acres (2 884 ha) are mostly 6–8 ft (2–3 m) in depth. The greatest recorded depth of peat is 13 ft (4 m) in this area.

Variations in the depths of peat over the underlying mineral deposits are characteristic of these landscapes. The surface of the deposits beneath the peat is gently undulating (Fig. 13.7). Where the sub-peat formations are highly calcareous some marked variations in soil pH can occur within one field, ranging from pH 8 near calcareous marl to pH 4 where the peat is deeper. Horizons influenced by a fluctuating water-table often contain gypsum, either as small crystals or as a whitish powder or grit filling worm channels or root holes. Fine white deposits of calcium sulphate also appear on exposed surfaces, especially in fallow fields, presumably as a result of capillary rise and subsequent evaporation of water.

Certain areas of the *Adventurers'* series are locally said to be 'drummy'. This local term indicates drained acid peatlands that appear to contain relatively large amounts

Table 13.1 Acreage and proportionate extent of the mapping units. (Source: Hodge and Seale, 1966)

Mapping unit	Acreage	Acreage	Percentage of area surveyed
Adventurers' series — shallow phase	8 180	15 320	11.1
Adventurers' series — deep phase	7 140		
Moulton complex		14 420	10.3
Wicken series		10 190	7.3
Wantage series	9 390	9 430	6.8
Wantage complex	40		
Swaffham Prior series	6 500	6 610	4.8
Swaffham Prior complex	110		
Milton series		6 470	4.7
Newmarket series		5 350	3.9
Burwell series	4 370	4 650	3.4
Burwell complex	280		
Peacock series		4 610	3.3
Clayhythe series		3 760	2.7
Midelney series		3 700	2.7
St Lawrence series		3 570	2.6
Stretham series		3 330	2.4
Landbeach series		3 300	2.4
Hanslope series		2 900	2.1
Freckenham series	2 200	2 630	1.9
Freckenham complex	430		
Bracks series		2 590	1.9
Soham series		2 560	1.9
Denchworth series		2 480	1.8
Aldreth series		2 410	1.7
Oakington series		2 360	1.7
Worlington series	1 820	2 230	1.6
Worlington complex	410		

Mapping unit	Acreage	Acreage	Percentage of area surveyed
Padney series		2 220	1.6
Disturbed Lode complex	2 140	2 180	1.6
Lode series	40		
Wilbraham series	1 780	1 900	1.4
Wilbraham complex	120		
Bottisham series		1 820	1.3
Block series		1 740	1.3
Cottenham series		1 400	1.0
Reach series	1 240	1 380	1.0
Reach complex	140		
Fordham series		910	0.7
Willingham series		900	0.7
Ashley series		730	0.5
Dullingham series		700	0.5
Earith series		570	0.4
Upware series		530	0.4
Icknield series		470	0.3
Chippenham series		440	0.3
Barway series		380	0.3
Chittering series		260	0.2
Isleham series		260	0.2
Redlodge series		200	0.1
Prickwillow series		170	0.1
Histon series		140	0.1
Spinney complex		100	0.1
Shell-marl phases		740	0.5

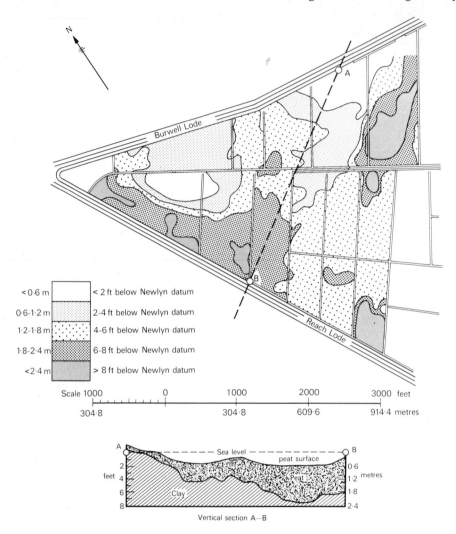

Fig. 13.7 Adventurers' Fen. Contour map of the Fen floor and cross-section. (Source: Hodge and Seale, 1966)

of ferric oxide. In soils of this kind structural elements in the subsoil at depths between 12–30 in (30–75 cm) have ferruginous coatings which give the horizon a dark reddish-brown tinge. Such horizons are always very acid and may have a hard consistency. From a land-use standpoint, these soils are difficult in that the 'drummy' layer is not easily re-wetted once it has dried out.

§13.5 LAND-USE OF LOWLAND PEATS AND WIND EROSION

When the land-use of the lowland peats of Britain is studied, it becomes apparent that there are great contrasts. In the Fens there are large fields under intensive agriculture where water-tables are kept low. In Somerset there are small fields used for grazing and the water-table is kept fairly high. Both areas have one thing in common, however — losses of peat soil. In the Fens the curse of wind erosion lies heavily on the land. In Somerset the peat extraction is likely to exhaust supplies of peat in an area of 5 900 acres (2 384 ha) near Glastonbury by about 2011.

The wind erosion of soils, or as it is colloquially termed 'blowing', has constituted a serious threat to arable farming in several areas. Although not confined to any one area of the country it is most marked in eastern England on the Fenland peats and adjoining light sands and sandy loams. The landscape here is open to the wind with few hedges. The arable rotation, of which potatoes and sugar beet form the basis, leaves the soil surface exposed, tilled and vulnerable. Crops like lettuce, carrots and sugar beet are particularly liable to damage from wind erosion, both before germination and at the seedling stage. The east of England is notable for long spells of dry, windy weather between the end of March and mid-May. It is at this time, particularly when the wind reaches an average speed of about 20 knots (10.2 m/sec), that a 'blow' can be expected.

Material can be moved by the wind in one of three processes: surface creep, saltation or deflation. In surface creep the movement is by rolling and sliding of larger particles along the surface. These large particles are too heavy to be lifted up by the wind and generally range from 0.5—3 mm in diameter. They are set in motion by the continuous impact of grains bouncing along the surface. Bouncing grains are those moving by saltation. They are generally 0.05—0.5 mm in diameter. The bouncing action begins when the particle is spun by wind at great speeds (200—1 000 revs/sec) and the air near the surface of the particle spins with the grain, thereby creating a partial vacuum above and a pressure beneath. These pressure changes lift the particle and the wind carries it forward. Eventually, the particle falls again and the process is repeated, leading to a bouncing effect (Chepil, 1958).

Sometimes the particles moved by surface creep and saltation are primary particles. In the Fens, however, they are often aggregates of clay or organic (peat) particles. Since aggregates contain some air space they are less dense and lighter than solid particles of similar size. Thus, for a given wind velocity aggregates will move more readily than primary particles of the same size.

Experiments have shown that particles smaller than 0.1 mm (i.e. smaller than fine silt) when occurring in a smooth bed are not moved even by strong winds. The reason for this is that these particles do not protrude above the non-turbulent layer of air which is near the ground surface (Fig. 13.8). In the average fenland field, however, there is a mixture of particles of differing sizes. If we assume that there is no protective cover of vegetation and that the soil is dry enough to blow (i.e. moisture tension is greater than 15 atmospheres) the erodible material is removed from the field. Eventually, non-erodible clods of earth become uncovered and with continued removal of the fine material the height of these clods, relative to the surface, becomes greater. Finally, the height of the clods is such that the surface is protected from further erosion. For any given wind velocity it has been found that the wider the spacing of

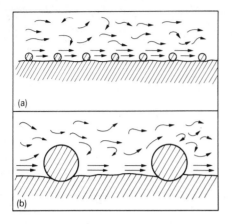

Fig. 13.8 Diagrammatic pattern of wind flow over a surface. Small particles in (*a*) do not protrude into the turbulent air and cannot be picked up by the wind. Particles in (*b*) absorb force of the wind and can be moved. (Source: Chepil, 1958)

the clods the greater is their height when movement stops. The relationship Distance between clods/Height of clods is found to be constant for a given wind velocity and range of size of movable particles in the soil. This constant, known as the critical surface roughness constant, varies between about 4 and 20 depending on the wind velocity and the size of the movable particles.

Thus, a field is at less risk immediately after ploughing when large clods are at the surface than after discing and rolling have provided a seed bed. When peat lands dry out in the spring, the surface has often been brought to a fine tilth for planting and, even where the plants are established, they are often not large enough to provide cover. A number of methods have been adopted to counter soil erosion (Sneesby, 1953, 1966), including planting of tree belts, strip cropping, claying (marling) at a rate of 200 tons per acre, spraying with water or a starch solution. As a means of wind-break protection a system based on willow hedges, single rows of barley and hessian screens has been used. The effect of a widespread hedge cover is to produce a micro-climate which is rather more humid than in the open fenland: this slows down the rate of drying out of the surface soil and accordingly reduces the risk of blowing. Barley rows planted in alternate rows with other crops serve to protect young crops from wind buffeting. Likewise hessian strips about 3 ft (1 m) high and spaced at 30 yard (30 m) intervals can protect soils and crops. Despite these measures considerable losses occur, as, for example, in the years 1943, 1956 and 1968. Accounts of the storm of 15–20 March 1968 report a number of losses (Robinson, 1968) (Fig. 13.9). Some 1 200 tons (1 219 tonnes) of soil were removed from Kesteven roadways. On one 200 acre (81 ha) farm in the Holton-le-moor area it was estimated that 3 000–4 000 tons of soil had been blown half a mile into the drains (Fig. 13.10). In Lindsey, clearance of drifted soil from fields cost £4 000 and clearing of infilled drains a further £6 000. In all, some 21 000 acres (8 498 ha) of land were affected.

As the peat wastes away the peat lands are changing their character. Already

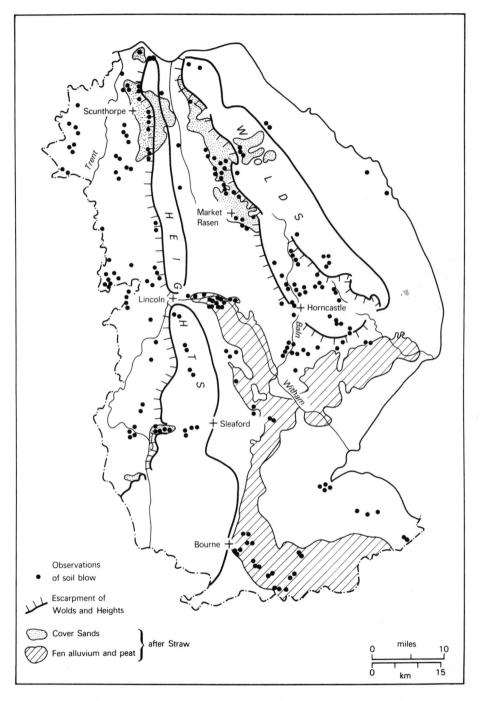

Fig. 13.9 Observations of soil 'blow' in Lincolnshire, March 1968. (Source: Robinson, 1968)

Fig. 13.10 A fen boundary ditch full almost to surface level with blown peat soil. The surface is so firm that one can walk across it without sinking into the blown material. (Photograph: N. J. Sneesby)

100 000 acres (40 400 ha) of black peat fen has either disappeared or become mixed with a high percentage of subsoil material. The farmers of the fens are now faced with the problems of farming the soil materials which formerly lay buried beneath the peat. These underlying materials are broadly of two types — (i) fine-grained (clays and silts), (ii) coarse-grained (sands and gravels). The clays are derived from marine or Oxford clay deposits and present two major problems. First, the clay surface beneath the peat is undulating and also the peat shrinks unevenly. As a result, if drains are placed at depths suitable to drain low spots, they are too low for effective drainage of high spots. Second, where clays occur within plough depth repeated ploughing at similar depths leads to a smeared surface in the clay. This surface — termed a plough pan — restricts downward and upward movement of moisture and may lead to rapid drying out of the surface.

In those areas where sands and gravels lie beneath peat the problems are largely concerned with lack of nutrients and moisture. Paradoxically, the situation is developing where irrigation is becoming necessary in areas which were once reclaimed from the water.

§13.6 BASIN AREAS AND THEIR SOILS

Apart from these problems there is the requirement to protect these low-lying lands from sea flooding since much now lies below sea level. The costs of such protective measures are high; for example, following floods in 1947 schemes costing £6½ million were introduced to build up banks and deepen channels to avoid river flooding. Nevertheless, the fen peats and silts are capable of producing such yields and crops that they represent first-quality land which justifies considerable capital expenditure.

At present the soil—landscape relationships mapped by the soil survey (Hodge and Seale, 1966) in the Fenland basin are as shown in Fig. 13.11. As can be seen, the *Adventurers'* series of the basins often lies between clay lowlands on which gley soils on clay substrata are dominant (*Denchworth, Wicken* and *St Lawrence* series). Interspersed with the peat soils there are areas in which organic soils are mixed with, or overlain by, mineral layers; *Padney* series (marl over peat); *Peacock* series — imperfectly or poorly drained clays derived from Gault or Jurassic clays; *Barway* series — calcareous loams from lake deposits; *Spinney* complex — humose gley soils over Corallian limestone (Table 13.2).

Table 13.2 Fenland basin soils. (Source: Hodge and Seale, 1966)

Major soil group	*Parent material*	*Series*
Alluvial gley soils	Clay over peat	Midelney
	Clay over loamy drift	Earith
Gley soils with clay substrata	Lake marl over clay or loamy drift	Barway
Peaty and humose gley soils	Loamy drift over Gault or Jurassic clays	Bracks
	Resorted Gault or Jurassic clays	Peacock
	Sandy and loamy drifts	Clayhythe
		Fordham
		Isleham
	Ferruginous coarse loams	Chittering
	Marly chalk (Chalk Marl)	Reach
	Chalky loamy drift	Wilbraham
	Lake marl over peat	Willingham
	Brashy limestone	Spinney
Calcareous immature soils	Lake marl over peat	Padney
	Disturbed Cambridge Greensand marl	Disturbed Lode complex
Basin peat soils	Peat	Adventurers'
	Peaty loam	Bottisham
	Peat with interbedded clay	Prickwillow

§13.7 EXPLOITATION OF PEAT BY CUTTING: THE INFLUENCE OF MAN ON PEAT LANDSCAPES

The Somerset peatlands, together with areas in Lancashire, Scotland and northern England, have been exploited for peat. Commercial production of peat in Somerset began in 1870 when it was used primarily for fuel and peat moss litter. At present, 90

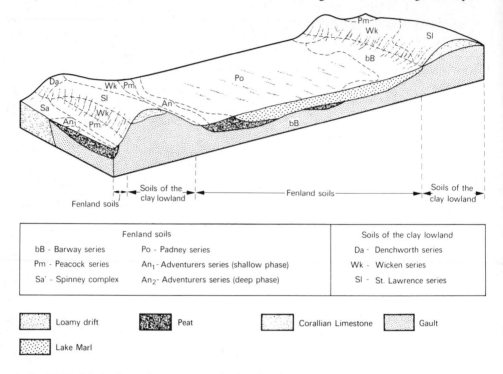

Fenland soils

Fenland soils		Soils of the clay lowland
bB - Barway series	Po - Padney series	Da - Denchworth series
Pm - Peacock series	An₁- Adventurers series (shallow phase)	Wk - Wicken series
Sa' - Spinney complex	An₂- Adventurers series (deep phase)	Sl - St. Lawrence series

Loamy drift   Peat   Corallian Limestone   Gault

Lake Marl

Fig. 13.11 Schematic section across Fenland and Upland at Soham Mere. (Source: Hodge and Seale, 1966)

per cent of the peat cut is used for horticulture. Output has risen considerably in the last two decades from 16 000 tons in 1954 to 63 000 tons in 1966 and 125 000 tons in 1973. One acre layer of peat provides 450 tons and it has been estimated that the quantity of peat extracted in the period 1966—86 will be 2 187 285 tons. Assuming four layers are extracted per acre the land requirement would be about 1 200 acres (Somerset County Planning Office, 1967) of the existing commercial peat area of about 5 900 acres. Estimates of the exhaustion of the deposit range from 45 years (on basis of rates quoted above) to 50—75 years assuming reserves of peat to be better and rates of extraction somewhat less.

There is at present no reclamation of peat turbaries for agriculture. It is estimated that 5—10 years must elapse after cutting before a restoration can be attempted. Even then it seems unlikely that it will be economic to clear the land for agriculture. As a result the flooded peat cuttings may be developed as amenity areas for boating and fishing and thus pass out of agricultural use.

The human record preserved in the peats is one of great interest, ranging as it does from lake village sites of prehistoric age to buried trackways that began in pre-history but in some cases continued into historic times (Godwin and Clapham, 1948). These regions still remain important as refuge areas for animals and plants but, as this section has sought to show, the natural landscape is now becoming a rarity and the natural soils likewise are fast disappearing.

It will be evident that the peat and clay lowlands present soil and land-use problems of a special kind. These problems are due to the interaction of man and landscape. In seeking to improve the land as a habitation and source of food, man has induced changes of an irreversible nature. It can be said that he has largely inverted the landscape so that areas which stood up have shrunk and some areas formerly depressions, i.e. river courses, now stand above the fields they once drained.

§13.8 INVERSION OF LANDSCAPE: THE CREEK RIDGES OF ROMNEY MARSH

The effect of land drainage on shrinkage of deposits is well seen on the reclaimed coastal marshlands of Romney Marsh. In this area a wide range of deposits have been laid down behind the shingle spits of Dungeness (Fig. 13.12). The stratigraphic sequence generally comprises:

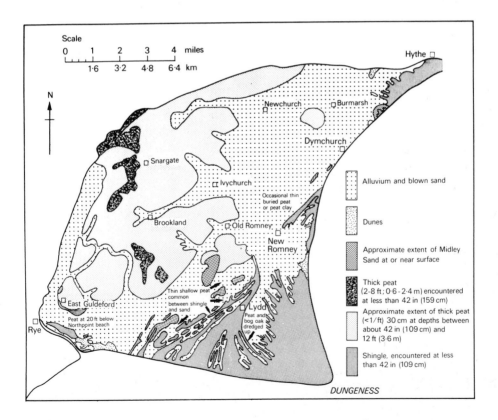

Fig. 13.12 The deposits and soil parent materials of Romney Marsh. (Source: Green, 1968)

(*a*) Young alluvium

(*b*) Peat

(*c*) Blue Clay

(*d*) Midley Sand

The surface deposits are dominated by marine alluvium (51 290 acres; 20 721 ha) and shingle deposits (8 540 acres; 3 522 ha). There is, however, some 990 acres (400 ha) of land where peat lies within 1 m of the surface. These areas in which peat forms a substrate are those in which shrinkage is most evident and the effects of differential shrinkage are clearly demonstrated where creek ridges occur in the landscape. The ridges are nearly symmetrical in cross-section and form complex dendritic systems. They are thought to correspond to a system of creeks cut into thick and extensive peat deposits. Subsequently clay, silt and sand were deposited over the landscape until the peat was buried and the creeks silted up (Green, 1968). The existing relief results from differential shrinkage of the peat after artificial drainage following enclosure. The sandy infill of the creek itself does not shrink following drainage, unlike the peat and clay deposits flanking the creek (Fig. 13.13*a*). The form of the ridge is, however, dependent on the depth of the creek in relation to the peat and clay deposits. Low creek ridges, such as those of the Lydd—Horsebones Bridge area are mainly found where the creeks eroded through thin peat into the sub-peat sediments. Many of these broad and low creek ridges also occur in areas where the peat rests on a relatively high floor where the base level of erosion was such that creeks were often cut some feet into Midley Sand.

In contrast there are creek ridges in The Dowels area of the Marsh which are well-defined and relatively narrow. These sharp ridges evolved from the infilling of deep creeks cut entirely in peat. A schematic illustration of the development of both narrow and broad creek ridges is given in Fig. 13.13*b*. Similar ridges occur in Holland where they are termed 'Kreekrugs' and the low-lying, irregularly shaped areas between are called 'Poels' (pools). Detailed examples of creek ridges and their influence on settlement are given by Green (1968). The soil series occurring on the creek ridges are grouped into seven soil associations on the basis of the nature and degree of development of the creek ridges and pools. An interesting feature of the ridges is that in some areas of the marsh they are associated with drinking holes for stock, whereas elsewhere drinking wells are not related to the ridges. In the former case water of low salinity is found only on the ridges and not in the pools, but elsewhere low salinity water is available on lowland as well as ridges. (See also §12.5, p. 230.)

The soils of Romney Marsh have been classified into the following classes:

(*a*) Calcareous soils on Marine Alluvium

(*b*) Decalcified soils on Marine Alluvium

(*c*) Soils on Old Beach Deposits

(*d*) Decalcified (Old) Marshland

(*e*) Calcareous (New) Marshland

(*f*) Miscellaneous Land Types

(a) Silted Creek

(b) Creek ridge

Sand [ ] Clay [≡] Peat [▨]

Fig. 13.13a Block diagrams of creek ridge development. (Source: Green, 1968)

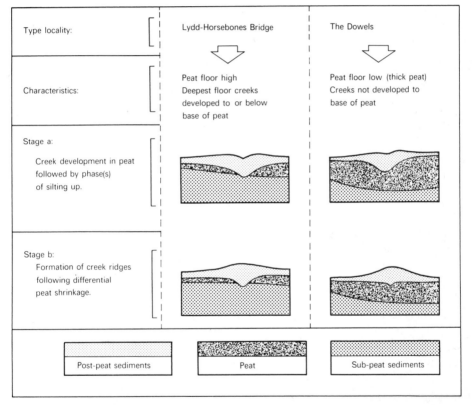

Fig. 13.13b Schematic illustration of creek ridge development in different localities. (Source: Green, 1968)

Few of the soils are more than 2 000 years old and there is a wide variation in age, the youngest parts of the Marsh having been reclaimed less than 100 years. In such circumstances such features as leaching, decalcification, development of soil structure, development of brown colours, organic matter distribution and texture differentiation can be used to separate the soils. Individual soil series can also be separated on the basis of drainage and texture (Table 12.3, p. 232; and Table 13.3).

Table 13.3 Relationship of drainage and mottling classes to soil series.
(Source: Green, 1968)

Drainage class	Mottling classes*	Soil series
Excessive	*d*	Beach Bank
		Dungeness
Well	*d*	Lydd
		Midley
Moderately well and imperfect	*b, c* and *a*	Newchurch
		Dymchurch
		Walland
		Brenzett
		Guldeford
		Ivychurch
	c, d and *a, b*	Agney
		Finn
		Romney
		Snargate
Imperfect	*a, b*	Greatstone
Poor	*a*	Dowels
		Fairfield
Very poor	*a*	Appledore

* Mottling
  (*a*) above 12 in   (*c*) between 15 and 24 in
  (*b*) between 12 and 15 in   (*d*) below 24 in

    The area of Romney Marsh is mainly noted for its sheep pastures but some diversification in the agriculture has taken place since 1946. Arable land and temporary grass (ley pastures) has increased and bulb growing, propagation of strawberry plants and beef production have been introduced locally.

FURTHER READING

Darby, H. C. (1969). *The Draining of the Fens*, Cambridge University Press.

Green, R. D. (1968). *Soils of Romney Marsh*, Bulletin No. 4, Rothamsted Experimental Station, Harpenden.

Williams, M. (1970). *The Draining of the Somerset Levels*, Cambridge University Press.

14   SOIL RESOURCES AND LAND CLASSIFICATION

One of the most pressing problems facing society today is the feeding of the people of the world. As far as soil resources are concerned this means that land should be used to its full potential for agricultural production. However, in urbanised countries, like Britain, there are rival claims for land-use, chiefly housing, industry, forestry and recreation. Moreover, the quality of soil resources is not the same all over Britain. There are distinct regional variations in the potential value of soils for agricultural production. The main content of this book discusses these regional variations but this chapter discusses some problems about the nature of resources, and soil resources in particular. It also considers some of the problems involved in decisions concerning how land should be classified.

§14.1 RESOURCE CONCEPTS

The concept of a resource is a complex one and therefore it is difficult to give one simple definition which embraces all aspects of the concept. One broad approach is that of O'Riordan (1971), who suggests that 'a resource is an attribute of the environment appraised by man to be of value over time within constraints imposed by his social, political and institutional framework'. The key point here is the appraisal by man of something to be of value. On this point Hunker (1964) has extended Zimmerman's (1951) idea of a resource as a culturally defined concept: 'a resource can no longer be conceived as a tangible object but as a functional relationship that exists between man's wants, his abilities and his appraisal of his environment' (Hunker, 1964). The point here is not so much that resources are significant in themselves, but that they become significant according to their appraisal by man. Having briefly considered a resource in general we can now focus our attention upon soil resources in particular.

14.1.1 Soil resources

The soil is like any other resource in that its significance depends upon its use by society. If man chose to live on a diet of one crop alone (for example, wheat) then all soils would be appraised in terms of their potential for growing that crop. If man chose to live on another food source (for example, sugar beet or, say, lichens) then, because each crop has its own particular requirements for optimum growth, each appraisal of soil resources would take a different viewpoint. However, man does not require soil to fulfil just one requirement, i.e. growing food. In practice soil resources have many demands placed upon them such as the growing of forest products and acting as site resources which are used for building and recreation. Furthermore the soil is an integral part of an ecosystem in which wild life exists. This is not only of aesthetic value to man but cannot be damaged or stressed without damaging the environment and the life it supports, including man.

Recently, because of the often conflicting demands which man places upon soil resources, there has been a growing awareness of the problems arising from an unrestrained and unplanned use of resources. There has been a growing consciousness of three main points of difficulty (O'Riordan, 1971):

1. Resources are not often plentiful or inexhaustible.

2. Cheap technological solutions for solving problems and for gaining the greatest benefit from resources are not always available.

3. The development of one resource is not usually independent of another.

The realisation of the falsity of assumptions which were implicit in the un-restrained use of resources has lead to an increasing awareness of the need for resource management concepts to be used.

At its simplest level resource management involves keeping the options open on the future development of resources: 'The goal should be to avert the thoughtless foreclosure of options' (Cloud, 1969). The definition by O'Riordan of resource management stresses that resources should be logically allocated in 'a process of decision making whereby resources are allocated over space and time according to the needs, aspirations and desires of man within the framework of his technological inventiveness, his political and social institutions and his legal and administrative arrangements'. It is a conscious process of allocation rather than a haphazard, competitive process. It takes into account physical, social, cultural, economic and institutional factors in an integrative way and it tries to plan predictively for the future rather than let things ramble on in a way summed up as the 'Vicious cycle of crisis and expedient action'.

14.1.2 Soil resource management

If the general themes from the resource management concepts discussed above are taken and explored for use in soil study then we can see that the rational allocation of soil resources for various uses could include the following considerations:

(*a*) Does the use give adequate economic returns?

(*b*) Does the use cater for social needs?

(c) Does the present, or proposed, use conserve the land for future (and possibly different) uses?

(d) Does the use maximise agricultural productivity?

(e) Does the use provide for consistently good crop yields?

An important problem in soil resource management is that consideration of all these points may give rise to conflicts in opinion. Earth scientists (for example, soil scientists, geomorphologists and agriculturalists) together with life scientists (botanists and zoologists) will provide assessments and evaluations of soil resources which may not always coincide with those of sociologists, economists, engineers or planners.

The need for *collectivity* in resource management, thereby acting for a group as a whole regardless of individual preferences, has been stressed by O'Riordan. However, the soil is largely managed by a collection of individuals (farmers and landowners) for agricultural use. Yet such use is often impinged upon by other uses such as recreation or housing. Thus amenity values may be in conflict with agricultural needs and economic values may be set against intangible assets such as visual beauty or wild-life habitats.

At the present time agriculture dominates the land-use of Britain and in 1965 it represented the use of 81.5 per cent of all land in the United Kingdom (Edwards and Wibberley, 1971). However, there is increasing demand for land for purposes other than agriculture; for example, recreation, forestry and urban growth. Thus Edwards and Wibberley have estimated changes in land-use likely to occur by the year 2000 if present trends continue (Table 14.1).

Table 14.1 **Present and projected land uses. (Source: Edwards and Wibberley, 1971)**

Land-use	1965		2000	
	Area 10^3 hectares	Per cent of total	Area 10^3 hectares	Per cent of total
Agricultural land	19 624	81.50	18 122	75.25
Urban land	2 043	8.49	2 745	11.43
Forestry and woodland	1 817	7.54	2 617	10.86

The significance of the loss of agricultural land can only be assessed by attempting to estimate the loss in productivity rather than the loss in area. Although less land may be available for agriculture the portion remaining may be capable of being used more intensively and more productively. For example, there was a loss of 1.2 per cent per year in 1959 but the loss in productivity only amounted to 0.55 per cent (Wibberley, 1959).

The society in which we live demands food, housing, timber, industry, recreation and transport. Each of these require land and so agricultural land is increasingly required for other purposes. It seems logical, therefore, that some assessment of land should be undertaken so that the poorer land is taken for non-agricultural purposes and the most productive land is retained in farming. In practice such assessments are difficult to make in an objective manner because it is difficult to set up a perfect scale of values for land. What seems to be a logical appraisal system for one group of people (for example, economists) may seem less satisfactory to other groups interested in political, social or aesthetic arguments. A particular dichotomy of opinions can be found at public inquiries on National Parks (Barkham, 1973). Often there are groups attributing a monetary value (to a resource) who believe in the strength of their position because they can quantify the benefits, and other groups who cannot place a monetary value on land but use aesthetic and intangible arguments and feel that they have the advantage of superior ideals.

There is also the dichotomy between the physical and social approaches to land classification and land planning. The physical approach, simply applied, would be to build houses on areas unfit for agriculture; for example, gravelly or marshy areas. This would leave the deep, well-drained and fertile soils for agricultural use. However, the actual location of urban growth is usually determined socially rather than physically as has been pointed out by Wibberley (1959). Distance from existing urban centres providing employment and service facilities is of prime consideration. In addition the worst agricultural land may also be the most costly on which to build houses or roads. Poor agricultural land may also be low-lying, damp and foggy thus bringing environmental penalties for settlement. Thus rationalisation of land-use problems is not a simple matter due to the many diverse issues involved.

§14.2 LAND CLASSIFICATION AND LAND CAPABILITY RATINGS

Regional planning problems demand that decisions be made concerning the value of land. This means that the characteristics of the land must be evaluated, for example soil type, slope, altitude, aspect, altitude, exposure, etc. Also some estimates of the potential use of the land, assuming the application of modern technology, is desirable.

The need for such information has led to the development of land classification schemes of various kinds together with crop productivity ratings. Vink (1960) has suggested that, 'all good land classification is based on a good soil classification'. Let us, therefore, examine the methods employed to classify land in Britain.

The first comprehensive land classification of Great Britain was initiated at the suggestion of the Advisory Maps Committee of the Ministry of Works and Planning (later the Ministry of Town and Country Planning). It was drawn from information collected between 1938 and 1942 for a special investigation carried out by the Land Utilisation Survey of Britain under the direction of Professor Sir Dudley Stamp. It was published in two sheets on a scale of 1:625 000 and divided land into ten categories of land (see Explanatory Text issued by Ordnance Survey, 1950).

Although this first land classification proved of outstanding value it became outdated as a result of post-war changes in farming practice and usage of land. As a result the Agricultural Land Service Research Group (MAFF) undertook to define the requirements for an up-to-date agricultural land classification system based on national standards but capable of application to small areas. They also undertook the task of preparing a new set of agricultural maps of a standardised kind. Surveying began in 1966 and maps are now available for all of England and Wales.

The Agricultural Land Service (ALS) scheme is aimed at identifying land with the greatest inherent value for crop production. In this context the inherent characteristics are those of soil, relief and climate. Other factors such as standard of management, adequacy of fixed equipment and farm structure are deemed less important in this respect because they are more susceptible to change. The physical limitations may operate in one or more of four principal ways:

(*a*) They affect the range of crops which can be grown.

(*b*) They influence the level of yield.

(*c*) They may determine the consistency of yield.

(*d*) They may considerably affect the cost of obtaining the yield.

Although flexibility of cropping is given considerable weight it does not outweigh ability to produce consistently high yields of a somewhat narrower range of crops. The classification places land into five grades: Grade I (minor or no limitations), Grade II (minor limitations), Grade III (land of average quality), Grade IV (land with severe limitations), Grade V (land of little agricultural value with very severe limitations). A summary of the characteristics of each grade is given and further information can be obtained from Agricultural Land Service Report No. 11, 1966.

AGRICULTURAL LAND CLASSIFICATION (Source: Ministry of Agriculture, Fisheries and Food, 1966)

Definition of the grades

Grade I

Land with very minor or no physical limitations to agricultural use. The soils are deep, well-drained loams, sandy loams, silt loams or peat, lying on level sites or gentle slopes and are easily cultivated. They retain good reserves of available water, either because of storage properties of the soil or because of the presence of a water-table within reach of roots, and are either well supplied with plant nutrients or highly responsive to fertilisers. No climatic factor restricts their agricultural use to any major extent.

Yields are consistently high on these soils and cropping highly flexible since most crops can be grown, including the more exacting horticultural crops.

Grade I land is often associated with the following deposits and situations:

Deposits and situations	*Soil characteristics*	*Soil series*
1. Brickearths	Well-drained, deep sandy loams, silt loams, loams or humose. Sometimes over clay loams without firm, very coarse structures	Hamble
2. Loamy river terrace deposits		—
3. Soft rocks weathering to medium textured soils		North Newton

4. Drained peat in lowland sites	Well-drained peat	Adventurers'
5. Reclaimed estuarine and marine deposits, e.g., in Lincolnshire, Kent (Romney Marsh), Yorkshire and Lancashire	Well-drained loams and clay loams	—
6. Sands and sandy loams with satisfactory water table control	(*a*) Deep loamy sands or humose loamy sands with water table supplying moisture to deep rooting plants	Sollom
	(*b*) Deep sandy loams or loamy sands overlying clayey sub-strata	Stockbridge

Grade II

Land with some minor limitations which exclude it from Grade I. Such limitations are frequently connected with the soil; for example, its texture, depth or drainage, though minor climatic or site restrictions, such as exposure or slope, may also cause land to be included in this grade.

These limitations may hinder cultivations or harvesting of crops, lead to lower yields or make the land less flexible than that in Grade I. However, a wide range of agricultural and horticultural crops can usually be grown, though there may be restrictions in the range of horticultural crops and arable root crops on some types of land in this grade.

Grade II land is often associated with the following deposits and situations:

Deposits and situations	*Soil characteristics*	*Soil series*
1. Red marls	Deep, well-drained and moderately well-drained silt loams over silty clay loams with firm, very coarse structures	Bromyard Worcester
2. Loess over or intermixed with clay-with-flints		Batcombe
3. Siltstones, sandstones and sandy drifts	(*a*) Moderately deep (12–18 in; 30–46 cm) sandy loams, loams or silt loams on rock	Munslow Shifnal
	(*b*) Sandy loams over loamy sands especially in wetter areas	Bridgnorth Newport
4. Chalk, chalk marl and limestone	Deep, well-drained loams	Wantage Aberford

5. Sandy drifts over clayey drifts (glacial sands over boulder clay)	Deep loamy sands over clays or clay loams, available water being increased by the low permeability of the subsoil	Rufford
6. Gravelly drift	Stony or gravelly sandy loams, loams or sandy clay loams	Baschurch Shillingford
7. Better drained boulder clay	Imperfectly drained soils with medium over fine textured horizons	Cottam

Grade III

Land with moderate limitations due to the soil, relief or climate, or some combination of these factors which restrict the choice of crops, timing of cultivations, or level of yield. Soil defects may be of structure, texture, drainage, depth, stoniness or water-holding capacity. Other defects, such as altitude, slope or rainfall, may also be limiting factors; for example, land over 400 ft (122 m) which has more than 40 in (101 cm) annual rainfall (45 in; 114 cm in north-west England, western Wales and the West Country) or land with a high proportion of moderately steep slopes (1 in 8 to 1 in 5) will generally not be graded above III.

The range of cropping is comparatively restricted on land in this grade. Only the less demanding horticultural crops can be grown and, towards the bottom of the grade, arable root crops are limited to forage crops. Grass and cereals are thus the principal crops and the land is capable of giving reasonable yields when judiciously managed and fertilised. In fact, some of the best quality permanent grassland may be placed in this grade where the physical characteristics of the land make arable cropping inadvisable.

Grade III land is often associated with the following deposits and situations:

Deposits and situations	Soil characteristics	Soil series
1. Boulder clays	Poorly drained medium to fine textured soils or imperfectly drained fine textured soils	Salop Hanslope
2. Better drained parts of clay vales, e.g., Lias, Oxford and London Clays (Vale of Avon and Thames valley)		Evesham Wicken
3. Estuarine and marine deposits, imperfectly or poorly drained	Imperfectly or poorly drained clays	Wentlloog
4. Oolitic and other limestones including chalk	Soils less than 10 in (25 cm) deep, of varying texture on soft or shattered rock	Sherborne Icknield

5. Red marls	Shallow, well-drained and moderately well-drained silt loams over silty clay loams with firm, very coarse structures	Bromyard Worcester
6. Stony deposits	Well-drained very stony soils of medium texture	Charity
7. Sandstones and sandy drifts	Sandy loams over loamy sands, especially in drier areas	Bridgnorth Newport

Grade IV

Land with severe limitations due to adverse soil, relief or climate, or a combination of these. Adverse soil characteristics include unsuitable texture and structure, wetness, shallow depth, stoniness or low water-holding capacity. Relief and climatic restrictions may include steep slopes, short growing season, high rainfall or exposure. For example, land over 600 ft (183 m) which has over 50 in (127 cm) annual rainfall or land with a high proportion of steep slopes (between 1 in 5 and 1 in 3) will generally not be graded above IV.

Land in this grade is generally only suitable for low-output enterprises. A high proportion of it will be under grass, with occasional fields of oats, barley or forage crops.

Grade IV land is often associated with the following deposits and situations:

Deposits and situations	Soil characteristics	Soil series
1. Land fringing upland and mountain rough grazings	—	—
2. Poorly drained areas of vales, e.g., Lias, Kimmeridge, Oxford and London clays (Vale of Avon and Thames valley)	Poorly or very poorly drained soils	Charlton Bank Denchworth Swanmore
3. Poorly drained areas of clay substrata, e.g., Culm Measures, Lacustrine clays	Poorly or very poorly drained soils	Tedburn Crewe
4. Sandstones and sandy or gravelly drifts (eastern England)	Excessively drained stony, loamy coarse sands and coarse sands with or without pans in drier areas	Freckenham Redlodge
5. Hard rocks weathering to shallow and/or stony soils	Soils less than 10 in (25 cm) deep on hard rock with low available water	—
6. Riverine and estuarine alluvium subject to regular flooding	Poorly drained soils of varying texture	—

Grade V

Land of little agricultural value with severe limitations due to adverse soil, relief or climate, or a combination of these. The main limitations include very steep slopes, excessive rainfall and exposure, poor to very poor drainage, shallow depth of soil, excessive stoniness, low water-holding capacity and severe plant nutrient deficiencies or toxicities. Land over 1 000 ft (305 m) which has more than 60 in (152 cm) annual rainfall or land with a high proportion of very steep slopes (greater than 1 in 3) will generally not be graded above V.

Grade V land is generally under grass or rough grazing, except for occasional pioneer forage crops.

Grade V land is often associated with the following deposits and situations:

Deposits and situations	*Soil characteristics*	*Soil series*
1. Flood plains of most major streams where pasture improvement is impossible	Very poorly drained soils	—
2. Undrained mossland or peat	Undrained peat or peaty soils	—
3. Mountains and upland rough grazings	Undrained peat, shallow soils, boulder dominated soils	Ynys and Hiraethog
4. Areas subject to disadvantages such as mining subsidence, smoke and other pollution, in consequence of industrial activities	—	—

Each map published in this series is at present on a scale of 1 : 63 360 and reports are being prepared for each map, thus providing a comprehensive commentary on the quality of land in Britain. The proportion of each of the grades for England and Wales as a whole and separately for each of the MAFF regions is given in Tables 14.2 and 14.3.

§14.3 THE SOIL SURVEY LAND-USE CAPABILITY CLASSIFICATION

A provisional land-use capability classification has been produced by the Soil Surveys of England and Wales and of Scotland. This classification is based, with some modifications, on the Land Capability Classification (Klingebiel and Montgomery, 1961) developed by the Soil Conservation Service of the United States Department of Agriculture.

The object of the classification is to present the results of soil surveys in a form which may be of more use to agricultural advisers, farmers, planners and other users.

Table 14.2 **Agricultural land classification. (Source: Ministry of Agriculture, 1974)**
Percentage of land in each grade — regional statistics
Proportions of the total land area of the MAFF regions in each of the Land Classification Grades

	Agricultural land					*Land primarily in urban use*	*Other land primarily in non-agricultural use*
	Grade I	Grade II	Grade III	Grade IV	Grade V		
England and Wales	2.3	11.8	39.6	16.0	11.3	8.5	10.5
Regions							
Wales	0.2	1.9	14.5	36.7	29.9	4.2	12.6
West Midlands	0.8	15.1	49.0	12.6	2.8	11.6	8.1
East Midlands	1.2	17.9	55.0	8.4	2.5	7.8	7.2
Eastern	9.1	27.6	40.3	5.7	0.1	7.8	9.4
Yorkshire and Lancashire	2.6	15.5	27.5	17.0	16.4	14.8	6.2
Northern	0.1	3.9	34.5	15.2	30.9	5.1	10.3
South-west	1.4	6.8	53.9	16.5	5.4	5.4	10.6
South-east	2.3	9.2	42.9	13.0	1.3	13.8	17.5

Table 14.3 **Proportions of the agricultural land area of the MAFF regions in each of the Agricultural Land Classification Grades. (Source: Ministry of Agriculture, 1974)**

	Grade I	*Grade II*	*Grade III*	*Grade IV*	*Grade V*
England and Wales	2.8	14.6	48.9	19.7	14.0
Regions					
Wales	0.2	2.2	17.5	44.1	36.0
West Midlands	1.0	18.8	61.1	15.7	3.4
East Midlands	1.4	21.1	64.7	9.9	2.9
Eastern	10.9	33.4	48.7	6.9	0.1
Yorkshire and Lancashire	3.2	19.7	34.8	21.5	20.8
Northern	0.1	4.6	40.8	18.0	36.5
South-west	1.7	8.0	64.2	19.6	6.5
South-east	3.4	13.4	62.3	18.9	2.0

Certain assumptions are made in the system adopted by the Soil Survey (Bibby and Mackney, 1969). It assumes a moderately high standard of management rather than present use. The capability classification may be changed by major reclamation schemes (e.g. pumping schemes) which permanently change the previous limitations in use. Minor changes, liable to regress with time, will not change the classification. The scheme also emphasises physical rather than chemical limitations, the latter being more easily overcome by modern farming technology.

Unlike the ALS scheme, the Soil Survey scheme has seven Land-Use Capability Classes. A summary of the characteristics of the classes is given below.

LAND-USE CAPABILITY CLASSES (Source: Soil Survey of England and Wales)

Class 1. *Land with very minor or no physical limitations to use*

Soils are usually well-drained deep loams, sandy loams or silty loams, related humic variants or peat, with good reserves of moisture or with suitable access for roots to moisture; they are either well supplied with plant nutrients or responsive to fertilisers. Sites are level or gently sloping and climate favourable. A wide range of crops can be grown and yields are good with moderate inputs of fertiliser.

Class 2. *Land with minor limitations that reduce the choice of crops and interfere with cultivations*

Limitations may include, singly or in combination, the effects of (1) moderate or imperfect drainage, (2) less than ideal rooting depth, (3) slightly unfavourable soil structure and texture, (4) moderate slopes, (5) slight erosion, (6) slightly unfavourable climate.

A wide range of crops can be grown though some root crops, and winter-harvested crops, may not be ideal choices because of difficulties in harvesting.

Class 3. *Land with moderate limitations that restrict the choice of crops and/or demand careful management*

Limitations may result from the effects of one or more of the following: (1) imperfect or poor drainage, (2) restrictions in rooting depth, (3) unfavourable structure and texture, (4) strongly sloping ground, (5) slight erosion, (6) moderately unfavourable to moderately severe climate.

The limitations affect the timing of cultivations and range of crops which are restricted mainly to grass, cereal and forage crops. While good yields are possible limitations are more difficult to overcome.

Class 4. *Land with moderately severe limitations that restrict the choice of crops and/or require very careful management practices*

Limitations are due to the effects of one or more of the following: (1) poor drainage difficult to remedy, (2) occasional damaging floods, (3) shallow and/or very stony soils, (4) moderately steep gradients, (5) slight erosion, (6) moderately severe climate.

Climatic disadvantages combine with other limitations to restrict the choice and yield of crops and increase risks. The main crop is grass, with cereals and forage crops as possible alternatives where the increased hazards can be accepted.

Class 5. *Land with severe limitations that restrict its use to pasture, forestry and recreation*

Limitations are due to one or more of the following defects which cannot be corrected: (1) poor or very poor drainage, (2) frequent damaging floods, (3) steep slopes, (4) severe risk of erosion, (5) severe climate.

High rainfall, exposure and a restricted growing season, prohibit arable cropping although mechanised pasture improvements are feasible. The land has a wide range of capability for forestry and recreation.

Class 6. *Land with very severe limitations that restrict use to rough grazing, forestry and recreation*

Of the following limitations one or more cannot be corrected: (1) very poor drainage, (2) liability to frequent damaging floods, (3) shallow soil, (4) stones or boulders, (5) very steep slopes, (6) severe erosion, (7) very severe climate.

The land has limitations which are sufficiently severe to prevent the use of machinery for pasture improvement. Very steep ground which has some sustained grazing value is included. On level or gently sloping upland sites wetness is closely correlated with peat or peaty or humose flush soils.

Class 7. *Land with extremely severe limitations that cannot be rectified*

Limitations result from one or more of the following defects: (1) very poorly drained boggy soils, (2) extremely stony, rocky or boulder strewn soils, bare rock, scree, or beach sand and gravels, (3) untreated waste tips, (4) very steep gradients, (5) severe erosion, (6) extremely severe climate.

Exposed situations, protracted snow cover and a short growing season preclude forestry though a poor type of rough grazing may be available for a few months.

LAND-USE CAPABILITY SUBCLASSES

Capability subclasses are divisions within capability classes based on the kinds of limitation affecting land-use; these are:

w   Wetness

s   Soil limitations

g   Gradient and soil pattern limitations

e   Liability to erosion

c   Climatic limitations

Wetness may result from a variety of causes including slowly permeable materials (e.g. clays), impermeable layers such as iron-pans, high water-tables or flushes by spring water and flooding by rivers.

Soil limitations include shallowness, which restricts root development and nutrient uptake and also reduces the water-holding capacity of the soil. Stoniness also limits water capacity and nutrient status and may have a severe economic penalty because of the effect of stones on cultivations and mechanised harvesting techniques. Soil texture and structure also play a role in limiting the use of some areas, e.g. weakly structured soils which show 'cappings' are often fine sandy or silty loams.

Gradient and soil pattern limitations are important in that with increasing angles of slope mechanised farming becomes more difficult (Curtis *et al.*, 1965). Where complex patterns occur the use of land may be limited by the occurrence of bad patches (e.g. bogs) in the fields.

Wind and water erosion are experienced in different parts of Britain (see Chapter 13 for discussion of wind erosion) and include marine erosion, gullies on slopes where overburning or overgrazing has been practised, and the formation of peat hags.

Climatic limitations have been assessed mainly in terms of water balance and temperature. Three major climatic groupings have been used based on R − PT and Tx where

R = average rainfall (mm)

PT = average potential transpiration (mm)

Tx = long-term average of mean daily maximum temperature

Using these properties three climatic groupings were defined in relation to land-use capability:

1. R − Pt < 100 mm and Tx > 15°C

    There are few, if any, limitations imposed on crop growth.

2. R − Pt < 300 mm and Tx > 14°C but excluding Group I.

    These climates are moderately unfavourable for crop growth and choice of crops.

3. R − Pt > 300 mm or Tx < 14°C

    These climates give rise to somewhat severe limitations on range of crops.

§14.4 CASE STUDIES IN LAND CLASSIFICATION

A few examples may help to show how land may be classified into one or another grade in practice. An important point to be borne in mind is that land may be assigned to a grade for a variety of reasons and that these reasons will change from region to region. For example, in an area of high rainfall a clay soil may be downgraded because it is moisture retentive. Yet in an area of lower rainfall this very factor of moisture retention is an asset.

1. Hull. Sheet 99 Agricultural Land Service (MAFF, 1970)
This map shows an area of soils derived from glacial and post-glacial deposits. There are no limitations placed on the use of these soils by gradient or climate. Thus the differences in land classification arise principally from soil characteristics such as texture, stoniness and drainage. The lighter textured (sandy loam) soils permit a wide range of cropping and are mostly placed in Class 2. The heavier textured (clay) soils are less easy to cultivate and are mainly assigned to Grade III. Much of this land in the middle grade is very productive even if its versatility for cropping is somewhat

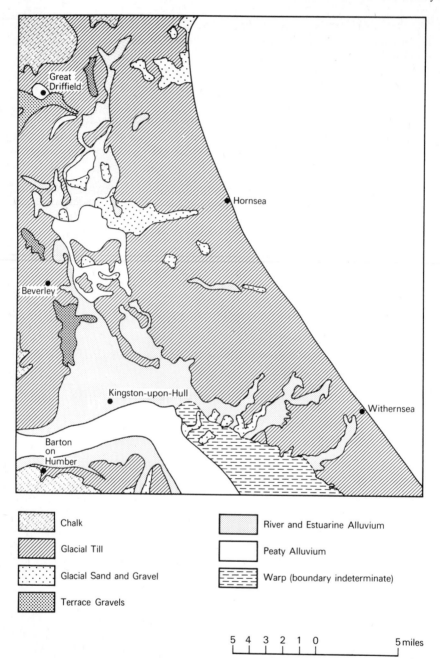

Chalk

Glacial Till

Glacial Sand and Gravel

Terrace Gravels

River and Estuarine Alluvium

Peaty Alluvium

Warp (boundary indeterminate)

5 4 3 2 1 0   5 miles

Fig. 14.1 Agricultural land classification. (Source: Ministry of Agriculture, Fisheries and Food map extract from Hull sheet 99. Report Map 3)

restricted. Some clay land suffering from poor drainage has been downgraded to Class 4 due to this deficiency. The area (in 1970) was mostly cropped with barley and wheat together with some potatoes. Some of the Class 2 land is capable of supporting horticultural crops but there is only limited horticultural use at present.

2. Beccles North Area. Sheet TM49, Soil Survey Land-Use Capability and Soil Maps. Soil Survey of England and Wales. Survey Record No. 1. 1970
A series of three maps, soils, land-use capability and drainage are published for this area and extracts of the land capability and soil map (scale 1 : 25 000) are shown in Fig. 14.2*a*, *b*. The extracts show a low-lying area including soils developed on fine textured mineral alluvium with saline subsoil (*Waveney* series), organic deposits (*Adventurers'* series and *Prickwillow* series), head and till deposits (*Hall, Ashley* and *Beccles* series), valley gravels and Pleistocene Sands and Gravels (*St Albans* and *Croxton* series).

Table 14.4 Soil groups and series

Soil group	Attributes	Series	Map symbol
Humus podzol	Acid sands with sub-surface colloidal humus accumulation	Redlodge	Ro
Brown earth	Ungleyed non-calcareous loam or sand soils	Hall, Freckenham, St Albans Croxton	hH, Fr SA CR
Brown earth with gleying	Non-calcareous loamy soils, slightly gleyed	Ashley	As
Brown calcareous soil with gleying	Moderately gleyed calcareous soil with impervious clay substrata	Hanslope	Hn
Surface-water gley soil	Moderately or intensely gleyed soils with impervious substrata and seasonal water-table, clay or with clayey subsoil	Aldeby, Beccles	aD, bW
Ground-water gley soil	Moderately to intensely gleyed soils with permanent water-table in the soil and substrata	Row, Gillingham	rO Gl
Peaty and humose gley soil	Intensely gleyed soils with permanent ground-water and peat or peat remnant surface	Isleham, Waveney	Id wE
Basin peat soil	Well-humified peat derived from fen plants; permanent ground-water	Adventurers', Prickwillow	An Pw
Raw peat soil	Poorly humified fen peat (Sedge-carr peat); permanent ground-water		

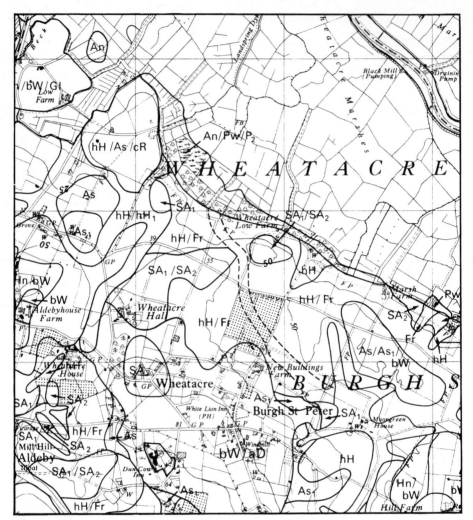

Fig. 14.2a Extract of soil map — north Beccles area (TM49). (Source: Corbett and Tatler, 1970)

The land quality in this small area ranges from Class 6w in the wet peaty areas in the vicinity of Wheatacre Low farm at the border of the Wheatacre marshes to Class 2s to the south near Burgh St Peter.

The *Hall, Ashley* and *Croxton* series all provide sandy loam or sandy clay loam soils which are moderately well-drained. These areas can be used for potatoes, sugar beet, tree and bush fruit, winter wheat and spring barley. In contrast, the *Waveney* series of organic silty clay loam over silty clay is unsuited to root crops or fruit crops. It is mainly used for wheat, barley and field beans and pump drainage is necessary. This series differs markedly from the *St Albans* series which is based on gravelly sand and is excessively drained so that it can suffer from severe droughtiness.

Fig. 14.2*b* **Extract of land capability map — north Beccles area. (Source: Corbett and Tatler, 1970)**

It will be evident that in this small area soil texture, stoniness and drainage varies considerably. These differences in turn affect soil properties such as available water capacity, base exchange capacity (nutrient status), porosity, aeration and soil temperatures.

3. Chalk Scarplands and Clay Vales. Ashford Sheet TR104. (Green and Fordham, 1973)

In the scarpland country of southern England contrasts of soil and quality are associated with topographic position. An example of an area near Ashford, Kent, shows a variety of physiographic regions. These range from downland consisting of plateau,

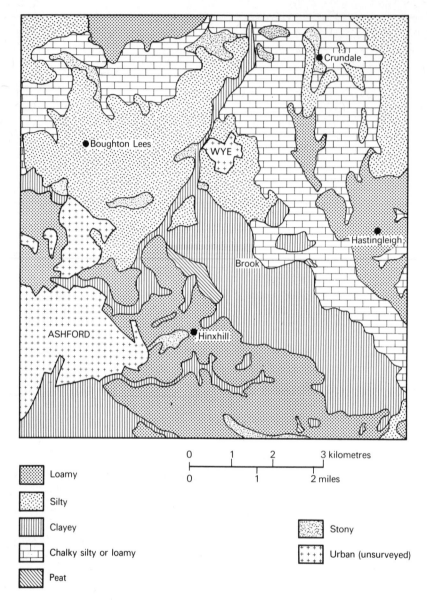

Fig. 14.3 **Soil texture and stoniness in an area near Ashford, Kent. (See Fig. 14.6 for a section across this area). (Source: Green and Fordham, 1973)**

escarpment, dry valleys and coombe slopes to undulating lowland and the Stour Valley. The undulating lowland can be sub-divided into Gault Vale, Lower Greensand Belt and Weald Clay Plain. Thus a wide range of geology and superficial drift deposits occur in this area producing variations in texture and stoniness (Fig. 14.2). Extracts from the soil maps and land classification maps of the area show the plateau, scarp and

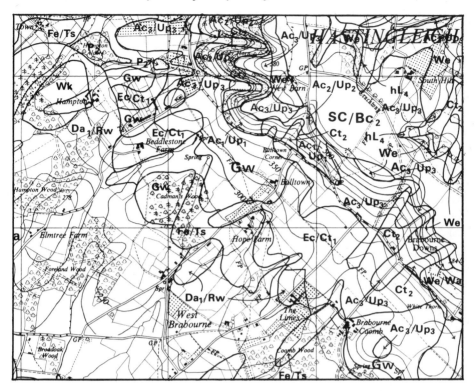

Fig. 14.4 Extract from a soil map — Ashford sheet (TR04). (Source: Green and Fordham, 1973)

undulating lowland (Figs 14.3 and 14.4). In the vicinity of Hastingleigh the scarpland gives rise to land classes 6gs (steep slopes and shallow soils) on the thin *Andover–Upton* soils comprising rendzinas and brown calcareous soils over thin silty drift and chalk. On the clay lowlands the *Denchworth* series a surface-water gley derived from Gault Clay provides Class 3w land. In contrast, the upland plateau carries loamy soils of the *Sheldwich* and *Batcombe* series based on Plateau Drift which can provide Class 3/2 land which may have some climatic limitations.

The general topographical relationships of the soils are illustrated in sectional form in Fig. 14.5 and the soil series and phases associated with different landscape elements are shown in Table 14.5.

§14.5 THE USES AND LIMITATIONS OF LAND CAPABILITY SURVEYS

The land classification map can be regarded as an information transfer from the earth scientists to the users (agronomists, economists, planners, etc.). In any such transfer there are problems concerning the relevance and the completeness of the data. Generally there is a loss of information because the scientist does not know precisely what

Table 14.5 Soils and landscape near Ashford, Kent. (Source: Green and Fordham, 1973)

	Landscape	Soil series and phases	Symbols
Plateau 500 ft (150 m)	Flat or gently sloping sites	Batcombe, Sheldwich, Maxted, Loamy Batcombe	Bc, SC mT
	Upper or middle slopes (11–20°)	Steep Andover, Steep Upton, Winchester, Wallop, Wye Downs	
	Very steep (20–30°) coombe sideslopes	Very steep Andover, Very steep Upton	
Downland — Escarpment, dry valleys and coombes	Gentle to moderate (<11°) valley side slopes including aprons, spurs and low benches	Rolling Andover, Rolling Upton, Charity, Shallow Coombe, Hamble, Stony Hamble	
	Dry valley and coombe footslopes and floors	Gore, Moderately deep Coombe, Deep Coombe	
	Gentle scarp footslopes with dissected benches	Shallow Coombe, Moderately deep Coombe, Gore, Eastwell	Ct, Gw, Ec
	250 ft (75 m) plateau	Batcombe	Bc
Gentle slopes to Stour floodplain	Plateau side slopes	Eastwell, Shallow Coombe, Winchester, Wallop, Moderately deep Coombe, Charity	We, Wa
Undulating lowland and Stour valley	Terrace and broad valley slopes	Charity, Hamble, Stony Hamble, Hook, Stony Hook, Park Gate	Cr, hL hK, Pz
	Hilltops, upper slopes and crests; shallow dry valley floors	Stony Denchworth, Rowsham, Stony Rowsham	Rw
Gault vale	Gentle to moderate slopes	Denchworth, Wicken, Glauconitic Denchworth, Glauconitic Rowsham	Da, Wk

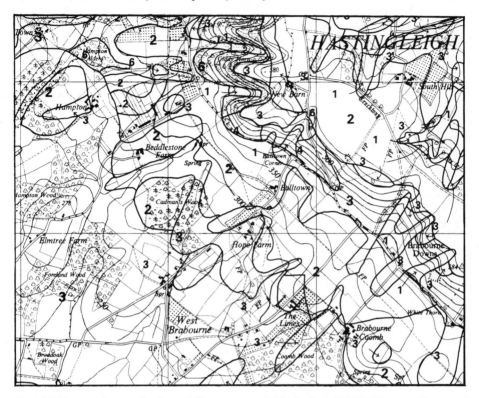

Fig. 14.5 Extract from a land capability map — Ashford sheet TR04. (Source: Green and Fordham, 1973)

data the user wants and, in some cases, the user may find it difficult to specify his requirements without knowledge of the range of data available (Young, 1972). There is further the serious problem that most classifications are general purpose classifications seeking to define categories of land in general terms. However, the requirements of different crops or different engineering structures are not the same. It may be, therefore, that the map is not sufficiently informative concerning particular uses (Gibbons, 1961). In these circumstances a special map may be required in which particular limitations are considered in the classification.

Soil is only one factor taken into account in the mapping of land capability. The role of soil data in land-use planning has been discussed by many authors including Young (1973), Mackney (1969), Mulcahy and Humphries (1967), Webster and Beckett (1968), Stephens (1953), Stewart (1968), Klingebiel and Montgomery (1966) and Bartelli *et al.* (1966). There are two main problems concerned with the use of soil data in land classification. First, better quantitative correlations between mappable (soil) characteristics and properties affecting yield are needed to give scientific expression to the soil factor in crop production. Such information could stem from experiments in which treatments are constant and soil property—yield relationships are examined

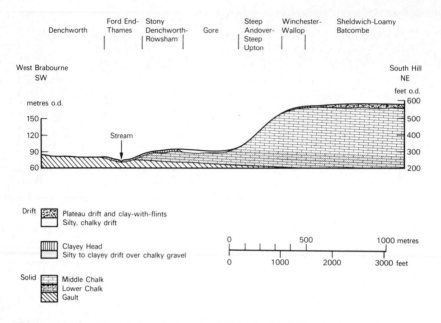

Fig. 14.6 Section through downland and undulating lowland. The dominant soil series occurring in different parts of the landscape are shown at the top of the figure. (Source: Green and Fordham, 1973)

(Mackney, 1969). Second, the user usually has to make decisions on a 'per field' basis and yet the land classification map normally shows units following soil boundaries. This problem may be overcome by rating the field according to the lowest grade found within it. Such an approach can be justified in that the limitations in part of a field (e.g. wetness) may be paramount in deciding land-use. Such an approach might, however, lead to lower output than is potentially available. Where this is the case a greater flexibility in the field boundaries will be sought. It is partly for this reason that the familiar hedgerows may disappear to allow more flexible use of land.

In practice the planning of the rational use of land is a complex problem requiring consideration of both physical and socio-economic factors. Nevertheless there must be an underlying assumption that existing soil resources will be conserved and where these soil resources are to be used for other purposes the poorest soils should be used first. In view of this it seems a paradox that land in Grades I and II should be used for urban development. However, such matters as accessibility, vandalism, landscape architecture and topography may all play a part in determining land-use. In many cases conflicts of interest arise which are difficult to resolve. Often the most difficult problems arise where the needs of people appear to be opposed to the requirements for nature conservation. In the following chapter several case studies have been chosen to illustrate the issues involved in making decisions about the interaction between soils and man.

FURTHER READING

Ministry of Agriculture, Fisheries and Food (MAFF) (1966). *Agricultural Land Classification*, Tech. Rep. Agric. Ld. Serv. No. 11.

Klingebiel, A. A. and Montgomery, P. H. (1961). *Land Capability Classification*, U.S. Dept. Agric. Soil Conserv. Serv. Agric. Handbk. No. 210.

MAFF (1968). *Agricultural Land Classification Map of England and Wales*, Explanatory Note, Agric. Ld. Serv.

15 SOME PROBLEMS OF MODERN FARMING

Some of the problems of modern farming, especially those concerning soil, are considered in this chapter. Themes currently under discussion by agriculturalists include the subjects of soil structure, soil drainage, eutrophication of inland waters and pollution by pesticides and herbicides. Although each of these is a very large topic in itself, and should merit an extensive treatment, the object of this discussion is to select topics and examples to emphasise the subtlety and complexity of the problems involved. Thus, the theme running through a rather diverse selection of topics is one of interdependence of factors in the agricultural ecosystem. The successful management of soils in Britain today involves a detailed understanding of the processes and factors involved, not only within the soil—crop system itself but also involving interactions within and between other systems; namely those of climate, drainage, geology, flora and fauna.

§15.1 SOIL STRUCTURE

One of the traditional aims of agriculturalists in Britain has been that of attempting to increase agricultural productivity by the addition of fertilisers in order to maintain soil nutrient levels (see for example Cooke, 1967). However, a parallel theme has emerged, that of the maintenance of a stable and well-developed soil structure. Structure management can be as important as fertility management in influencing crop productivity.

Structure management is important because of the relationship between the structure and the pore space in a soil. The existence of pore spaces between the solid particles of the soil is vital to drainage and aeration of the soil and hence to the satisfactory growth of plant roots. The size, shape and frequency of these pore spaces is determined by the way in which the solid particles of the soil are grouped together, or aggregated, into soil structures. For these reasons soil structure plays an important part in determining available water capacity and aeration in soils.

Soil structure can be viewed in two ways; static and dynamic. When seen in a static sense the important factors are simply the size, shape and interconnectivity of the soil pore spaces. The deepest, most interconnected pores allow for the greatest penetration of plant roots into the soil. The largest pores aid aeration and the rapid transport of water while the smallest pores act to retain water and reduce permeability (Table 15.1).

Table 15.1 The relationship between pore diameter and pore function. (After Brewer, 1964)

Pore type	Size range (μm)	Function
Macropore	Over 75	Rapid transport of water and air
Mesopore	75–30	Reservoir of water
Micropores	Less than 30	Water unavailable to plants

It is evident from Table 15.1 that soils should possess both macro- and meso-pores for healthy plant root growth, with ample water and air.

When soil structure is viewed in a dynamic sense it is evident that the pore space characteristics of any one soil are not necessarily constant over time. In this context it is the *stability* of the pores that becomes important. The pore stability will depend upon the structure stability and this in turn depends in part upon soil texture, in part upon the presence of cementing agents and in part upon soil management practices.

If the soil contains a large proportion of swelling clays (chiefly montmorillonite) and humus it will undergo shrinkage when water is removed by drainage or evaporation. As a result macro-scale tension cracks may appear, separating the soil into large structural units with fissures in between. In this way, although pore size may initially control water permeability, drying out may result in a temporary increase in permeability.

As far as plant growth is concerned the important factor is that the soil pore space characteristics are strongly influenced by the stability of the soil structure when wetted. The presence of earthworms, chemical cements, inorganic colloids (particles of less than 0.001 mm diameter), humus and plant roots themselves all act to encourage the formation of soil structures stable in water.

The study of soil structure stability under cultivation (MAFF, 1970; Cooke and Williams, 1971) has lead to the knowledge that not all soils have inherently stable structures. The principal bonding agents are organic, clay and iron colloids.

It is believed that soils with a high silt content may be prone to instability because silt lacks clay or organic colloids. In Britain the areas where structure instability problems tend to occur are shown in Fig. 15.1. The significant parent materials are the silts, shales and clays. Clay soils may be prone to structure instability because of their inherently poor drainage, i.e. frequent wetting. Soil structures are generally at their least stable when wet. This is chiefly because the colloidal bonds which act to bind the soil particles may be partly soluble and are therefore more effective when the soil is relatively dry. One of the other important factors influencing the stability of soil structures is that of the length of time in which the soil has been cultivated. Under agricultural use soil structures may be compressed or deformed by pressure from farm

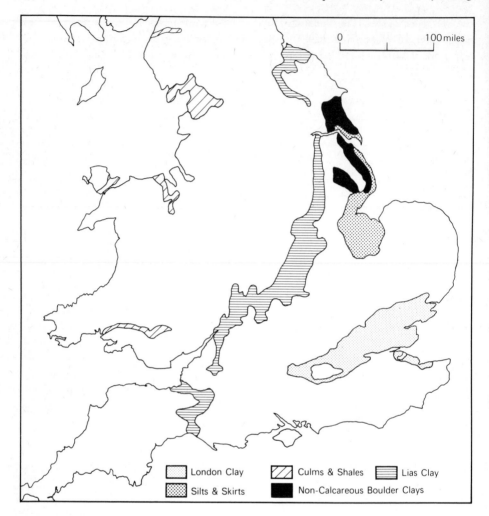

Fig. 15.1 Areas prone to soil structure deterioration under intensive cultivation. (Source: *The Times*, **23 November 1970)**

machinery or from grazing animals. This action of compression naturally reduces the size and amount of pore spaces in the soil and therefore, as may be deduced from Table 15.1, drainage and aeration are impeded and root growth is hindered. Soils which have been uncultivated and have been under grassland show a greater structure stability when cultivated than those which have been under continuous cultivation. This is chiefly due to the increase in organic matter content which takes place when the soil is under grass. The effects of humus, earthworms, cementing agents, texture and cultivation have been investigated in research work on structure stability which is discussed below. The important factors influencing structure stability are biological activity, biological, chemical and physical bonding and cultivation policy.

15.1.1. Biological chemical and physical factors

The general significance of earthworms to soil structure has been known ever since the pioneering work of Darwin (1882). More recent work has attempted to quantify the effects of earthworm activity upon soil structure. An interesting example is the work of Van de Westeringh (1972), who compared the soil structures in two sets of soil profiles; one set had abundant earthworms, the other set was worm free. In all other respects the soils were virtually identical. The worm-free soils had been treated with pesticides (copper oxychloride or DNOC) which killed the worms. These soils showed not only a marked decrease in the incorporation of organic matter into the soil (giving a noticeably thicker organic 'O' horizon) but also a markedly lower number of pore spaces of a specific size. The relationship between the presence of worms and the numbers of visible pores of a specific size is shown in Table 15.2.

Table 15.2 The relationship between presence of earthworms and pore spaces. (Source: Van de Westeringh, 1972)

| Soil depth (in) | (cm) | 1. No worms present | | 2. Worms present | |
| | | Pores | | Pores | |
		Over 0.16 in (4 mm) dia.	0.08—0.16 in (2—4 mm) dia.	Over 0.16 in (4 mm) dia.	0.08—0.16 in (2—4 mm) dia.
0—3.94	0—10	0	3	8	19
3.94—7.87	10—20	1	1	7	15
7.87—11.81	20—30	4	3	8	9
11.81—15.75	30—40	7	4	1	5

As can be seen from the table, there is a striking loss of porosity concomitant with the loss of earthworms; this feature being most marked in the upper layers of the soil. Overall, the structures in the topsoil of the worm-free profile were sub-angular, blocky and cloddy but in the control profile where worms were present there was a good porous crumb structure more amenable to plant growth (Fig. 15.2). One of the most important aspects of this research is, however, that the structures in Profile I (worm free) were far less stable in water than those in Profile II where the worm species *Allolobophora caliginosa*, *A. rosea* and *A. chlorotica* were present together with some individuals of *Lumbricus terrestris*. The research demonstrates that the absence of earthworms decreases soil porosity and leads to instability of the soil structures.

The importance of the binding action of humus and micro-organisms on soil structure stability was demonstrated by Swaby (1949, 1950). He undertook a number of experiments whereby soils were inoculated with fungi, bacteria or actinomycetes while the control soil sample was sterile. The structures of the control sample were unstable and broke down easily when wetted but the inoculated soils were far more stable, especially where the fungal mycelia were present. Not only do the fungal mycelia help to bind the particles together, but the decomposition products of the fungi were important cementing agents. It was also evident from his research that the boundary between inorganic and organic factors tends to break down, for Swaby

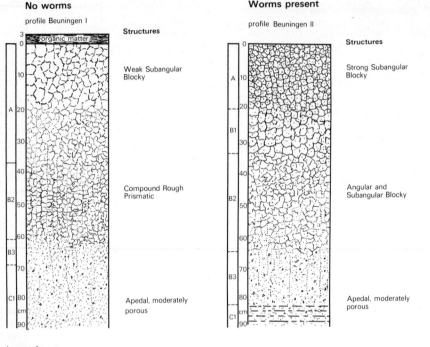

Fig. 15.2 Soil horizons and soil structures in profiles without and with worms. (Source: Van de Westeringh, 1972)

concluded that the soil crumbs may be held together by a calcium—humate 'cement' and by sesquioxides as well as by fungal mycelia. These conclusions were later supported by the work of Arca and Weed (1968) on the effects of iron oxide and clays on soil aggregates and porosity. Iron may be especially important because under anaerobic, waterlogged conditions (such as may be found in poorly drained fields) the iron is chemically reduced to the ferrous state which is soluble. However, when oxidised (as may occur if the soil is drained) the iron changes to the ferric state which is relatively insoluble and tends to precipitate. This precipitate of ferric oxide can act as a cement between soil particles.

15.1.2 Policy: cultivation and management

The report of the Ministry of Agriculture, Fisheries and Food (1970) — *Modern Farming and the Soil* — emphasised that cultivation should not take place when soils have a moisture content higher than field capacity. Only when the soil is below field capacity does the soil have sufficient strength, stability and consistence for the structure to withstand the weight of farm machinery. This also applies to the stocking of fields, as the hooves of grazing animals can exert considerable pressures capable of damaging soil structure when the soil is wet. It is only when the soil moisture content is at a level lower than field capacity that the formation and hardening of organic and inorganic cements leads to an increase in structure stability.

The consequences of cultivation when wet can be readily seen in a soil profile (Fig. 15.3). A smeared layer or plough pan may be visible at between 5 and 12 in (15—30 cm) depth. Pan formation can occur at even shallower depths if a field is overstocked with grazing animals when the soil is too wet. Clearly, when smearing and structure collapse take place under pressure the vertical penetration of water, and indeed plants roots and air, is hindered. It is thought that under these anaerobic conditions the binding organic agents may be degraded and also the cementing ferric iron may be reduced to the more soluble ferrous, both having a further deleterious effect on soil structure.

Several methods can be adopted to avoid structure deterioration. Firstly, the policy of ploughing only when the soil water content is lower than field capacity is the most important. The *timeliness of cultivation* is, therefore, a crucial issue. However, the variability of the British weather may mean that the actual field situation may fall short of the optimum conditions for long periods. If a particularly wet spring occurs the farmer may feel obliged to cultivate the soil in order to sow seed early enough even if the soil is too wet to avoid damage to structure.

A second technique for structure improvement is to improve the drainage of fields by the increased use of field drains in areas prone to structure collapse. Thirdly, organic matters, lime or soil conditioners may be applied to add stability. Soil conditioners (Weakly, 1960) are synthetic organic polymers which help to aggregate soil particles and thus stabilise soil structures. Lime and organic matter aid structure stabilisation not only directly, but also by promoting the growth of soil organisms which are important in stabilising soil structures. Fourth, and lastly, as far as cultivation with tractors is concerned, cage wheels may be used in addition to the normal tractor tyres. This helps to distribute the weight of the tractor more evenly, causing less soil compaction. The compressive and smearing action of a normal tractor tyre passing over the soil (Fig. 15.4) can be compared with the effect of an auxiliary cage (Fig. 15.5).

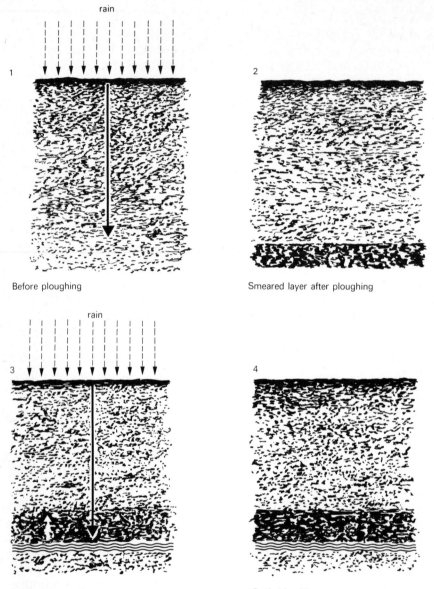

Fig. 15.3 Formation of plough pans in unstable structures. (Source: Ministry of Agriculture, Fisheries and Food, 1970b)

While it is possible to employ these technological solutions to the problems of soil structure stability it is also clear that an assessment of the soil's liability to structure deterioration is helpful so that structure deterioration may be avoided in the first place. It is very necessary to stress this point because it might be easy to gain the

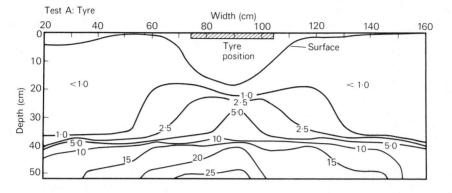

Fig. 15.4 Distribution of cone resistance (bar) after passage of test equipment Test A: Tyre. Contours show the compression of the soil under the tyre track and the sideways displacement of surface soil. (Source: Soane, 1973)

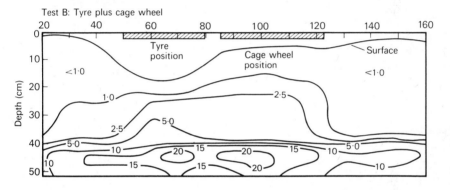

Fig. 15.5 Distribution of cone resistance (bar) after passage of test equipment Test B: Tyre plus cage wheel. Contours show that the spreading of the load results in less compaction. (Source: Soane, 1973)

impression that modern farming necessarily harms the soil. This is not the case; it is simply that some soils are more suited to intensive agriculture than others. Thus it may be possible to cultivate, for example, a loamy soil or sandy loam soil at a high moisture content without serious damage. However, in a silt soil, at a similar moisture content, serious damage may result because a silt may not necessarily have the load-bearing properties that sandy soils have. Indeed, some farmers in Britain report that they can use heavy crawler tractors in early spring without structure damage while nearby farmers report that cultivation at this time is impossible without causing structure deterioration. Local variations in soil type may well be responsible for this situation.

Clearly, knowledge of the load-bearing properties of soil types under different moisture conditions is a crucial step in the prevention of structure deterioration. Work at Rothamsted Experimental Station has recently concentrated upon the measurement of structure stability (Cooke and Williams, 1971) by a study of the decrease in the amount of pore space (increase in bulk density) upon the slaking of structures with water. While more research is needed on this topic it is becoming clear that there is no cure for the heaviest of lands having structures which are unstable in water, drain badly and compact easily. These should simply be put under pasture and grazed with care. At the other end of the spectrum there are robust loam and sandy loam soils where structure deterioration presents little or no problem. Many soils, however, require careful drainage, and the timeliness of cultivation is of prime importance.

§15.2 SOIL DRAINAGE AND THE EUTROPHICATION OF WATER SUPPLIES

Although in the context of soil structure problems the extension of soil drainage schemes may be beneficial, they may present problems in respect of the quality of water supply. We may here emphasise our general theme of the subtlety and complexity of the problems involved and also point out that although action may be taken to tackle one set of problems this action may well create other problems. In the case of land drainage it is thought that the increased level of nutrients (eutrophication) in inland waters may be causally connected with improvements in drainage and with fertiliser applications.

In recent years attention has been drawn to the increase of nitrates in water supplies, its possible sources from agricultural fertilisers and its possible effects on human health (Tomlinson, 1970; Cooke, 1972). Although there is no doubt that an increase of nitrates in streams is occurring in some places in Britain (Fig. 15.6) and this is coincident with an increased use of fertilisers, the mechanisms of cause and effect are by no means clear or proven as Edwards (1976) and Green (1973) have pointed out. It appears that the problem may not be a simple one of increased nitrate application to the fields leading directly to increased nitrate levels in water. Changes in

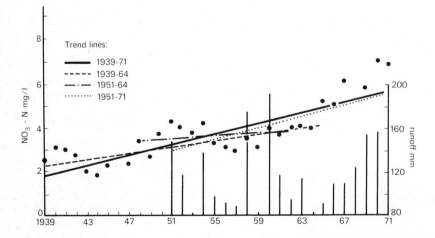

Fig. 15.6 Mean annual nitrate nitrogen concentration and run-off, River Stour. (Source: Edwards, 1976)

drainage and climate may also be involved. Green has suggested that on the one hand an increase in soil drainage may simply be leading to an increased leaching of soil nutrients, including nitrogen, while on the other hand the noticeable increase (since 1965) of intensive rainfall may have flushed more fertilisers out from the soil. The argument is that periodic, intensive rainfall has a more marked effect on the leaching of soil nutrients than rain falling in small amounts evenly spaced throughout time.

A further line of evidence comes from the observations of Edwards of the Yorkshire River Authority. He shows that not only have nitrates increased, but other constituents have as well, especially the highly soluble carbonates. The key point is that these elements are not being supplied as fertilisers and therefore it can be argued that some other factor is important besides application of fertilisers (Table 15.3). Furthermore, it is not economic for farmers to waste money on excessive and increasing doses of fertiliser.

Table 15.3 **Increases in soluble constituents in inland waters. (After Edwards, 1976)**

Nitrates (N mg/l)		*Non-carbonate hardness* (HCO_3^- mg/l)	
1939	*circa* 2	1948	110
1971	*circa* 5*	1972	140

* The WHO maximum acceptable figure for human health is 10 mg/l.

Green makes the point that a useful contrast can be made between East Anglia and the Hampshire Basin. In the former nitrates have increased in streams in recent years; this coincides with increased fertiliser use and also with increased field drainage. However, in the latter area there has been no great increase in field drainage and neither has there been any great increase in nitrates in the streams, even though the rate of fertiliser application has been comparable in the two areas. This evidence is, of course, circumstantial and the coincidence of trends does not prove cause and effect; other factors may well be important. Nevertheless one message is very clear. The relationship of eutrophication to farm fertilisers is complex and involves many factors. The statement that heavy use of fertilisers directly causes water pollution of water supplies is over-simplistic and ignores many subtle links in a ramifying chain of events that tend to typify so many environmental problems.

§15.3 THE AGRICULTURAL ECOSYSTEM: PESTICIDES AND HERBICIDES

When studying certain aspects of modern farming and soil problems a very useful viewpoint to adopt is that of the *agricultural ecosystem*. By doing so we focus our attention upon a number of important themes. Firstly, the functional and ecological inter-relationships that exist between plants and domesticated animals on the one hand and the physical environment on the other. Secondly, we may focus upon the continuity of the flow of materials and energy in an ecosystem dominated by man's activities. Thirdly, one may perceive that one action rarely has just one reaction. Multiple consequences and side-effects are common. We have already touched upon some of these themes but they are especially important when discussing the problems of pesticides and herbicides.

The agricultural use of pesticides and herbicides has underlined, in a unique and striking way, the interdependence of ecosystems. The dramatic and well-publicised side-effects throughout food chains so eloquently expressed by Carson (1963) shows that pesticides applied to crops, seed and soil may rapidly find their way through hydrological and biological systems to be distributed throughout the world. The initial realisation of these facts had a dramatic and emotional impact so that in more recent years a great deal of scientific research has been undertaken. This seeks to minimise the harmful side-effects and to maximise the benefits that accrue from the use of pesticides and herbicides.

It appears that pesticides are here to stay. Modern agriculturalists maintain that it is no longer economic to farm competitively without them. It is undoubted that they are helping considerably in increasing the world's food supply. But they can constitute a serious pollutant.

Pollution in this case needs careful definition: it may either be a chemical in the wrong place (at any concentration) or it may be in the right place but at the wrong concentration. Drainage is often an important factor in getting pollutants into the wrong place. If a chemical is non-biodegradable it may be transported in mobile soil water and thence into streams. Additionally concentration may occur. While a dose may be applied that is safe for a particular acreage of field one of the functions of a drainage system is to collect and integrate soil waters. Thus chemicals may be collected by field drains and ditches into streams leading to much higher concentrations than originally applied.

It is important to attempt to quantify these losses in order to attempt to estimate a safe dosage for a field. Such quantitative experiments have, for example, been under-taken in America (United States Department of Agriculture, 1971) on the use of Dieldrin. The pesticide was applied at a rate of 5 lb/acre and some 30 per cent of it was immediately lost by direct volatilisation into the atmosphere during application. Three per cent was lost in the first year in eroded topsoil and 0.1 per cent was lost in drainage waters (at concentrations of 20 to 30 p.p.m.). Of the remainder, slow volatili-sation from the soil occurred at a rate which, by projection, would take between 10–12 years for the pesticide to disappear. This logarithmic rate of decomposition has also been noted by other workers; for example, Edwards (1966).

Clearly the lessons which can be learnt from such experiments are that the direct volatilisation of pesticides at the time of spraying should be minimised because this appears to be one of the chief ways that the chemicals may be generally distributed into the environment. Furthermore topsoil erosion, with its removal of absorbed pesticide, should be controlled and the more rapid decomposition of the pesticide in the soil should be encouraged.

On the question of decomposition in the soil, the mechanisms and rates are now becoming more fully understood, especially as a result of the work being carried out at

Rothamsted Experimental Station, and by chemical manufacturers. Probably one of the most satisfactory results to come out of this research is that chemicals have been discovered which can be decomposed by soil organisms, but which are nevertheless effective in their intended role of pest and weed eradication. From other research it appears that not only may soil micro-organisms simply decompose some chemicals they may, in fact, be using them as a food source. For example, the bacterium *Coryne-bacterium simplex* appears to be able to use the herbicide 'DNOC' as a source of carbon and nitrogen.

Because much of the breakdown in soils is bacterial the factors of temperature and moisture are important if rapid breakdown is to be encouraged. In a warm, moist soil, where bacterial growth is encouraged, the decomposition of a chemical may take only a few months whereas it may persist for over a year in a dry soil. Furthermore, the incorporation of organic matter in the soil appears to minimise the harmful effects of many chemicals; the chemicals being bound into the organic structures.

A further factor is that different chemicals have inherently different persistence characteristics. From the work of Hocombe *et al.* (1966) we know that the herbicide chloroprophon persists for less than 5 weeks but simazine may persist for over 60 weeks. In the latter case, such a relatively long life time means that accumulation would take place from year to year as not all of each year's dosage would have decomposed by the time of the next application. The relative persistences of some herbicides is shown in Table 15.4.

Table 15.4 Persistence of some herbicides (compiled from several sources)

Herbicide	Normal field dose (kg/ha)	Persistence (weeks)
Simazine	1	Over 60
Linuron	2	25–60
Chloroprophon	1.5	Less than 5
Dichlorobenil	1.5	25–60

What recommendations can be put forward as a result of this research? One might concern the type of soil management. The maintenance of a high soil organic matter content is clearly beneficial to the growth of micro-organisms as is the maintenance of optimum water and temperature conditions where possible. Thus drainage or irrigation together with the addition of farmyard manure may be useful management techniques.

A second approach is by improving the technical aspects of application and use. This approach is probably more important than that of soil management because it may prevent problems occurring in the first place. As a guiding rule it should be stated that a chemical should be applied and be effective when and where it is required and should decompose when and where it is not required. Thus non-persistent chemicals which decompose rapidly, preferably within a year, are desirable. One recent technical advance is the use of chemicals which are readily decomposable but which are encapsulated and are released only slowly from the capsule throughout the growing season. If these capsules are drilled into the soil next to the seed (rather than employing an overall spray) then many of the problems of environmental pollution which have in the past resulted from the application of pesticides and herbicides may be avoided in the future. Pesticides and herbicides should be agriculturally effective without being ecologically damaging and it appears that modern research is making this possible.

§15.4 MODERN FARMING AND SOIL PROBLEMS: CONCLUSION

From the wide ranging nature of this chapter it is clear that it is very necessary to take an ecological view of these problems. Soil conditions, drainage, water supply, pesticides, herbicides, flora, fauna and farming practice all interact in the natural habitat. It is not enough to consider soil conditions *per se* nor simply soil conditions and farming practice. Such a simplistic approach is insufficient in terms of achieving profitable farming in the long term. Moreover it is woefully inadequate in terms of the broader environmental issues. It is possible to have two objectives, first maximisation of agricultural output; and second, protection of the environment, and its flora and fauna. In reality, these two objectives are, in the long term, interdependent.

In the next chapter we shall attempt to show by case studies how the inherent nature of soils may have an effect on man and how these problems can be handled to obtain the maximum benefit from the land. These case studies are chosen from a wide range of possible examples. Those selected deal with the following topics:

1. Reclamation of derelict land.
2. Problems of soils and human and animal health.
3. Hedges and conservation in lowland areas.
4. Competing land-use and pressures on soils of upland areas.

FURTHER READING

The report entitled *Modern Farming and the Soil* (Ministry of Agriculture, Fisheries and Food, 1970) is essential reading. The Annual Reports of Rothamsted Experimental Station regularly have articles of interest, for example, Cooke and Williams, 1971.

16 SOILS AND MAN: CASE STUDIES

§16.1 RECLAMATION OF DERELICT LAND

This case study is derived from work by the University of Newcastle upon Tyne (1971) on the reclamation of pit heaps (mining spoil). This is an interesting example because basic, theoretical knowledge on plant colonisation and rock weathering have been used in an applied sense. It is of general interest because it is one aspect of the attempts being made to maximise the use of limited land resources in Britain today. Clearly, in the natural state, the capability of derelict land is very low yet the capability may be considerably improved for various uses if reclamation is carried out.

It is possible to identify two rather separate topics within the subject of land reclamation. One is the reclamation of an area for visual amenity: the area should simply be made to look pleasant and be aesthetically acceptable. The other is reclamation for agricultural production and this is rather more difficult and the chances of success are less easily forecast.

The physical conditions of the spoil parent material impose very severe restrictions on plant growth. The shales that constitute the soil-forming material are usually unweathered, not having been exposed on the surface of the earth for very long, and they provide infertile soils. It should be emphasised that rock weathering and soil fertility are complementary: it is weathering which releases the nutrients into the soil. However, when the rock does begin to weather, the products may not be entirely beneficial for plant growth at the outset. The weathering product of shale tends to be almost pure kaolinite clay which has a low cation exchange capacity and is liable to puddling when wet. The chemical products tend to be very acid. The acidity is derived from iron sulphides present in the shales which weather in the presence of water to produce sulphuric acid. A typical pH profile through a weathered pit heap is shown on Table 16.1.

Very large quantities of lime are required to raise the pH of the topsoil to a level suitable for plant growth. For example, at the pH of 2.5 (depth 3–8 in; 10–22 cm) an application of 12 tons/acre (5 tonnes/hectare) of lime is needed in order to raise the soil pH to around pH 6–7. At a pH of 3.7, the surface layer, which has already been limed to some extent, requires a further 2.5 tons/acre (1 tonne/hectare). (pH is a

Table 16.1 Soil profile in weathered pit heap. (Source: University of Newcastle upon Tyne, 1971)

Depth (in)	(cm)	pH	Explanation
0–3.9	(0–10)	3.7	Limed topsoil
3.9–8.6	(10–22)	2.5	Highly acid weathered rock
8.6–14.5	(22–37)	4.4	Weathering rock
14.5–20.7	(37–52)	5.4	Weathering rock
20.4–34.6	(52–88)	7.4	Unweathered rock
34.6+	(88+)	7.5	Unweathered rock

measurement on a logarithmic scale and therefore the difference in acidity between pH 2 and pH 3 is far greater than that between pH 3 and pH 4 and therefore far more lime is needed per unit of pH the lower the measurement is down the pH scale.)

Extreme physical conditions, especially of soil acidity and infertility, coupled with extremes of drought and wetness (at the top and foot of slopes respectively) together with extremes of temperature on the uninsulated surface, make this an unpromising environment for colonisation with vegetation. One method of improvement would obviously be to add large amounts of lime, together with other fertilisers, and to clothe the pit heap with some kind of stabilising material (e.g. nylon fibres, straw or hemp nets). Seed could then be applied to establish vegetation. The difficulty with this approach is that it would be very expensive.

An alternative approach is to attempt to plant species which would tolerate the existing harsh conditions. This involves the study of the vegetation which occurs naturally on older, mature pit heaps. Using knowledge of which species can tolerate the difficult physical conditions it is possible to vegetate the heaps by sowing seeds of these naturally occurring tolerant species.

In other words the first approach is to modify the soil conditions to suit the desired vegetation and the second is to use the more tolerant species suitable for the soil. The first approach is necessary where it is desired to grow a more demanding agricultural crop. The second would be sufficient for reclamation on grounds of visual amenity alone. In the latter case it may also be useful to breed an experimental specialised ecotype of say, grass, in order to clothe the heap in green and thereby remove a local eyesore.

The natural colonisers of the pit heaps include some mosses and some grasses such as the cocksfoot (*Dactylis glomerata*), the false oat (*Arrhenatherum elatius*), white bent (*Agrostis tenuis*) as well as some tree species; for example, Birch (*Betula*), Hawthorn (*Crataegus*), Alder (*Alnus*) and shrubs such as Gorse (*Ulex*) and the Bramble (*Rubus*). These species can be planted for visual amenity reclamation, especially when some of the sulphate has been leached out and therefore the acidity is not extreme. The older, leached heaps are termed mature or receptive for planting.

The financial return from those heaps planted with the species mentioned above will be very limited, both in terms of grazing and of timber. Agriculturally useful grasses and economic softwoods may, however, be planted after some soil treatment. In the case study the useful grasses fescue (*Festuca rubra*) and rye (*Lolium perenne*) were planted together with the nitrogen-fixing white clover (*Trifolium repens*) and also a mixture of grasses occurring in the natural succession. Furthermore, for timber, not only trees of the natural succession were planted but also coniferous species, especially the pines *Pinus nigra* and *P. contorta*. It was found that *P. nigra* was the most successful, with the highest rate of survival and an adequate timber production. In this way useful agricultural land may be gained from derelict land, productive in terms of grazing and also of timber. Regular checks are necessary on the grasses and cattle for any toxic elements which may be present in the pit, a point elaborated upon in the next section. Thus, in summary, the procedure outlined below allows a derelict area to be reclaimed and put to good use by raising its capability:

1. Study the natural succession of vegetation to suggest which species to plant.

2. Introduce species of economic use.

3. Carry out small-scale experiments with seed mixtures.

4. Carry out soil testing of parent material.

5. Ensure conditions are suitable for desired seed mixtures.

6. Monitor results and check for possible side effects.

7. Recommendations for further plots.

This procedure does, however, involve considerable effort and expenditure which is not possible in all cases. It is only when increasing pressure is put upon land-use that this kind of reclamation will become increasingly worthwhile.

§16.2 PROBLEMS OF SOILS AND HUMAN AND ANIMAL HEALTH

Soils may be locally polluted by heavy metals and trace elements which may have a deleterious effect on human health if crops or livestock products from these areas are consumed by man. In some cases, for example, next to old lead mines, the problems are well known and obvious. But in other cases the problem may be a very subtle one, difficult to prove and difficult to eliminate. In the study of the relationships between human health and soils one method of approach is to compare maps of human disease and mortality (for example, from cancer) with maps of the distribution of trace elements (for example, lead or copper). Maps produced by the Institute of Geological Sciences show trace element concentrations in soils and rocks in Britain. There is, however, a major shortcoming to this approach in that the coincidence of map patterns is not proof, in itself, of cause and effect. Furthermore, the food eaten by any one individual will have usually come from very diverse sources, not just from the local area.

In Britain there are some well recognised areas where problems with trace elements occur. It is known that soils developed on marine black shales (mostly of Carboniferous age) can be high in trace elements which may be injurious to animal health. Molybdenum is an example of such an element in the black shales of south-west

England, the Welsh borderlands and in the north-west of England. A high molybdenum content in soils has the effect of depressing the availability of copper. This has an effect on livestock, mostly by causing deformities of their bone structures. Cattle with deformed limbs and splayed hooves can be seen, for example, on the Somerset Levels on the *Evesham* series, a gleyed calcareous soil developed on Lias clays (Findlay, 1965) where the pastures giving rise to this condition are termed 'teart' pastures. Mild symptoms appear at concentrations greater than 3 p.p.m. molybdenum in the herbage and in severe cases the concentration may rise to 20 p.p.m. On the Mendip Hills, Somerset, and also in the Peak District of Derbyshire, lead occurs in soils derived from lead veins in the Carboniferous Limestone, the mining spoil having been distributed by mining activity. The disturbed ground is termed 'gruffy ground' on Mendip (Findlay, 1965). Lead is also found distributed by mining activity in central and south Wales especially in alluvial deposits downstream of mining areas. Thus pastures many miles from lead veins may be contaminated through the transport of a lead-rich sediment in rivers and the deposition of this sediment during floods.

Frequently igneous rocks show a higher concentration of potentially harmful trace elements than do sedimentary rocks. The sedimentary rocks, such as sandstones, have already been through weathering cycles in geological times and may often be leached of these elements but this is not always the case. In rocks like shales and clays the leaching products may have accumulated in earlier weathering phases in geological time. The relative concentrations in different parent materials are shown in Table 16.2.

Table 16.2 Concentrations of trace elements in soils on various parent materials in Britain (after data from the Institute of Geological Sciences). (Figures in p.p.m.)

Element	Serpentine	Granite	Shales	Sandstone
Cobalt	80	2	20	3
Chromium	3 000	5	200	30
Copper	20	10	10	10
Manganese	3 000	700	1 000	200

Although the degree of causality in the links between the occurrence of trace elements and human health is difficult to establish, with farm animals the relationships are a little clearer. Sheep, for example, appear to find a concentration of above about 30 p.p.m. copper in their body tissue toxic but much higher concentrations of zinc can readily be tolerated. Horses appear to be more liable to lead poisoning than sheep or cattle as they tend to ingest some soil as well as grass. This may lead to lead poisoning because lead is not very mobile and tends to remain in the soil rather than move upwards into the vegetation.

It might be thought that the vigour of the plant would be a guide to toxicity. However, healthy plant growth need not be a sign that a plant is safe for consumption by man. Experiments on lettuces have shown that even with as much as 1 000 p.p.m. cadmium in the plant tissues, plant growth is unimpaired but this concentration would be toxic to man.

That apparent associations exist between soils and human health is shown by some research workers (for example, Armstrong, 1962, 1964; Sauer *et al.*, 1963; Hill *et al.*, 1973). Statistical tests have suggested that soil contents of organic matter, zinc and cobalt were positively and significantly correlated with mortality from stomach cancer, but again it is difficult to be sure of the actual causal links. As Howe (1959) has pointed out, it is possible that it is more valid to study the relationship between disease and water quality, rather than with soils. This is because an individual usually drinks water from one identifiable source, while vegetables and meat can come from a multiplicity of possible sources and possible soils. Certainly research in this subject is a challenging one because of the complexity of the factors involved.

One may note that man-made pollution of soils can lead to toxic concentrations of trace elements as well as toxicity due to the weathering of rocks. For example, lead, cadmium, mercury and chromium can be produced as a waste product of industrial manufacture. Fumes released into the atmosphere can reach the land as dry fallout or be washed out by rainfall. Another example of possible man-made contamination arises from the increasing pressure for the use of sewage sludge on farmland. This is seen as a more desirable alternative than the pollution of inland waters with organic matter. But the sludge may contain between 20 and 100 p.p.m. copper, and as the organic matter is oxidised in the soil by the soil micro-organisms then the copper will be concentrated and left as a residue in the soil. The concentrations may then rise to a toxic level reaching figures of 1 000—5 000 p.p.m. Lead may also be found in sewage sludge (1 000—2 000 p.p.m.) but it is relatively insoluble and is therefore less liable to be taken up by plants than the more mobile zinc and copper. It is clear that great caution should be exercised in the application of sewage sludge to fields. Monitoring, not only the trace element concentration of the original sludge, but also the soils subsequent to application while the sludge is being decomposed, should be carried out.

It is useful to distinguish between those elements which are essential and are therefore only a pollutant at high concentrations and those elements which are not essential and are a pollutant at any level. In the first group, copper and zinc are important plant nutrients. However, nickel, lead, cadmium, mercury and chromium are not essential and are therefore pollutants at whatever levels they may be found. At levels around 2 p.p.m. copper is deficient and plant growth will be hindered; around 20 p.p.m. is normal and up to about 100 p.p.m. is the approximate maximum concentration tolerated. Above 100 p.p.m. the concentrations may be deleterious to plant growth and livestock and constitute a possible human health hazard (though it should be emphasised that different crops and different animals possess different tolerances to the higher concentrations). Examples of ranges, averages and general toxicities are shown in Table 16.3.

Where trace elements reach toxic concentrations the most effective remedy is liming. This has the effect of raising the pH of the soil towards alkalinity and as most of the trace elements show a greater solubility in acid conditions the elements are rendered largely insoluble and immobile in the soil. Furthermore, many toxic elements may be adsorbed relatively harmlessly in the organic matter of the soil and so main-

Table 16.3 Trace element concentrations in soils and possible toxicities (after data from the Institute of Geological Sciences)

Element	Average*	Normal range	Toxic concentrations†
Essential			
Copper	20	2–1 000	1 000–5 000
Zinc	3 000	60–2 000	2 000–50 000
Non-essential			
Nickel	40	10–1 000	1 000–8 000
Lead	10	0.5–200	1 000–2 000
Cadmium	0.1	0.5–1.0	100–1 000
Mercury	0.2	0.1–0.3	9–30
Chromium	100	90–110	110–8 000

* Figures in p.p.m.
† Recorded at specific localities.

taining the organic matter levels in soils is an additional management procedure which can limit risk of uptake by plants and animals.

The problems of soils and human health are not necessarily clearcut. The causal links are not always proven and thus the definition of the problems and the precise remedy in specific cases may be elusive. Some soil resources may be put at risk for long periods by trace element pollution; in other cases the contamination may be slight and treatment such as liming may provide a remedy so that the soil resource may be used with profit. In the absence of clearly defined causal links it is perhaps best to emphasise as far as possible the hazards associated with trace elements and to monitor the side effects of industry and of the disposal of sewage sludge.

§16.3 HEDGE REMOVAL AND CONSERVATION

The arguments for the removal of hedges in parts of Britain are strong. They include the maximisation of the use of limited soil resources and an improvement in economic efficiency of farms. On the other hand such action leads to problems of wind erosion and hedge removal is not welcomed by conservationists concerned to preserve habitats for wildlife.

The arguments supporting the removal of hedges can be summarised as follows:

1. A tall hedge can cast shadows which may restrict the rate of crop growth next to the hedge because of the reduced insolation.

2. Hedgerows require maintenance which is costly, both in terms of time and money.

3. Hedges can harbour vermin and disease.

4. The enlargement of fields lessens the time incurred through stopping and turning machinery at the edges of fields.

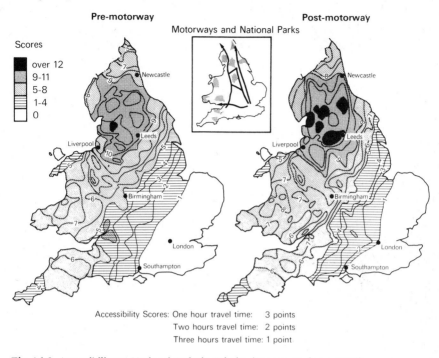

Accessibility Scores: One hour travel time: 3 points
Two hours travel time: 2 points
Three hours travel time: 1 point

Fig. 16.3 Accessibility to National Parks in Britain. (Source: Jackson, 1970)

are arranged. The difficulty with alternative routes is that of getting individuals to follow them, but under skilful management this may be achieved by use of well-placed signs and walking aids.

It may be concluded that under natural conditions the soils on steep slopes and peaty areas may be inherently prone to movement and erosion by mass movement, slope wash and piping. However, this instability is increased under pressure from trampling by people or by grazing animals. This feature of inherent instability is a characteristic which has to be taken into account when planning the utilisation of upland soils.

The questions of soil instability and recreation pressures highlight some of the modern problems of the use of upland soil resources. However, the traditional use of these areas has been sheep farming and the infertility of the soils and slow growth of the vegetation have placed limitations upon productivity. It is because of this low productivity potential that the value of sheep farming may be challenged by alternative uses.

Someone once described the Plynlimon range in mid-Wales as 'miles and miles of damn all' (Edmonds, 1973). He continues: 'These days have passed. The hiker, forester, conservationist and water engineer are now strong competitors to the hill farmer with his sheep'. Edmonds describes how artificial feed blocks have recently been used to supplement the diet of heather and grass (mainly *Nardus* and *Molinia*) for sheep on a farm near Aberystwyth. This innovation has turned 'miles and miles of damn all' into

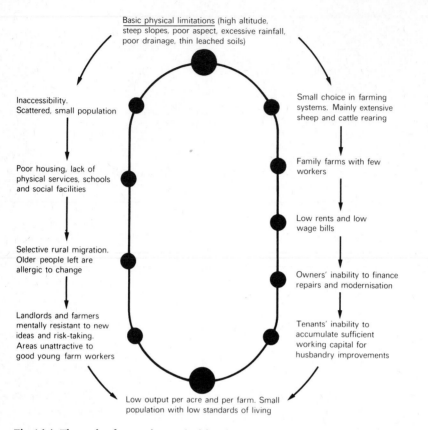

Basic physical limitations (high altitude, steep slopes, poor aspect, excessive rainfall, poor drainage, thin leached soils)

Inaccessibility. Scattered, small population

Poor housing, lack of physical services, schools and social facilities

Selective rural migration. Older people left are allergic to change

Landlords and farmers mentally resistant to new ideas and risk-taking. Areas unattractive to good young farm workers

Small choice in farming systems. Mainly extensive sheep and cattle rearing

Family farms with few workers

Low rents and low wage bills

Owners' inability to finance repairs and modernisation

Tenants' inability to accumulate sufficient working capital for husbandry improvements

Low output per acre and per farm. Small population with low standards of living

Fig. 16.4 The cycle of events in marginal farming areas. (Source: Wibberley, 1959)

useful profitable grazing. The use of feed blocks during the winter does not mean that the area can be stocked with more and more sheep but that the lamb productivity of existing flocks can be greatly increased so as to make the sheep farming a more economic alternative to other non-agricultural uses. Using the feed blocks the average weight per ewe was raised from 60—70 lb (27—38 kg) live weight and 95 per cent of the lambs survived, instead of 60—65 per cent as previously.

The challenge that the hill farmer is facing is to make the raising of sheep a viable economic alternative to the other rival claims upon a hill area in the face of harsh climatic conditions and difficult and infertile soil conditions. Traditionally the farmers feel a loyalty to their heritage of the land, however difficult it may be, and are reluctant to see alternative uses in many cases. In 1959 Wibberley wrote of the physical, economic and social difficulties facing hill farmers. He cited high altitude, steep slopes, poor aspect, excessive rainfall, poor drainage and thin leached soils as the severe physical limitations the farmer had to face. The socio-economic factors of inaccessibility, rural—urban migration and the lack of economic incentives all acted to enforce a low output per acre and per farm which could only support a small population with

low standards of living (Fig. 16.4). This low standard of living resulted in little investment which in turn restricted any improvement in the productivity. While the vicious
circle described by Wibberley still holds in some areas, especially in the remoter areas,
the realisation of the non-agricultural potential of the upland areas has presented land-
use planners with a choice of uses. However, the use of land often demonstrates that
there are at least two sides to every issue. Water engineers may be using areas which are
seen as agriculturally unprofitable for water storage. On the other hand they may
flood agriculturally useful land in the valley bottoms and alter the visual amenity of
the area. Similarly, forestry meets an economic demand but may take the land from
other potential users and alters the visual impression of an area. Thus, each use of
upland soil resources has its arguments for and against it. Each has its potential benefit
and possible potential damage. Rational allocation of land could perhaps take into
account the balance of the factors such as present social and economic demand, the
conservation of land resources for the future and the stability and productivity
potential of the land resource.

Two alternative themes emerge for the future use of upland areas in Britain. One
is the improved use of grassland for grazing — the traditional use — and the other is the
allocation of sectors of an area for a particular use or group of uses.

Seeding of grassland is a way in which the pasture may be improved for sheep.
The sheep tend to selectively eat the more palatable grasses (such as *Festuca* and
Agrostis) leaving the less palatable grasses (e.g. *Nardus*). A very slow deterioration of
the quality of grazing tends to result from this practice, leaving *Nardus* as the
dominant grass. Furthermore plants such as *Eriophorum* and *Sphagnum* are of very
little value for grazing while *Calluna* and *Molinia* have but limited value. Grassland
reclamation schemes are now being tried in some of the hill farms in Wales where
fertilising with lime and phosphate is followed by treatment of the existing vegetation
by mowing or burning or both. Rotovators are used to form a seed bed in small local
patches rather than wholesale ploughing which can lead to soil erosion. Seeding
follows and the desired plants planted in the rotovated patches are encouraged by the
application of the fertilisers so that they soon gain a foothold. Plants such as rye grass
(a forage crop) and clover (which fixes nitrogen) are used in order to improve the
productivity of the pastures. These schemes cost up to about £60 per acre and such
investment is not always possible. In some areas reclamation is physically impossible
and it may be more productive to afforest the area. In such cases mixed forestry and
sheep farming has been carried out. The forest areas are carefully planned to provide
sheltered areas for sheep in bad weather while still allowing easy access to the grazings.

Present trends in land-use changes are illustrated in the study of the North York
Moors National Park by Statham (1972). In this area large-scale changes have occurred
where open moorland, formerly used for sheep grazing and grouse shooting, has been
afforested and cultivated. Of the area of open moors existing in 1933 about one-half
was cultivated or under forest by 1971. In the remaining areas Statham considers that
forestry has the same potential as the other traditional uses and a choice, therefore,
cannot be based on productivity factors alone. It is suggested that conservation areas
should be preserved and recreational uses should be developed in view of an increasing

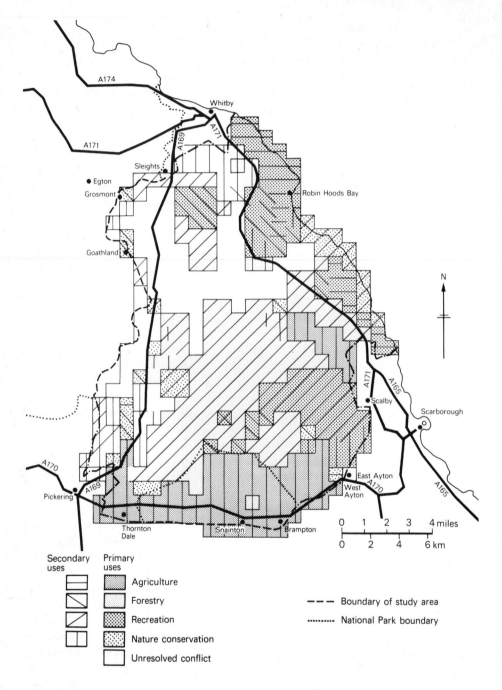

Fig. 16.5 North York Moors conservation project. Optimum uses ideal competitive situation. (Source: Statham, 1972)

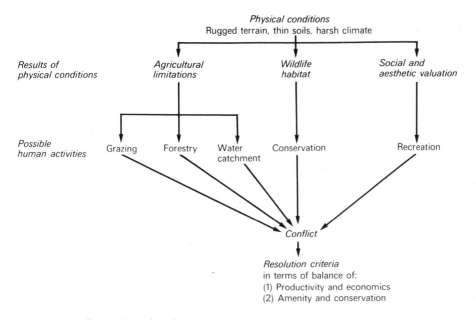

Fig. 16.6 **Planning problems in upland areas — physical conditions.**

demand for leisure areas. In the areas designated as 'conflict areas' by Statham the issue of land-use may be resolved by determining what the area is best suited for in terms of economic productivity. This consideration is used, however, in combination with assessments of which areas are the most outstanding in terms of amenity, recreation and conservation. Using these criteria Statham arrives at a zoned scheme, such as that illustrated in Fig. 16.5.

In conclusion, it can be suggested that either we may take land resources as they are and allocate them in zones or sectors or uses according to their potential agricultural productivity without investment in land improvements, keeping the least suitable land for recreation, conservation or forestry, or we may act to improve the existing resource by capital investment on such items as re-seeding and fertilisers. Where capital is available the latter is likely to prove profitable but where it is not available the former is likely to form a better solution as shown in Fig. 16.6.

There is likely to be a conflict of interest between agriculture, forestry, water catchment, hunting, fishing, conservation and recreation. It is to be hoped that planners of the future will take soil conditions into account when decisions are being made about the uses of land. The evaluation of land capability can form a basis for the rational allocation of land resources but it by no means provides the final answer. The planner should be aware of the physical processes which operate in upland environments so that he may gauge the effects which any one course of action may have upon the environment. He should also be sensitive to social and economics needs and to the arguments for the conservation of wild life and habitat. Thus, using the potential of the soil for agriculture as a starting point, it is to be hoped that all the other factors may be balanced in arriving at rational decisions for the allocation of land — leading to uses of our land resources which will be the most beneficial at the present time and for future generations.

APPENDIX 1   Methods and terms used in profile descriptions
(Source: Corbett and Tatler, 1970)

The terminology used in soil description is defined in the *Field Handbook of the Soil Survey of Great Britain* (Hodgson, 1974). A brief outline of the more important of these terms is given below.

Site characteristics

The site is described under locality, map reference, vegetation and relief, including aspect, elevation and slope.

Depth and clarity of horizons

The depth of horizon boundaries are measured in cm from the surface of the mineral soil, and where they fluctuate widely the range of variation is noted. Horizon boundaries may be described as *even*, if at the same depth across the face; *undulating*, if upward or downward projections are wider than their depth; or *irregular*, if pockets are deeper than their width.

The clarity is described as *sharp*, if the transition zone is less than 2.5 cm wide; *narrow*, if there is a gradual transition through more than 5 cm.

Colour

This is described by comparing the colour of a moist soil fragment with the Munsell Soil Color Charts. According to the system of notation, each colour may be considered as a resultant of three variables: the *hue* indicating its relationship to the spectral colours yellow, red or blue; the *value* its lightness or darkness, and the *chroma* the strength or departure from a neutral colour of the same value. Thus the hue 10 YR, the value 5 and the chroma 6 are combined to give a notation 10 YR 5/6; the colours are grouped under standard names, the name 'yellowish-brown', for example, covering the notations 10 YR 5/4, 10 YR 5/6 and 10 YR 5/8.

Many soil horizons, particularly those which are incompletely weathered or subjected to seasonal waterlogging, are variecoloured, and the self-explanatory terms mottled and micro-speckled are then used. Mottles are described in terms of contrast (*faint, distinct, prominent*), abundance (*few, common, many*) and size.

Texture

Texture refers to the particle-size distribution of inorganic soil material which passes a 2 mm sieve. Standard methods of mechanical analysis are used in the laboratory, but in the field a soil is assigned to a texture class (Fig. 1.2) by estimating the proportions of sand (2.0−0.5 mm), silt (0.05−0.002 mm) and clay (less than 0.002 mm) particle-size grades in a small sample of moist soil worked between finger and thumb. In assessing the texture of surface horizons, allowance has to be made for the influence of organic matter, significant amounts of which tend to make both sandy and clayey soils feel more silty.

Organic matter status

With increasing amounts of organic matter, it becomes more difficult and less useful to assess the texture of the mineral fraction, and soil materials are classed according to their estimated organic matter content, estimates for representative samples being checked by laboratory determinations of organic carbon (Table A.1).

Table A.1 Humose and organic soil materials*

	Organic carbon per cent	Organic matter (Org. C % × 1.7) per cent	Approximate equivalent per cent loss on ignition
Peat	>25	43	50
Peaty loam	15—25	26—43	30—50
Humose mineral soil	7.5—15	13—26	15—30

* Developed from Pizer *et al.* 1961.

Soil materials containing between 7.5 and 15 per cent organic carbon are designated by the prefix humose, together with the textural class of the mineral fraction (e.g. humose clay, humose sandy loam). Materials containing between 15 and 25 per cent organic carbon are regarded as organic and are classed as peaty.

Stoniness

The following broad classes of stoniness are recognised:

Very rare stones	<1% by volume
Slightly stony	1—5% by volume
Stony	5—20% by volume
Very stony	20—50% by volume
Extremely stony	50—75% by volume

In addition indications of the size, shape and nature of the stones are given. Size is reported in accordance with the following categories:

Gravel	2 mm to 1 cm
Small stones	1—5 cm
Medium stones	5—10 cm
Large stones	10—20 cm
Very large stones	20 cm.

Structure

The ultimate particles of soil are arranged to form aggregates or peds with characteristic shapes and sizes, separated by voids or planes of separation. Structure is defined by the degree of aggregate development and by the size and shape of the peds present.

The following terms are used in assessing the distinctness of structure.

Structureless — no observable peds, massive if coherent and single grain if non-coherent.
Weak — indistinct peds; when disturbed breaks into much unaggregated material.
Moderate — well-formed peds; little unaggregated material when disturbed.
Strong — peds distinct in place; remains aggregated when disturbed.

The basic types of *peds* are defined as follows:

Platy — vertical axis much shorter than horizontal.
Prismatic — vertical axis longer than horizontal, vertical faces well defined, vertices usually angular; *columnar* reserved for prisms with rounded tops.
Blocky — dimensions of peds are of the same order, enclosed by plane or curved surfaces that are casts or moulds formed by faces of adjacent peds; sub-divided into *angular* and *sub-angular*.
Granular — small, sub-rounded or irregular peds without distinct edges or faces, usually hard and relatively non-porous.
Crumb — soft, porous, granular aggregates, like bread crumbs.

Each type is divided into size classes, from very fine or very thin to very coarse or very thick.

Roots

An estimate is made of the abundance and size of roots, which are also classified either as woody, fibrous, fleshy or rhizomatous.

Soil drainage classes

The relationship of soil drainage class to water regime and soil morphology are illustrated in Fig. A.1. The following terms are used to describe classes:
Well drained (*excessive*). The soil is coarse textured or very shallow, with low available water capacity and is saturated only during and after heavy rain. Surplus water is removed very rapidly. Any water table is well below the soil horizons.

Profiles are generally brownish, yellowish or reddish in colour and are free of mottling.
Well drained. The soil is rarely saturated in any horizon within the upper 90 cm except during or immediately after heavy rain.

The profiles are usually free of mottling throughout. Sub-surface horizons are generally brownish, reddish or yellowish.
Moderately well drained. Some part of the soil within the upper 90 cm is saturated for short periods in winter or after heavy rain but no horizon above 60 cm remains saturated for more than a month during the year.

Colours typical of well-drained soils in similar materials are usually dominant in

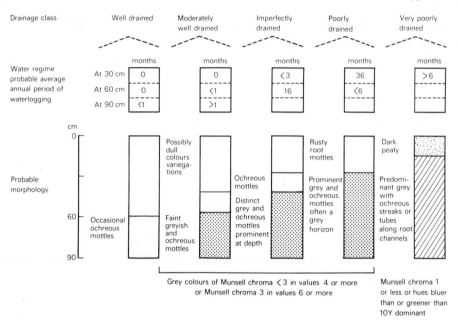

Fig. A.1 Drainage class, water regime and soil morphology. (Source: Corbett and Tatler, 1970)

the profile, but may be slightly lower in chroma, especially on ped faces, and faint to distinct ochreous or grey mottling may occur below 60 cm.

Imperfectly drained. Some part of the soil within the upper 60 cm is saturated for several months but not for a major part of the year.

The colours of sub-surface horizons are commonly lower in chroma and/or yellower in hue than in well-drained soils in similar materials, greyish or ochreous mottling is usually distinct at 45 cm and may become prominent below that depth. There is usually little or no evidence of gleying to a depth of 30 cm.

Poorly drained. The soil is saturated for at least half the year within the upper 60 cm, but the upper 30 cm is free of saturation during most of the growing season.

The profiles normally show evidence of strong gleying. A horizons are usually appreciably darker and/or greyer in colour than in well-drained soils in similar materials, and contain rusty mottles. A horizon with matrix colours of low chroma, or with much ochreous and greyish mottling, normally occurs at 30 cm or less. Grey colours are prominent on ped faces in fissured clayey soils or in the matrix of weak structured soils.

Very poorly drained. Some part of the soil is saturated at less than 30 cm for at least half the year. Some part of the soil within the upper 60 cm is permanently saturated.

The profiles usually have peaty or humose surface horizons, and the sub-surface horizons have colours with low (near neutral) chroma, and yellowish to bluish hues.

HORIZON NOMENCLATURE
Organic and organic—mineral surface horizons

L   Undecomposed litter.
F   Partially decomposed litter.
H   Well decomposed humus layer, low in mineral content.
A   Mixed mineral—organic layer.
Ap   Ploughed layer of cultivated soils.
Ah   A humose horizon.
Ag   A horizon with rusty mottling subject to periodic waterlogging.

Mineral sub-surface horizons

E   Eluvial horizon, depleted of clay and/or sesquioxides.
Ea   Bleached ash-like horizon in podzolised soils.
Eb   Brown (paler when dry) friable weakly structured horizon depleted of clay.
B   Altered horizon distinguished from the overlying A or E horizons and the underlying less altered C, by colour and/or structure, or by illuvial concentrations of the following materials denoted by suffixes.
     t   Illuvial clay, characteristic of brown earths (*sols lessivés*).
     h   Illuvial humus, characteristic of podzols.
     fs   Illuvial iron and aluminium (sesquioxides)
C   A horizon that is little altered, except by gleying and is either like or unlike the material in which overlying horizons have developed. (Where two or more distinct depositional horizons occur in the lower part of the profile they are designated C1, C2, etc.)

Bg
Eg   }   Mottled (gleyed) horizon subject to waterlogging; where gleying is only weakly expressed the suffix (g) is used.
Cg

Bfeg }
Cfeg }   In gley soils, horizons of maximum iron accumulation.

A/C }
B/C }   Horizons of transitional or intermediate character.

Horizons in soils derived from peat

Aop   Ploughed layer of peaty soil or peat.
Bo   Decomposed sub-surface horizon distinguished by its structure.
Co   Little altered peat substratum.

Lithologic discontinuities

Many soils have composite profiles with layers inherited from stratified parent materials, as indicated by significant change in particle size distribution (particularly in grades coarser than $2\mu m$) or mineralogy. Roman numerals are prefixed to the horizon designations to indicate such lithological discontinuities. The uppermost layer is not numbered (the number I being understood); the first contrasting layer is numbered II; others, if present, are numbered consecutively with depth.

APPENDIX 2   Original soil classification for England and Wales
(After Avery, in Clarke, 1940)

I. SOILS OF THE BROWN EARTH GROUP

Three characteristics form the basic definition of the normal brown earths:

1.   The soil has free drainage throughout the profile.

2.   There is no vertical differentiation of silica and sesquioxides in the clay fraction.

3.   There is no natural free $CaCO_3$ in the soil horizon.

Other morphological and chemical features may vary. Thus the soil may be of any colour, but this colour is more or less uniform throughout the profile; the degree of acidity may vary widely.

The virgin soils are usually characterised by an accumulation of leaf litter on the surface, which is underlain by mull humus. Under cultivation the surface is altered, added bases may be present, and to this extent arable soils will differ from the normal brown earths.

The Brown Earth group is divided into soils of low base status and soils of high base status. Soils with a high base status are only slightly acid, and become neutral with depth; they are derived from base-rich parent materials. Those of low base status have a tendency to acidity throughout the profile.

Sub-types of the Brown Earth group

(a) *Creep or colluvial soils.* This group is dependent on topography for its development. In morphological and chemical characteristics it is the same as the normal brown earth soils.

(b) *Brown earths and gleyed B and C horizons.* The soils of this sub-type are the same as the normal brown earths except for a suggestion of gleying in the lower horizons. This gleying is no more than an occasional bluish or rusty mottling. The effect may be due to rare rises in ground-water or to a slight impedance in drainage.

(c) *Leached soils from calcareous parent materials.* These soils are characterised by a red-brown colour and a condition of base unsaturation. They may be quite acid, and if $CaCO_3$ is present it is in the form of hard lumps. Organic matter is light in colour, but is not necessarily low. Secondary $CaCO_3$ may occur at the base of the B horizon, or in the parent material.

II. SOILS OF THE PODZOL GROUP

The chief morphological characteristics of normal podzolised soils are:

1. The presence of a bleached (grey) layer under the surface raw humus.

2. The yellow to rusty coloured accumulation layer which follows.

   The chemical characteristics are found in the differentiation of the silica and sesquioxides of the clay fraction.

   Under cultivation the surface raw humus is absent. Arable soils may show the typical grey and rusty layers, or these may be almost entirely obliterated. All transitions occur, but so long as the clay fraction shows differentiation of the silica and sesquioxides, such soils are included in the Podzol group.

Sub-types of the Podzol group:

(*a*) *Slightly to strongly podzolised soils.* These depend on the thickness of the bleached horizon.

(*b*) *Concealed podzols.* The soil has a raw humus surface layer but no bleached layer. The translocation of sesquioxide is proved by the changes in the silica–sesquioxide ratio.

(*c*) *Peaty podzolised soils.* In these the raw humus has developed into peat. They may vary from a slightly to strongly podzolised condition (for definition of peat see later).

(*d*) *Podzolised soils with gleying.* These are essentially podzolised soils in the upper layers, but exhibit signs of impedance by gleying in the B or C horizons.

(*e*) *Truncated podzols.* Here the surface soil has the characters of a B horizon. Under grass vegetation the iron colours are washed by humus.

III. SOILS OF THE GLEY GROUP

The characteristic of gleying is the presence of greenish, bluish-grey, rusty or yellowish spots or mottling:

1. *Surface-water gley soils.* In these the excessive water is on the surface and produces gleying in the surface horizons. In the lower horizons gleying progressively decreases, or may be absent altogether.

2. *Ground-water gley soils.* The surface of such soils may be dry, at least seasonally and often permanently, with little or no gleying. Gleying is essentially present in the lower layers. This group includes soils with slow percolation, not necessarily occurring only in depressions.

Sub-types of ground-water gley soils:

(*a*) *Gley podzolised soils.* These soils have a raw humus surface and a bleached A horizon. The B horizon is thin or absent. Gleying occurs below this level.

(*b*) *Peaty gley podzolised soils.* Essentially similar to 'gley podzolised soils' but peat replaces raw humus.

(*c*) *Peaty gley soils.* These soils are completely gleyed and carry a peaty surface.

(*d*) *Gley calcareous soils.* These are characterised by a grey colour and a moderately high organic matter content. Calcium carbonate occurs throughout the profile and increases with depth; the soil is base saturated. There is little change in the silica—sesquioxide ratio down the profile. Secondary calcium carbonate often occurs in the form of concretions, especially in the lower layers. Gleying is shown by the presence of bluish, greenish, rusty or yellow spots and mottling.

IV. SOILS OF THE CALCAREOUS GROUP

These soils are developed from calcareous parent materials, contain primary calcium carbonate in the soil horizons and are base saturated.

Sub-types of the Calcareous group:

(*a*) *Grey calcareous soils* (Rendzina type). Under natural vegetation these soils show a very dark surface horizon, a high content of organic matter and well-developed crumb structure. There is no differentiation of the silica—sesquioxide ratio down the profile. Calcium carbonate increases in amount with depth until the parent material is reached. Secondary deposition of calcium carbonate may occur. Under arable cultivation organic matter is lower, calcium carbonate is higher and the soils may be pale grey or almost white in colour. Crumb structure is less pronounced.

(*b*) *Red and brown calcareous soils.* These are formed on hard limestone and do not occur on the chalk. They are shallow, being characterised by a red or brown colour and by the presence of fragmentary calcareous rock. The organic matter content is usually low and the silica—sesquioxide ratio constant throughout the profile. Secondary desposition of calcium carbonate may occur in the parent material.

(*c*) *Calcareous soils with gleyed B and C horizons.* These soils, in the upper layers, are similar to either sub-types (*a*) or (*b*), but show slight gleying in the lower horizons.

V. SOILS OF THE ORGANIC GROUP

The soil character is determined by the presence of twenty or more centimetres or waterlogged organic matter, termed Peat. There are two groups:

1. *Basin Peat.* Soligenous in origin, i.e. formed under the influence of excessive or stagnant ground-water.

2. *Moss Peat.* Ombrogenous in origin, i.e. formed under the influence of heavy rainfall and low summer temperature.

Basin Peat. The main development forms of this group are as follows:

a1. *Fen* (including Carr). This is formed under the influence of calcareous or base-rich ground-water. Transition phase is grass-moor, etc.

a2. *Raised Moss.* This is ombrogenous as a result of accumulation of a1 above ground-water level.

b1. *Acid Low Moor.* This is formed under the influence of drainage from acid or base-poor rock and soils, e.g. podzolised surface or raised moss.

b2. *Raised Moss.* As in a2.

Note. Moss peats are predominantly ombrogenous since they develop under conditions of high rainfall on a substratum lying above ground-water level. The ultimate form of these is 'Raised Moss' and may develop over any organic soil when it grows above ground-water influence.

Moss peat covering a region is termed 'blanket moss' and is to be regarded as climatic in the pedological sense.

Sub-type of b2. Hill Peat

This is a variety of Blanket Moss formed on hill tops and slopes which varies from the main type in distribution and character and is therefore to be mapped separately as 'Hill Peat'.

VI. UNDIFFERENTIATED ALLUVIUM GROUP

Owing to the great variety of soils which may be encountered in a comparatively small area of alluvium, some surveyors do not attempt to differentiate them. In such cases the soils are allocated to this group. Where alluvial flats are extensive careful survey will be worth while. In this case the different series identified will be allocated to one of the other five groups.

APPENDIX 3 Summary of USDA Soil Survey staff classification (1960) (7th Approximation) (After Bridges, 1970)

1. *Entisols*. Weakly developed (generally azonal) soils
   - with features of gleying
   - with strong artificial disturbance
   - on alluvial deposits
   - with sand or loamy sand texture
   - other Entisols (e.g. lithosols, some regosols)

   Aquents
   Arents
   Fluvents
   Psamments
   Orthents

2. *Vertisols*. Cracking clay soils
   - usually moist
   - dry for short periods
   - dry for a long period
   - usually dry

   Uderts
   Usterts
   Xererts
   Torrerts

3. *Inceptisols*. Moderately developed soils, not in other orders
   - with features of gleying
   - on volcanic ash
   - in a tropical climate
   - with an umbric epipedon
   - other Incepitsols (e.g. most brown earths)

   Aquepts
   Andepts
   Tropepts
   Umbrepts
   Ochrepts

4. *Aridisols*. Semi-desert and desert soils
   - with an argillic horizon
   - other soils of dry areas (e.g. grey desert soils)

   Argids
   Orthids

5. *Mollisols*. Soils of high base status with a dark A horizon
   - with albic and argillic horizons
   - with features of gleying
   - on highly calcareous parent materials
   - others in cold climates
   - others in humid climates
   - others in sub-humid climates
   - others in sub-arid climates

   Albolls
   Aquolls
   Rendolls
   Borolls
   Udolls
   Ustolls
   Xerolls

6. *Spodosols*. Soils with a spodic horizon (e.g. podzols)
   - with features of gleying
   - with little humus in spodic horizon
   - with little iron in spodic horizon
   - with iron and humus

   Aquods
   Ferrods
   Humods
   Orthods

7. *Alfisols*. Soils with an argillic horizon and moderate to high base content
   - with features of gleying
   - others in cold climates
   - others in humid climates

   Aqualfs
   Boralfs
   Udalfs

    — others in sub-humid climates   Ustalfs
    — others in sub-arid climates   Xeralfs

8. *Ultisols*. Soils with an argillic horizon and low base content
    — with features of gleying   Aquults
    — with a humose A horizon   Humults
    — in humid climates   Udults
    — others in sub-humid climates   Ustults
    — others in sub-arid climates   Xerults

9. *Oxisols*. Soils with an oxic horizon or with plinthite near the surface
    — with features of gleying   Aquox
    — with a humose A horizon   Humox
    — others in humid climates   Orthox
    — others in drier climates   Ustox

10. *Histosols*. Organic soils (sub-orders not finalised)

APPENDIX 4 New soil classification for England and Wales
(After Avery, 1973)

Major group	Group	Sub-group
Lithomorphic (A/C) soils Normally well-drained soils with distinct, humose or organic topsoil and bedrock or little altered unconsolidated material at 30 cm or less	*Rankers* With non-calcareous topsoil over bedrock (including massive limestone) or non-calcareous uncon-solidated material (excluding sand)	Humic ranker Grey ranker Brown ranker Podzolic ranker Stagnogleyic (fragic) ranker
	Sand-rankers In non-calcareous, sandy material	Typical sand-ranker Podzolic sand-ranker Gleyic sand-ranker
	Ranker-like alluvial soils In non-calcareous recent alluvium (usually coarse textured)	Typical ranker-like alluvial soil Gleyic ranker-like alluvial soil
	Rendzinas Over extremely calcareous non-alluvial material, fragmentary limestone or chalk	Humic rendzina Grey rendzina Brown rendzina Colluvial rendzina Gleyic rendzina Humic gleyic rendzina
	Pararendzinas Over moderately calcareous non-alluvial (excluding sand) material	Typical pararendzina Humic pararendzina Colluvial pararendzina Stagnogleyic pararendzina Gleyic pararendzina
	Sand-pararendzinas In calcareous sandy material	Typical sand-pararendzina

Major group	Group	Sub-group
	Rendzina-like alluvial soils In calcareous recent alluvium	Typical rendzina-like alluvial soil Gleyic rendzina-like alluvial soil
Brown soils Well drained to imperfectly drained soils (excluding Pelosols) with an altered sub-surface (B) horizon, usually brownish, that has soil structure rather than rock structure and extends below 30 cm depth	*Brown calcareous earths* Non-alluvial, loamy or clayey, with friable moderately calcareous sub-surface horizon	Typical brown calcareous earth Gleyic brown calcareous earth Stagnogleyic brown calcareous earth
	Brown calcareous sands Non-alluvial, sandy, with moderately calcareous sub-surface horizon	Typical brown calcareous sand
	Brown calcareous alluvial soils In calcareous recent alluvium	Typical brown calcareous alluvial soil Gleyic brown calcareous alluvial soil
	Brown earths (sensu stricto) Non-alluvial, non-calcareous loamy, with brown or reddish friable sub-surface horizon	Typical brown earth Stagnogleyic brown earth Gleyic brown earth Ferritic brown earth Stagnogleyic ferritic brown earth
	Brown sands Non-alluvial, sandy or sandy gravelly	Typical brown sand Gleyic brown sand Stagnogleyic brown sand Argillic brown sand Gleyic argillic brown sand
	Brown alluvial soils Non-calcareous in recent alluvium	Typical brown alluvial soil Gleyic brown alluvial soil

Major group	Group	Sub-group
	Argillic brown earths Loamy or loamy over clayey, with sub-surface horizon of clay accumulation, normally brown or reddish	Typical argillic brown earth Stagnogleyic argillic brown earth Gleyic argillic brown earth
	Paleo-argillic brown earths Loamy or clayey, with strong brown to red sub-surface horizon of clay accumulation attributable to pedogenic alteration before the last glacial period	Typical paleo-argillic brown earth Stagnogleyic paleo-argillic brown earth
Podzolic soils Well drained to poorly drained soils with black, dark brown or ochreous sub-surface (B) horizon in which aluminium and/or iron have accumulated in amorphous forms associated with organic matter. An overlying bleached horizon, a peaty topsoil, or both, may or may not be present	*Brown podzolic soils* Loamy or sandy, normally well drained, with a dark brown or ocheous friable sub-surface horizon and no overlying bleached horizon or peaty topsoil	Typical brown podzolic soil Humic brown podzolic soil Paleo-argillic brown podzolic soil Stagnogleyic brown podzolic soil Gleyic brown podzolic soil
	Gley-podzols With dark brown or black sub-surface horizon over a grey or mottled (gleyed) horizon affected by fluctuating ground-water or impeded drainage. A bleached horizon, a peaty topsoil or both may be present	Typical (humus) gley-podzol Humo-ferric gley-podzol Stagnogley-podzol Humic (peaty) gley-podzol

Major group	Group	Sub-group
	Podzols (sensu stricto) Sandy or coarse loamy, normally well drained, with a bleached horizon and/or dark brown or black sub-surface horizon enriched in humus and no immediately underlying grey or mottled (gleyed) horizon or peaty topsoil	Typical (humo-ferric) podzol Humus podzol Ferric podzol Paleo-argillic podzol Ferri-humic podzol
	Stagnopodzols With peaty topsoil, periodically wet (gleyed) bleached horizon, or both, over a thin ironpan and/or a brown or ochreous relatively friable sub-surface horizon	Ironpan stagnopodzol Humus-ironpan stagno-podzol Hardpan stagnopodzol Ferric stagnopodzol
Pelosols Slowly permeable non-alluvial clayey soils that crack deeply in dry seasons with brown, greyish or reddish blocky or prismatic sub-surface horizon, usually slightly mottled	*Calcareous pelosols* With calcareous sub-surface horizon	Typical calcareous pelosol
	Argillic pelosols With sub-surface horizon of clay accumulation, normally non-calcareous	Typical argillic pelosol
	Non-calcareous pelosols Without argillic horizon	Typical non-calcareous pelosol
Gley soils With distinct, humose or peaty topsoil and grey or grey-and-brown mottled (gleyed) sub-surface horizon altered by reduction, or reduction and segregation, of iron caused by periodic or permanent saturation by water in the presence	1. Gley soils without a humose or peaty top-soil, seasonally wet in the absence of effective artificial drainage.	
	Alluvial gley soils In loamy or clayey recent alluvium affected by	Typical (non-calcareous) alluvial gley soil

Major group	*Group*	*Sub-group*
of organic matter. Horizons characteristic of podzolic soils are absent	fluctuating ground-water	Calcareous alluvial gley soil Pelo-alluvial gley soil Pelo-calcareous alluvial gley soil Sulphuric alluvial gley soil
	Sandy gley soils Sandy, permeable, affected by fluctuating ground-water	Typical (non-calcareous) sandy gley soil Calcareous sandy gley soil
	Cambic gley soils Loamy or clayey, non-alluvial, with a relatively permeable substratum affected by fluctuating ground-water	Typical (non-calcareous) cambic gley soil Calcaro-cambic gley soil Pelo-cambic gley soil
	Argillic gley soils Loamy or loamy over clayey, with a sub-surface horizon of clay accumulation and a relatively permeable substratum affected by fluctuating ground-water	Typical argillic gley soil Sandy-argillic gley soil
	Stagnogley soils Non-calcareous, non-alluvial, with loamy or clayey, relatively impermeable sub-surface horizon or substratum that impedes drainage.	Typical stagnogley soil Pelo-stagnogley soil Cambic stagnogley soil Paleo-argillic stagnogley soil Sandy stagnogley soil
	2. Gley soils with a humose or peaty topsoil normally wet for most of the year in the absence of effective artificial drainage	

Major group	Group	Sub-group
	Humic-alluvial gley soils In loamy or clayey recent alluvium	Typical (non-calcareous) humic-alluvial gley soil Calcareous humic-alluvial gley soil
	Humic-sandy gley soils Sandy, permeable, affected by high ground-water	Typical humic-sandy gley soil
	Humic gley soils (sensu stricto) Loamy or clayey, non-alluvial, affected by high ground-water	Typical (non-calcareous) humic gley soil Calcareous humic gley soil Argillic humic gley soil
	Stagnohumic gley soils Non-calcareous, with loamy or clayey, relatively impermeable sub-surface horizon or substratum that impedes drainage	Cambic stagnohumic gley soil Argillic stagnohumic gley soil Paleo-argillic stagnohumic gley soil Sandy stagnohumic gley soil
Man-made soils With thick man-made topsoil or disturbed soil (including material recognisably derived from pedogenic horizons) more than 40 cm thick	*Man-made humic soils* With thick man-made topsoil *Disturbed soils* Without thick man-made topsoil	Sandy man-made humus soil Earthy man-made humus soil
Peat soils With a dominantly organic layer at least 40 cm thick formed under wet conditions and starting at the surface or within 30 cm depth	*Raw peat soils* Permanently waterlogged and/or contain more than 15 per cent recognisable plant remains within the upper 20 cm	Raw oligo-fibrous peat soil Raw eu-fibrous peat soil Raw oligo-amorphous peat soil Raw eutro-amorphous peat soil

Major group	Group	Sub-group
	Earthy peat soils	
	With relatively firm (drained) topsoil, normally black, containing few recognisable plant remains	Earthy oligo-fibrous peat soil
		Earthy eu-fibrous peat soil
		Earthy oligo-amorphous peat soil
		Earthy eutro-amorphous peat soil
		Earthy sulphuric peat soil

BIBLIOGRAPHY

Ackermann, K. J. and Cave, R. (1967). 'Superficial deposits and structures, including landslip, in the Stroud district, Gloucestershire', *Proc. Geol. Assoc.*, **78**, 567—86.

Acland, Sir F. (1918). *Final Report of the Forestry Sub-Committee Ministry of Reconstruction* (Cd. 8881), HMSO, London.

Anon (1897). *Official Guide to the Great Eastern Railway*, Cassell, London.

Arca, M. N. and Weed, S. B. (1968). 'Soil aggregation and porosity in relation to contents of free iron oxide and clay', *Soil Sci.*, **101**, 164—70.

Arkell, W. J. (1947*a*). *The Geology of Oxford*, Clarendon Press, Oxford.

Arkell, W. J. (1947*b*). 'The geology of the Evenlode Gorge', *Proc. Geol. Assoc.*, **58**, 87—114.

Arkell, W. J. (1971). *The Jurassic System of Great Britain*, Clarendon Press, Oxford.

Armstrong, R. W. (1962). 'Cancer and soil: Review and counsel', *Prof. Geogr.*, **14**, 7—13.

Armstrong, R. W. (1964). 'Environmental factors involved in studying the relationship between soil elements and disease', *Amer. J. Public Health*, **54**, 1536—44.

Astbury, A. K. (1958). *The Black Fens*, Golden Head Press, Cambridge.

Avery, B. W. (1955). 'The soils of the Glastonbury District of Somerset', *Mem. Soil Surv. Gt. Brit.*, Soil Survey, Harpenden.

Avery, B. W. (1958). 'A sequence of beechwood soils on the Chiltern Hills, England', *J. Soil Sci.*, **9**, 210—24.

Avery, B. W. (1964). 'The soils and land use of the district around Aylesbury and Hemel Hempstead', *Mem. Soil Surv. Gt. Brit.*, Soil Survey, Harpenden.

Avery, B. W. (1969). 'Problems of soil classification', in J. G. Sheals (ed.), *The Soil Ecosystem*, Systematics Association, London, 9—17.

Avery, B. W. (1973). 'Soil classification in the soil survey of England and Wales', *J. Soil Sci.*, **24**, 324—38.

Baird, W. W. and Tarrant, J. R. (1972). 'Vanishing hedgerows', *Geogrl. Mag.*, **44**, 545—51.

Baldwin, M., Kellogg, C. E. and Thorp, J. (1938). 'Soil classification', in *Soils and Men, US Dept. Agric. Yearbook*, 979—1001.

Ball, D. F. (1960). 'The soils and land use of the district around Rhyl and Denbigh', *Mem. Soil Surv. Gt. Brit.*, Soil Survey, Harpenden.

Ball, D. F. (1963). 'Soils and land use of the district around Bangor and Beaumaris', *Mem. Soil Surv. Gt. Brit.*, Soil Survey, Harpenden.

Ball, D. F. (1966). 'Late glacial scree in Wales', *Biul. peryglac.*, **15**, 151—63.

Ball, D. F. (1967). 'Stone pavements in soils of Caernarvonshire, North Wales', *J. Soil Sci.*, **18**, 103—8.

Ball, D. F. and Goodier, R. (1968). 'Large sorted stone-stripes in the Rhinog Mountains, North Wales', *Geografiska Annaler*, **50**, Ser A. 54—9.

Barker, T. (1858). 'On water meadows as suitable for Wales and other mountain districts', *J. Bath. W. sth Cties' Assoc.*, **6**, 267—86.

Barkham, J. P. (1972). 'Recreation and environment in part of the Lake District National Park', *Fieldworker*, Dec. 1971.

Barkham, J. P. (1973). 'Recreational carrying capacity: A problem of perception', *Area*, 5(3), 218—22.

Barry, T. A. (1954). 'Some considerations affecting the classification of the bogs of Ireland and their peats', *First Int. Peat Congress, Dublin*, B2, 8—16.

Bartelli, L. J. *et al.* (1966). *Soil Surveys and Land Use Planning*, Soil Science Society of America, Madison.

Baver, L. D. (1938). 'Ewald Wollny — a pioneer in soil and water conservation research', *Proc. Soil Sci. Soc. Amer.*, 3, 330—3.

Baver, L. D. (1956). *Soil Physics*, Wiley, New York, 488.

Beckett, P. H. T. and Webster, R. (1971). 'Soil variability: A review', *Soils and Fertilizers*, 34, 1—15.

Beckinsale, R. P. (1970). 'Physical problems of Cotswold rivers and valleys', *Proc. Cotteswold Nat. Field Club*, 31, 184—95.

Beckinsale, R. P. and Smith, K. W. (1953). 'Some morphological features of the valleys of the North Cotswolds: the Windrush and its tributaries', *Proc. Cotteswold Nat. Field Club*, 31, 184—95.

Bibby, J. S. (1973). 'Land capability', in J. Tivy (ed.), *The Organic Resources of Scotland*, Oliver and Boyd, Edinburgh, 51—65.

Bibby, J. S. and Mackney, D. (1969). 'Land use capability classification', *Tech. Monogr. Soil Surv. Gt. Brit.*, 1.

Bie, S. W. and Beckett, P. H. T. (1970). 'The Costs of Soil Survey', *Soils and Fertilizers*, 33, 203—17.

Bloomfield, C. (1949). 'Gleying', *Soils and Fertilizers*, 32, 1—3.

Bloomfield, C. (1951). 'Experiments on the mechanism of gley formation', *J. Soil Sci.*, 2, 196—211.

Bloomfield, C. (1953). 'A study of podzolisation. Parts I—II. I. The mobilization of iron and aluminium by Scots pine needles. II. The mobilisation of iron and aluminium by the leaves and bark of *Agathis australis* (Kauri)', *J. Soil Sci.*, 4, 5—23.

Bloomfield, C. (1954). 'A study of podzolisation: Parts III, IV and V', *J. Soil Sci.*, 5, 39—56.

Bloomfield, C. (1955). 'A study of podzolisation: Part VI. The immobilization of iron and aluminium', *J. Soil Sci.*, 6, 284—92.

Bloomfield, C. (1957). 'The possible significance of polyphenols in soil formation', *J. Sci. Food Agric.*, 8, 389—92.

Bower, M. M. (1961). 'The distribution of erosion in Blanket Peat Bogs in the Pennines', *Trans. Inst. Brit. Geogr.*, 29, 17—30.

Bown, C. J. (1973). 'The soils of Carrick and the country round Girvan', *Mem. Soil Surv. Gt. Brit.*, Soil Survey, Harpenden.

Brewer, R. (1964). *Fabric and Mineral Analysis of Soil*, Wiley, New York.

Bridges, E. M. (1961). 'Aspect and time in soil formation', *Agriculture, Lond.*, 68, 258—63.

Bridges, E. M. (1966). 'The soils and land use of the district north of Derby', *Mem. Soil Surv. Gt. Brit.*, Soil Survey, Harpenden.

Bridges, E. M. (1970). *World Soils*, Cambridge University Press.

Briggs, D. J. and Courtney, F. M. (1972). 'Ridge-and-trough topography in the north Cotswolds', *Proc. Cotteswold Nat. Field Club*, 36, 94—103.

Buckman, H. O. and Brady, N. C. (1974). *The Nature and Properties of Soils*, Macmillan.

Burnham, C. P. (1970). 'The regional pattern of soil formation in Great Britain', *Scott. Geogr. Mag.*, 86, 25—34.

Butler, B. E. (1958). 'The diversity of concepts about soils', *J. Aust. Inst. agric. Sci.*, 24, 14—20.

Butler, B. E. (1959). 'Periodic phenomena in landscapes as a basis for soil studies', *Soil Publ., CSIRO, Aust.*, 14.

Carson, R. (1963). *Silent Spring*, Penguin.

Charlesworth, J. K. (1928). 'The South Wales end moraine', *Quart. J. geol. Soc., Lond.*, **85**, 335—58.

Chatwin, C. P. (1961). *British Regional Geology: East Anglia and Adjoining Areas*, HMSO, London.

Chepil, W. S. (1958). 'Soil conditions that influence wind erosion', *Technical Bulletin, US Dept. Agric.* No. 1185.

Clarke, G. R. (1940). *Soil Survey of England & Wales Field Handbook*, Clarendon Press, Oxford.

Clarke, G. R. (1957). *The Study of Soil in the Field*, Oxford University Press.

Clayden, B. (1964). 'Soils of the Middle Teign Valley District of Devon', *Bull. Soil Surv. Gt. Brit.*, Soil Survey, Harpenden.

Clayden, B. (1971). 'Soils of the Exeter District', *Mem. Soil Surv. Gt. Brit.*, Soil Survey, Harpenden.

Clayden, B. and Manley, D. J. R. (1964). 'The Soils of the Dartmoor Granite', in I. G. Simmons (ed.), *Dartmoor Essays*, Devon Association Advanced Science, 117—40.

Clayton, K. M. (1957). 'Some aspects of the glacial deposits of Essex', *Proc. Geol. Assoc.*, **68**, 1—21.

Clayton, K. M. (1960). 'The Landforms of parts of southern Essex', *Trans. Inst. Brit. Geogr.*, **28**, 55—74.

Clawson, M. and Held, R. B. (1965). *Soil Conservation in Perspective*, Prentice-Hall.

Clements, F. E. (1916). 'Plant succession, an analysis of the development of vegetation', *Publs Carnegie Instn*, 242.

Cloud, P. E. (1969). Preface in: *Resources and Man*, National Academy of Sciences, W. H. Freeman, San Francisco.

Cooke, G. W. (1967). *The Control of Soil Fertility*, Crosby-Lockwood, London.

Cooke, G. W. and Williams, R. J. B. (1971). 'Problems with Cultivations and Soil Structure at Saxmundham', *Rep. Rothamsted Exp. Stn.*, 1971 (2), 122—42.

Cooke, G. W. (1972). 'Fertilisers and society', *Proc. Fertil. Soc.*, No. 121.

Cope, D. W. (1972). 'Soils in Gloucestershire I: Sheet SO82 (Norton)', *Soil Surv. Record*, Soil Survey, Harpenden.

Cope, D. W. (1975). 'Soils in Wiltshire: Sheet SUO3 (Wilton)', *Soil Surv. Record*, Soil Survey, Harpenden.

Corbett, W. M. (1973). 'Breckland Forest Soils', *Spec. Surv. Soil Surv. Gt. Brit.*, Soil Survey, Harpenden.

Corbett, W. M. and Tatler, W. (1970). 'Soils in Norfolk: Sheet TM49 (Beccles North)', *Soil Surv. Record*, Soil Survey, Harpenden.

Coulson, C. B., Davies, R. I. and Lewis, D. A. (1960). 'Polyphenols in plant, humus and soil', *J. Soil Sci.*, **11**, 20—44.

Country Landowners Association and the National Farmers Union (1967). *Reclamation in Exmoor National Park*, Joint Statement published by authors.

Courtney, F. M. (1972). 'Variation within the Sherborne Soil Mapping Unit', unpublished M.Sc. thesis, University of Bristol.

Courtney, F. M. (1973). 'A taxonometric study of the Sherborne soil mapping unit', *Trans. Inst. Brit. Geogr.*, 58, 113—24.

Courtney, F. M. (1974). 'New soil classification: was it worth it?', *Area*, 6, 205—6.

Courtney, F. M. and Findlay, D. C. (1976 in press). 'Soils in Gloucestershire: Sheet SP12 (Stow-on-the-Wold)', *Soil Surv. Record*, Soil Survey, Harpenden.

Crabtree, K. (1968). 'Pollen analysis', *Science Progress*, 56, 83—101.

Crampton, C. B. (1959). 'Analysis of heavy minerals in certain drift soils of Yorkshire', *Proc. Yorks. Geol. Soc.*, 32, 69—81.

Crampton, C. B. (1961). 'An interpretation of the micro-minerology of certain Glamorgan soils: the influence of ice and wind', *J. Soil Sci.*, 12, 158—71.

Crampton, C. B. (1963). 'The development and morphology of iron pan podzols in Mid and South Wales', *J. Soil Sci.*, 14, 282—302.

Crampton, C. B. (1972). 'Soils of the Vale of Glamorgan', *Mem. Soil Surv. Gt. Brit.*, Soil Survey, Harpenden.

Crampton, C. B. and Taylor, J. A. (1967). 'Solifluction terraces in South Wales', *Biul. Peryglac.*, 16, 15—36.

Crawford, D. V. (1956). 'Microbiological aspects of podzolisation', *6th Int. Congr. Soil Sci.*, C, 197—202.

Crompton, A. (1961). 'A brief account of the soils of Yorkshire, *J. Yorks, Grassland Soc.*, 3, 27—35.

Crompton, A. and Matthews, B. (1970). 'Soils of the Leeds District', *Mem. Soil Surv. Gt. Brit.*, Soil Survey, Harpenden.

Crompton, E. (1951). 'Soils of North-West England', *Scientific Horticulture*, 10, 169—77.

Crompton, E. (1952). 'Some morphological features associated with poor soil drainage', *J. Soil Sci.*, 3, 277—89.

Crompton, E. (1953). 'Grow the soil to grow the grass: some pedological aspects of marginal land improvement', *Agriculture, Lond.*, 60, 301—8.

Crompton, E. (1960). 'The significance of the weathering/leaching ratio in the differentiation of major soil groups with particular reference to some very strongly leached brown earths on the hills of Britain', *7th Int. Congr. Soil Sci.*, 4, 406—12.

Crompton, E. (1966). 'The Soils of the Preston district of Lancashire', *Mem. Soil Surv. Gt. Brit.*, Soil Survey, Harpenden.

Crompton, E. and Osmond, D. A. (1954). 'The Soils of the Wem district of Shropshire', *Mem. Soil Surv. Gt. Brit.*, Soil Survey, Harpenden.

Curtis, L. F. (1965). 'Soil Erosion on Levisham Moor, North York Moors. Institute of British Geographers Symposium', *Occasional Publication, British Geomorphological Research Centre*, 2, 19—21.

Curtis, L. F. (1968). 'Surface patterns on Godney Moor, Somerset', *Proc. Bristol Naturalists Society*, 31, 415—20.

Curtis, L. F. (1971). 'Soils of Exmoor forest', *Spec. Surv. Soil Surv. Gt. Brit.*, 5, Soil Survey, Harpenden.

Curtis, L. F. (1973). 'The application of photography to soil mapping from the air', in J. Cruise and A. A. Newman (eds), *Photographic Techniques in Scientific Research*, Vol. 1, Academic Press, 57—110.

Curtis, L. F. (1974). 'A study of the soils of Exmoor Forest', unpublished Ph.D. thesis, University of Bristol.

Curtis, L. F. (1975). 'Landscape periodicity and soil development', in R. Peel, M. Chisholm and P. Haggett (eds), *Processes in physical and human geography*, Heinemann, 247—65.

Curtis, L. F., Doornkamp, J. C. and Gregory, K. J. (1965). 'The description of relief in field studies of soils', *J. Soil Sci.*, 16, 16—30.

Curtis, L. F. and James, J. H. (1959). 'Frost heaved soils of Barrow, Rutland', *Proc. Geol. Assoc.*, 70, 310—14.

Curtis, L. F. and Trudgill, S. (1974). 'The measurement of soil moisture', *Tech. Bull. British Geomorphological Research Group*, 13.

Damann, A. W. H. (1962). 'Development of hydromorphic humus podzols and some notes on the classification of podzols in general', *J. Soil Sci.*, 13, 92—7.

Darby, H. C. (1969); *The Draining of the Fens* (2nd edn), Cambridge University Press.

Darwin, C. (1882). *Vegetable Mould and Earthworms: the formation of vegetable mould through the action of worms with observation on their habits*, John Murray, London.

Davies, D. B. (1973). 'Soil stabilisation in agricultural practice', *J. Sci. Fd. Agric.*, 24, 1313—14.

Davis, W. M. (1899). 'The drainage of cuestas', *Proc. Geol. Assoc.*, 16, 75—93.

Davis, W. M. (1909). 'The valleys of the Cotteswold Hills', *Proc. Geol. Assoc.*, 21, 150—2.

De Bakker, H. and Schelling, J. (1966). *Systeem van Bodemclassificatie voor Nederland. De Hogere Niveans* (with English summary), Pudoc, Wageningen.

Dimbleby, G. W. (1952). 'The historical status of moorland in north-east Yorkshire', *New Phytologist*, 51, 349—54.

Dimbleby, G. W. (1961*a*). 'Transported material in the soil profile', *J. Soil Sci.*, **12**, 12—22.

Dimbleby, G. W. (1961*b*). 'Soil pollen analysis', *J. Soil Sci.*, **12**, 1—11.

Dimbleby, G. W. (1962). 'The development of British heathlands and their soils', *Oxf. For. Mem.*, **23**.

Dines, H. G., Holmes, S. C. A. and Robbie, J. A. (1954). 'Geology of the country around Chatham', *Mem. geol. Surv. U.K.*

Donahue, R. L., Schickluna, J. C. and Robertson, L. S. (1971). *Soils: An Introduction to Soils and Plant Growth*, Prentice-Hall.

Duchaufour, P. (1959). *La dynamique du sol foresteir en climat atlantique*, Presses Universitaires, Laval.

Duchaufour, P. (1970). *Precis de Pedologie* (3rd edition), Masson, Paris.

Dury, G. H. (1960). 'Misfit streams: problems in interpretation, discharge and distribution', *Geogr. Rev.*, **50**, 219—42.

Dylik, J. (1966). In D. L. Linton and R. S. Waters (eds), 'The Exeter Symposium: Discussion', *Biul. peryglac.*, **15**, 133—49.

Edmonds, D. T., Painter, R. B. and Ashley, C. D. (1970). 'A semi-quantitative hydrological classification of soils in North-East England', *J. Soil Sci.*, **21**, 256—64.

Edmonds, H. (1973). 'How to turn "miles and miles of damn all" into profitable grazing', *Livestock Farming*, Oct., 1973, 37—9.

Edwards, A. M. and Wibberley, G. P. (1971). 'An Agricultural Land Budget for Britain 1965—2000', Wye College (University of London), School of Rural Economics and Related Studies, *Studies in Rural Land Use No. 10*.

Edwards, A. M. C. (1976). 'Long term changes in the water quality of agricultural catchments', *Symposium on Applied Physical Geography, Inst. Brit. Geogr.*, January 1974.

Edwards, C. A. (1966). 'Insecticide residues in soils', *Residue Review*, **13**, 83—132.

Emerson, W. W. (1959). 'The structure of soil crumbs', *J. Soil Sci.*, **10**, 235—44.

Exmoor Society (1966). *Can Exmoor Survive: A technical assessment*, Exmoor Society.

Faegri, K. and Iversen, J. (1964). *Textbook of Pollen Analysis*, Blackwell Scientific Publications, Oxford.

Farnham, R. S. and Finney, H. R. (1965). 'Classification and properties of organic soils', *Advances in Agronomy*, **17**, 115—62.

Finch, T. F. (1971). 'Soils of County Clare', *Soil Surv. Bull., Ireland*, An Foras Taluntais, Dublin.

Finch, T. F. and Ryan, P. (1966). 'Soils of Co. Limerick', *Soil Surv. Bull., Ireland*, An Foras Taluntais, Dublin.

Finch, T. and Synge, F. M. (1966). 'The drifts and soils of West Clare and the adjoining parts of Counties Kerry and Limerick', *Irish Geogr.*, **5**, 161—72.

Finch, V. C. and Trewartha, G. T. (1942). *Elements of Geography*, McGraw-Hill.

Findlay, D. C. (1965). 'Soils of the Mendip District of Somerset', *Mem. Soil Surv., Gt. Brit.*, Soil Survey, Harpenden.

Findlay, D. C. (1970). *Making 1 : 25,000 soil maps* (mimeographed), Soil Survey, Harpenden.

Findlay, D. C. (1976). 'The soils of the Southern Cotswolds and surrounding country', *Mem. Soil Surv. Gt. Brit.*, Soil Survey, Harpenden.

Fitzpatrick, E. A. (1956*a*). 'Progress report on the observations of periglacial phenomena in the British Isles', *Biul. peryglac.*, **4**, 99—115.

Fitzpatrick, E. A. (1956*b*). 'An indurated soil horizon formed by permafrost', *J. Soil Sci.*, **7**, 248—54.

Fitzpatrick, E. A. (1958). 'An introduction to the periglacial geomorphology of Scotland', *Scottish Geographical Magazine*, **74**, 28—36.

Fitzpatrick, E. A. (1964). 'The Soils of Scotland', Ch. 3, in J. H. Burnett (ed.), *The Vegetation of Scotland*, Oliver and Boyd, London.

Fordham, S. J. and Green, R. D. (1973). 'Soils in Kent II: Sheet TR35 (Deal)', *Soil Surv. Record*, Soil Survey, Harpenden.

Forestry Commission (1963). *Forestry in England* (booklet), HMSO.

Foth, H. D. and Turk, L. M. (1972). *Fundamentals of Soil Science* (5th edn), Wiley.

Fox, Sir Cyril (1932). *The Personality of Britain — its Influence on Inhabitant and Invader in Prehistoric and Early Historic Times*, National Museum of Wales, Cardiff.

Fraser, G. K. (1943). 'Peat deposits of Scotland', *D.S.I.R. Geol. Survey of Great Britain*. Wartime Pamphlet No. 36.

Fraser, G. K. (1954). 'Classification and nomenclature of peat and peat deposits', *First Int. Peat Congress*, Dublin B2, 1—8.

Furness, R. R. (1971). 'Soils in Cheshire I: Sheet SJ65 (Crewe West)', *Soil Surv. Record*, Soil Survey, Harpenden.

Gallois, R. W. (1965). 'The Wealden District (4th edn), *British Regional Geology*, HMSO, London.

Galloway, R. W. (1961a). 'Solifluction in Scotland', *Scottish Geographical Magazine*, **77**, 75—87.

Galloway, R. W. (1961b). 'Ice wedges and involutions in Scotland', *Biul. periglac.*, **10**, 169—93.

Gardiner, M. J. and Ryan, P. (1962). 'Relic soil on limestone on Ireland', *Irish J. Agric. Res.*, **1**(2), 181—8.

Gardiner, M. J. and Ryan, P. (1964). 'The Soils of Co. Wexford', *Bull. Soil Surv., Ireland*, An Foras Taluntais, Dublin.

Gibbons, F. R. (1961). 'Some misconceptions about what soil surveys can do', *J. Soil Sci.*, **12**, 96—100.

Glentworth, R. (1954). 'Soils of the country round Banff, Huntly and Turriff', *Mem. Soil Surv. Gt. Brit.*, Soil Survey, Harpenden.

Glentworth, R. and Dion, H. G. (1950). 'The association or hydrologic sequence in certain soils of the podsolic zone of north-east Scotland', *J. Soil Sci.*, **1**, 35—49.

Glentworth, R. and Muir, J. W. (1963). 'The soils of the country round Aberdeen, Inverurie and Fraserburgh', *Mem. Soil Surv. Gt. Brit.*, Soil Survey, Harpenden.

Gloyne, R. W. (1960). 'Wind as a factor in hill climates', Symposium: *Hill Climates and Land usage with special reference to the Highland Zone of Britain*, Univ. College of Wales, Aberystwyth, 23—30.

Godwin, H. (1941). 'Studies in the post-glacial history of British vegetation: VI. Correlations in the Somerset Levels', *New Phytologist*, **40**, 108—32.

Godwin, H. (1955). 'The botanical and geological history of the Somerset Levels', *Advancement of Science*, **12**, 319—22.

Godwin, H. (1956). *The History of the British Flora*, Cambridge University Press.

Godwin, H. and Clapham, A. R. (1948). 'Studies in the post-glacial history of British vegetation: IX. Prehistoric trackways in the Somerset Levels', *Philosophical Transactions, Royal Society*, Series B, **233**, 249—73.

Great Eastern Railway (1897). *Official Guide*, Cassells, London.

Green, R. D. (1968). 'Soils of Romney Marsh', *Bull. Soil Surv. Gt. Brit.*, **4**, Soil Survey, Harpenden.

Green, R. D. and Fordham, S. J. (1973). 'Soils in Kent — I', *Soil Surv. Record, 14*, Soil Survey, Harpenden.

Green, F. H. W. (1973). 'Aspects of the changing environment: some factors affecting the aquatic environment in recent years', *J. Environ. Management*, **1**, 377—91.

Grigal, D. F. and Arneman, H. F. (1969). 'Numerical classification of some forested Minnesota soils', *Proc. Soil Sci. Soc. Amer.*, **33**, 433—8.

Hall, B. R. and Folland, C. J. (1970). 'Soils of Lancashire', *Bull. Soil Surv. Gt. Brit.*, **5**, Soil Survey, Harpenden.

Hallsworth, E. G., Costin, A. B. and Gibbons, F. R. (1953). 'Studies in pedogenesis in New South Wales: VI. On the classification of soils showing features of podzol morphology', *J. Soil Sci.*, **4**, 241—56.

Harrod, T. R. (1971). 'Soils in Devon I: Sheet ST10 (Honiton)', *Soil Surv. Record*, 9, Soil Survey, Harpenden.

Hawkins, A. B. and Kellaway, G. A. (1971). 'Field meeting at Bristol and Bath with special reference to new evidence of glaciation', *Proc. Geol. Assoc.*, **82**, 267–91.

Henin, S., Feodoroff, A., Gras, R. and Monnier, G. (1960). *Le Profil Cultural: Principles de Physique du Sol*, Societe d'Editions des Ingenieurs, Agricoles, Paris.

Hill, M. J. *et al.* (1973). 'Bacteria, nitrosamines and cancer of the stomach', *Br. J. Cancer*, **28**, 562–7.

Hills, R. C. (1968). 'Infiltration measurement using cylinders as a method of assessing the influence of land management and soil type on the occurrence of overland flow', unpublished Ph.D. thesis, University of Bristol.

Hocombe, S. D., Holly, G. K. and Parker, C. (1966). 'The persistence of some new herbicides in the soil', *Proc. 8th Brit. Weed Control Conf.*, 605–13.

Hodge, C. A. H. and Seale, R. S. (1966). 'The soils of the district around Cambridge', *Mem. Soil Surv. Gt. Brit.*, Soil Survey, Harpenden.

Hodgson, J. M. (1967). 'Soils of the West Sussex Coastal Plain', *Bull. Soil Surv. Gt. Brit.*, 3, Soil Survey, Harpenden.

Hodgson, J. M. (1972). 'The soils of the Ludlow district', *Mem. Soil Surv. Gt. Brit.*, Soil Survey, Harpenden.

Hodgson, J. M. (1974). 'Soil Survey Field Handbook', *Tech. Monogr. Soil Surv. Gt. Brit.*, 5.

Hodgson, J. M., Catt, J. A. and Weir, A. H. (1967). 'The origin and development of Clay-with-flints and associated soil horizons on the South Downs', *J. Soil Sci.*, **18**, 85–102.

Hodgson, J. M., Rayner, J. H. and Catt, J. A. (1974). 'The geomorphological significance of Clay-with-flints on the South Downs', *Trans. Inst. Brit. Geogr.*, 61, 119–29.

Hopp, H. and Slater, C. S. (1948). 'Influence of earthworms on soil productivity', *Soil Sci.*, **66**, 421–8.

Howe, G. M. (1959). 'The geographical distribution of disease with special reference to cancer of the lung and stomach in Wales', *Brit. J. Preventive and Social Medicine*, 13, 204–10.

Hull, E. (1855). 'Physical geography and Pleistocene phenomena of the Cotteswold Hills', *Quart. J. geol. Soc., Lond.*, 11, 477–96.

Hunker, H. L. (ed.) (1964). *Introduction to World Resources*, Harper and Row, New York.

Jackson, R. (1970). 'Motorways and National Parks in Britain', *Area*, 4, 26–9.

Jarvis, M. G. (1973). 'Soils of the Wantage and Abingdon District', *Mem. Soil Surv. Gt. Brit.*, Soil Survey, Harpenden.

Jarvis, R. A. (1968). 'Soils of the Reading District', *Mem. Soil Surv. Gt. Brit.*, Soil Survey, Harpenden.

Jarvis, R. A. (1973). 'Soils in Yorkshire II: Sheet SE60 (Armthorpe)', *Soil Surv. Record*, Soil Survey, Harpenden.

Jeanson, C. and Monnier, G. (1965). 'Studies on the stability of soil structure: influence of moulds and soil fauna', in E. G. Hallsworth and D. V. Crawford (eds), *Experimental Pedology*, Butterworths, London.

Jenny, H. (1941). *Factors of Soil Formation*, McGraw-Hill, London.

Kay, F. F. (1934). 'A soil survey of the eastern part of the Vale of the White Horse', *Bull. Fac. Agric. Hort. Univ. Reading*, No. 48.

Kay, F. F. (1939). 'A soil survey of the strawberry district of south Hampshire', *Bull. Fac. Agric. Hort. Univ. Reading*, No. 52.

Kay, F. F. (1940). Unpublished survey around Caistens and Partridges Farms, Hampshire. In F. H. W. Green (ed.), *Hampshire, the Land of Britain*, part 89, Reports of the Land Utilisation Survey of Britain, Geographical Publications Ltd, London.

Kellaway, G. A. (1971). 'Glaciation and the stones of Stonehenge', *Nature, Lond.*, 233, 30–5.

King, D. W. (1969). 'Soils of the Luton and Bedford District', *Spec. Surv. Soil Surv. Gt. Brit.*, Soil Survey, Harpenden.

Kittredge, J. (1948). *Forest Influences*, McGraw-Hill, New York.

Klingebiel, A. A. and Montgomery, P. H. (1966). 'Land capability classification', *Agricultural Handbook*, No. 210, Soil Conservation Service, US Dept Agric.

Kubiena, W. L. (1953). *The Soils of Europe*, Murby, London.

Lamb, H. H. (1961). 'Climatic change within historical time as seen in circulation maps and diagrams', *Ann. New York Acad. Sci.*, 95, 124−61.

Leeper, G. W. (1956). 'The classification of soils', *J. Soil Sci.*, 18, 77−80.

Loveday, J. (1962). 'Plateau deposits of the southern Chiltern Hills', *Proc. Geol. Assoc.*, 73, 83−102.

Low, A. J. (1955). 'Improvements in the structural state of soils under leys', *J. Soil Sci.*, 6, 179−99.

Lowdermilk, W. C. (1930). 'Influence of forest litter on run-off percolation and erosion', *J. Forestry*, 28, 474−91.

Luchshev, A. A. (1940). 'Precipitation under the forest canopy', *Trudȳ vses. nauchno-issled. Inst. les. Khoz*, 18, 113−48 (in H. L. Penman, 1963).

Lucy, W. C. (1872). 'The gravels of the Severn, Avon and Evenlode and their extension over the Cotswold Hills', *Proc. Cotteswold Nat. Field Club*, 5, 71−142.

Lundquist, J. (1962). 'Patterned ground and related frost phenomena in Sweden', *Sver. geol. Unders. Afh.*, Ser. C., No. 583.

MacDermot, E. T. (1911). *The History of the Forest of Exmoor*, The Wessex Press. (Reprinted by David and Charles, Newton Abbot, 1973.)

Mackereth, F. J. H. (1965). 'Chemical investigations of lake sediments and their interpretation', *Proc. Royal Soc.*, B, 161, 295−309.

Mackney, D. (1961). 'A podzol development sequence in oakwoods and heath in central England', *J. Soil Sci.*, 12, 23−40.

Mackney, D. and Burnham, C. P. (1964). 'The soils of the West Midlands', *Bull. Soil Surv. Gt. Brit.*, Soil Survey, Harpenden.

Mackney, D. and Burnham, C. P. (1966). 'The soils of the Church Stretton District of Shropshire', *Mem. Soil Surv. Gt. Brit.*, Soil Survey, Harpenden.

Mackney, D. (1969). 'The agronomic significance of soil mapping units', in J. G. Sheals (ed.), *The Soil Ecosystem*, Systematics Association, London, 55−62.

Manil, G. (1956). 'Aspects dynamiques du profil pedologique', *6th Int. Congr. Soil Sci.*, E, 439−41.

Manil, G. (1963). 'Profil chimique, solum biodynamique et autres caracteristiques ecologiques du profil pedologique', *Sci. Sol.*, 1, 31−46.

Manley, G. (1945). 'The effective rate of altitudinal change in temperate Atlantic climates', *Geogrl Rev.*, 35, 408−17.

Manley, G. (1947). 'Snow cover in the British Isles', *Met. Mag.*, 76, 1−8.

Manley, G. (1961). 'Late and Post Glacial climatic fluctuations and their relationships to those shown by the instrumental record of the past 300 years', *Ann. New York Acad. Sci.*, 95, 162−72.

Marshall, W. (1809). *Review and Abstract of County Reports to the Board of Agriculture*. Vols. I−V. York: Longman *et al.* (Reprinted David and Charles, Newton Abbot, 1968.)

Matthews, B. (1971). 'Soils in Yorkshire I: Sheet SE65 (York East)', *Soil Survey Record*, 6, Soil Survey, Harpenden.

Miles, R. (1967). *Forestry in the English Landscape*, Faber and Faber, London.

Miers, R. H. and Thomasson, A. J. (1971). 'Land drainage', Ch. 5 in A. J. Thomasson (1971) op. cit.

Miller, R., Common, R. and Galloway, R. W. (1954). 'Stone Stripes and other surface features of Tinto Hill', *Geogrl J.*, 120, 216−19.

Milne, G. (1936). 'A soil reconnaissance journey through parts of Tanganyika Territory', published in 1947 in *J. Ecology*, 35, 192−265.

Ministry of Agriculture, Fisheries and Food (1966). 'Agricultural Land Classification', *Agricultural Land Service Report*, 11.

Ministry of Agriculture, Fisheries and Food (1970). *Land Classification Survey Sheet 99. Holderness*, Min. Agr. Fish. Food. (Explanatory Notes accompany map.)

Ministry of Agriculture, Fisheries and Food (1970). *Modern Farming and Soil*, HMSO.

Ministry of Agriculture, Fisheries and Food (1974). 'Agricultural Development and Advisory Service, Land Service', *Agricultural Land Classification of England and Wales*. (Presentation and exhibition, London, 1974.)

Mitchell, B. D. and Jarvis, R. A. (1956). 'The soils of the country round Kilmarnock', *Mem. Soil Surv. Gt. Brit.*, Soil Survey, Harpenden.

Mitchell, G. F. (1960). 'The Pleistocene history of the Irish Sea', *Advmt Sci.*, 17, 313—25.

Mitchell, G. F., Penny, L. F., Shotton, F. S. and West, R. G. (1973). *A Correlation of Quaternary Deposits in the British Isles*, Geological Society, London.

Moore, P. D. and Bellamy, D. J. (1973). *Peatlands*, Elek Science, London.

Moore, P. D. and Chater, E. H. (1969). 'Studies in the vegetational history of mid-Wales. 1. The post-glacial period in Cardiganshire', *New Phytologist*, 68, 183—96.

Morgan, R. P. C. (1971). 'A morphometric study of some stream valley systems on the English chalklands', *Trans. Inst. Brit. Geogr.*, 54, 33—44.

Muir, A. (1934). 'Soils of the Tiendland State Forest: a preliminary survey', *Forestry*, 8, 25—55.

Muir, A. (1961). 'The podzol and podzolic soils', *Adv. Agron.*, 13, 1—56.

Mulcahy, M. J. and Humphries, A. W. (1967). 'Soil classification, soil surveys and land use', *Soils and Fertilizers*, 30, 1—8.

Musgrave, G. W. (1947). 'The quantitative evaluation of factors in water erosion — a first approximation', *J. Soil and Water Conserv.*, 2, 133—8.

Nordbye, D. J. and Campbell, D. A. (1951). 'Preliminary infiltration investigations in New Zealand', *C.R. Ass. Int. Hydrologie Sci.*, Brussels, 2, 98—101.

Norris, J. M. (1970). 'Multivariate methods in the study of soils', *Soils and Fertilizers*, 33, 313—18.

O'Riordan, T. (1971). *Perspectives on Resource Management*, Pion, London.

Orwin, C. S. and Sellick, R. J. (1970). *The Reclamation of Exmoor Forest*, David and Charles, Newton Abbot.

Palmer, J. (1966). 'Landforms, drainage and settlement in the Vale of York', in S. R. Eyre and G. R. J. Jones (eds), *Geography as Human Ecology*, Edward Arnold, London.

Palmer, J. and Neilson, R. A. (1962). 'The origin of granite tors on Dartmoor, Devonshire', *Proc. Yorks. geol. Soc.*, 33, 315—40.

Parizek, E. J. and Woodruffe, J. F. (1957). 'Description and origin of stone layers in soils of the United States', *Journ. Geol.*, 65, 24—34.

Pearsall, W. H. (1950). *Mountains and Moorlands*, Collins, London.

Peltier, L. C. (1950). 'The geographic cycle in periglacial regions as it is related to climatic geomorphology', *Annals. Assoc. Amer. Geogr.*, 40(3), 214—36.

Penman, H. L. (1963). 'Vegetation and hydrology', *Technical Communication, Commonwealth Bureau of Soils*, 53.

Pepper, D. M. (1973). 'A comparison of the "Argile a Silex" of Northern France with the "Clay-with-flints" of Southern England', *Proc. Geol. Assoc.*, 84, 331—52.

Perelman, A. I. (1955). 'Riady migratsie khimicheskikh elementov v kope vyvetrivania', *Dokl. Akad. Nauk SSSR*, 103, 669—72.

Perrin, R. M. S. (1955). 'Studies on pedogenesis', unpublished Ph.D. thesis, University of Cambridge.

Perrin, R. M. S. (1956). 'The nature of "Chalk Heath" soils', *Nature, Lond.*, 178, 31—2.

Perrin, R. M. S., Davies, H. and Fysh, M. D. (1974). 'Distribution of late Pleistocene aeolian deposits in eastern and southern England', *Nature, Lond.*, 248, 320—4.

Polynov, B. B. (1937). *The Cycle of Weathering* (translated by A. Muir), Thos. Murby, London.

Pizer, N. H., Wright, H. A., Caldwell, T. H., Hargrave, J., Burgess, G. R., Cory, V. and Boyd, D. A. (1961). 'A study of the peat fenlands with particular reference to potato manuring', *J. Agric. Sci., Camb.*, 56, 197—211.

Radford, C. A. R. (1952). 'Prehistoric settlements on Dartmoor and the Cornish Moors', *Proc. Prehist. Soc.*, 18, 55—84.

Ragg, J. M. (1960). 'The soils of the country round Kelso and Lauder', *Mem. Soil Surv. Gt. Brit.*, Soil Survey, Harpenden.

Ragg, J. M. (1973). 'Factors in soil formation', in J. Tivy (ed.), *The Organic Resources of Scotland*, Oliver and Boyd, Edinburgh, 38—50.

Ragg, J. M. and Clayden, B. (1973). 'The classification of some British soils according to the comprehensive system of the United States', *Tech. Monogr. Soil Surv. Gt. Brit.*, Soil Survey, Harpenden.

Raistrick, A. (1933). 'The correlation of glacial retreat stages across the Pennines', *Proc. Yorks. geol. Soc.*, 22, 199—214.

Rayner, J. H. (1966). 'Classification of soils by numerical methods', *J. Soil Sci.*, 17, 79—92.

Reid, C. (1903). 'Geology of the country around Chichester', *Mem. geol. Surv. U.K.*

Richardson, L. (1929). 'The country around Moreton in Marsh', *Mem. geol. Surv. U.K.*

Richardson, L. and Sandford, K. S. (1960). 'Ditchford gravel pit near Stratton-on-Fosse, Gloucestershire, and the occurrence of a mammoth tooth', *Proc. Cotteswold Nat. Field Club*, 33, 172—6.

Robinson, D. N. (1968). 'Soil erosion by wind in Lincolnshire, March, 1968', *East Midland Geographer*, 4, 351—62.

Robinson, G. W. (1949). *Soils: Their Origin, Constitution and Classification*, Thos. Murby, London.

Robinson, K. L. (1948). 'The soils of Dorset', in R. Good (ed.), *A Geographical Handbook of the Dorset Flora*, Dorset Natural History and Archaeological Society.

Rogowski, A. S. and Kirkham, D. (1962). 'Moisture, pressure and formation of water-stable soil aggregates', *Proc. Soil Sci. Soc. Amer.*, 26, 213—16.

Rudge, T. (1807). *General view of the Agriculture of the County of Gloucester*, Board of Agriculture.

Ruhe, R. V. (1959). 'Stone lines in soils', *Soil Sci.*, 87, 223—31.

Ruhe, R. V. (1960). 'Elements of the soil landscape', *7th Int. Congr. Soil Sci.*, 4, 165—70.

Russell, E. W. (1973). *Soil Conditions and Plant Growth* (10th edn), Longman, London.

Salisbury, E. J. (1925). 'Note on the edaphic succession in some dune soils with special reference to the time factor', *J. Ecol.*, 13, 322—8.

Salter, P. J. and Williams, J. B. (1967). 'The influence of texture on the moisture characteristics of soils: IV. A method of estimating the available water-capacity of profiles in the field', *J. Soil Sci.*, 18, 174—81.

Sarkar, P. K., Bidwell, O. W. and Marcus, L. F. (1966). 'Selection of characteristics for numerical classification of soils', *Proc. Soil Sci. Soc. Amer.*, 30, 269—72.

Sauer, H. I. *et al.* (1963). 'Cardiovascular disease mortality patterns in Georgia and North Carolina', *Public Health Reports, Washington*, 81, 455—65.

Savigear, R. A. (1960). 'Slopes and hills in West Africa', *Zeitschrift für Geomorphologie Supplbd*, 1, 156—71.

Schnitzer, M. (1969). 'Relations between fulvic acid, a soil humic compound and inorganic soil constituents', *Proc. Soil Sci. Soc. Amer.*, 33, 75—81.

Seddon, B. (1965). 'Prehistoric climate and agriculture: a review of recent Palaeo—Ecological investigations', in J. A. Taylor (ed.), *Climatic Change with Special Reference to Wales*, University of Aberystwyth. Memorandum No. 8, 19—32.

Simmons, I. G. (1964). 'An ecological history of Dartmoor', in I. G. Simmons (ed.), *Dartmoor Essays*, 191—215, Devon Association for Advanced Science.

Simonson, R. W. (1968). 'Concept of soil', *Adv. Agron.*, 20, 1—47.

Simpson, J. B. (1932). 'Stone polygons on Scottish Mountains', *Scott. Geogr. Mag.*, 48, 37—8.

Sissons, J. B. (1967). *The Evolution of Scotland's Scenery*, Oliver and Boyd, Edinburgh.

Smith, G. D. (1965). 'Lectures on soil classification', *Pedowgie*, 4, 5—135.

Smithson, F. (1953). 'The micro-minerology of North Wales soils', *J. Soil Sci.*, 4, 194—210.

Sneesby, N. J. (1953). 'Wind erosion and the value of shelter belts', *Agriculture, Lond.*, 60, 263—71.

Sneesby, N. J. (1966). 'Erosion control on the Black Fens', *Agriculture, Lond.*, **73**, 391–4.

Soane, B. D. (1973). 'Techniques for measuring changes in the packing state and cone resistance of soil after the passage of wheels and tracks', *J. Soil Sci.*, **24**, 311–23.

Soil Survey Staff (1960). *Soil Classification: A Comprehensive System: 7th approximation*, US Government Printing Office, Washington DC.

Somerset County Planning Office (1967). *Peat in Central Somerset: A Planning Study*, County Planning Office, Taunton.

Sparks, B. W. and West, R. G. (1972). *The Ice Age in Britain*, Methuen, London.

Stagg, M. (1973). 'Storm runoff in a small catchment in the Mendip Hills', unpublished M.Sc. thesis, University of Bristol.

Stahlfelt, M. G. (1944). 'The water consumption of spruce and its influence on the water regime of the soil', *K. Lantbr Akad. Tidskr.*, **83**, 425–505.

Statham, D. C. (1972). 'Land use changes in the moorlands of the North York Moors National Park', *Centre for Environmental Studies, University Working Paper*, 16.

Stephens, C. G. (1953). 'Soil Surveys for Land Use Development', *F.A.O. Agricultural Studies*, 20.

Stephens, N. (1966). 'Some Pleistocene deposits in North Devon', *Biul. peryglac.*, **15**, 103–14.

Stewart, G. A. (ed.) (1968). *Land Evaluation*, CSIRO/UNESCO Symposium, Macmillan.

Stobbe, P. C. and Wright, J. R. (1959). 'Modern concepts of the genesis of podzols', *Proc. Soil Sci. Soc. Am.*, **23**, 161–4.

Sturdy, R. G. (1971). 'Soils in Essex I: Sheet TQ59 (Harold Hill)', *Soil Surv. Record*, Soil Survey, Harpenden.

Swaby, R. J. (1949). 'The influence of humus on soil aggregation', *J. Gen. Microbiol.*, **3**, 236–54.

Swaby, R. J. (1950). 'The influence of humus on soil aggregation', *J. Soil Sci.*, **1**, 182–94.

Swindale, L. D. and Jackson, M. L. (1956). 'Genetic processes in some residual podzolised soils of New Zealand', *6th Int. Congr. Soil Sci.*, E, 233–9.

Tansley, A. G. (1953). *The British Islands and Their Vegetation*, Cambridge University Press.

Tavernier, R. J. F. and Smith, G. D. (1957). 'The concept of Braunede (brown forest soil) in Europe and the United States', *Adv. Agron.*, **9**, 217–89.

Taylor, J. A. (1958). 'Growing season as affected by aspect and soil texture', *Symposium: The Growing Season*, University College of Wales, Aberystwyth, 1–8.

Taylor, B. J., Price, R. H. and Trotter, F. M. (1963). 'Geology of the country around Stockport and Knutsford', *Mem. Geol. Surv. U.K.*

Taylor, J. A. and Tucker, R. B. (1970). 'The peat deposits of Wales: An inventory and interpretation', *3rd Int. Peat Congr.*, 163–73.

Ten Khak-Mun, V. (1973). 'Participation of micro-organisms in the formation of manganese–iron concretions in soils of the brown forest zone of the Far East', *Soviet Soil Science*, **5**, 96–9.

Thomasson, A. J. (1969). 'Soils of the Saffron Walden District', *Spec. Surv. Soil Surv. Gt. Brit.*, Soil Survey, Harpenden.

Thomasson, A. J. (1971). 'Soils of the Melton Mowbray District', *Mem. Soil Surv. Gt. Brit.*, Soil Survey, Harpenden.

Thomasson, A. J. and Avery, B. W. (1963). 'The soils of Hertfordshire', *Trans. Hertfordshire Natural History Society*, **25**, 247–63.

Thomasson, A. J. and Avery, B. W. (1970). 'The soils of Hertfordshire', *Spec. Surv. Soil Surv. Gt. Brit.*, Soil Survey, Harpenden.

Toleman, R. D. L. (1974). 'Silvicultural problems and their relationship to upland soils in Wales', *Welsh Soils Discussion Group*, University of Aberystwyth, 1–10.

Tomlinson, T. E. (1970). 'Trends in nitrate concentrations in English rivers in relation to fertiliser use', *Water Treat. Exam.*, **19**, 277–89.

Trafford, B. D. (1970). 'Field drainage', *J. Royal Agric. Soc. Engl.*, **31**, 129–52.

Troll, C. (1958). *Structure Soils, Solifluction and Frost Climates of the Earth*, US Army Snow, Ice and Permafrost Research Establishment. Translation 43, 121 pp.

Turner, J. (1964). 'The anthropogenic factor in vegetational history. I. Tregarron and Whixall mosses', *New Phytologist*, **63**, 73–90.

United States Department of Agriculture (1971). *Annual Report*, US Govt. Printing Office, Washington DC.

USDA Soil Survey Staff (1951). *Soil Survey Manual. US Dept. Agric. Handbook* 18, Govt. Printer, Washington.

University of Newcastle upon Tyne (1971). *Proceedings of Derelict Land Symposium* (2 vols).

Van de Westeringh, W. (1972). 'Deterioration in soil structure in worm free orchard soils', *Pedobiologia*, 12, 6—15.

Vink, A. P. A. (1960). *Planning of soil surveys in land development*, International Institute of Land Reclamation and Improvement.

Vancouver, C. (1808). *General View of the Agriculture of the County of Devon*, Board of Agriculture.

Washburn, A. L. (1956). 'Classification of patterned ground and review of suggested origins', *Bull. geol. Soc. Amer.*, 67, 823—66.

Washburn, A. L. (1973). *Periglacial Processes and Environments*, Edward Arnold, London.

Waters, R. S. (1961). 'Involutions and ice-wedges in Devon', *Nature, Lond.*, 189, 389—90.

Waters, R. S. (1964). 'The Pleistocene legacy to the geomorphology of Dartmoor', in I. G. Simmons (ed.), *Dartmoor Essays*, Devon Association for Advanced Science.

Waters, R. S. (1965). 'The geomorphological significance of Pleistocene frost action in south-west England', in J. B. Whittow and P. D. Wood (eds), *Essays in Geography for Austin Miller*, Reading University, 39—57.

Waters, R. S. (1966). 'The Exeter symposium', *Biul. peryglac.*, 15, 123—49.

Watt, A. S. and Jones, E. W. (1950). 'The ecology of the Cairngorms. Part I: The environment and the altitudinal zonation of the vegetation', *J. Ecol.*, 36, 283—304.

Weakly, H. E. (1960). 'Effect of HPAN Soil conditioners on runoff, erosion and soil aggregation', *J. Soil and Water Conserv.*, 15, 169—71.

Webster, R. (1968). 'Fundamental objections to the 7th Approximation', *J. Soil Sci.*, 19, 354—66.

Webster, R. and Beckett, P. H. T. (1968). 'Quality and usefulness of soil maps', *Nature, Lond.*, 219, 680—2.

Weir, A. H., Catt, J. A. and Madgett, P. A. (1971). 'Postglacial soil formation in the loess of Pegwell Bay, Kent (England)', *Geoderma*, 5, 131—49.

West, R. G. (1968). *Pleistocene Geology and Biology*, Longman, London.

West, R. G. and Donner, J. J. (1956). 'The glaciations of East Anglia and the East Midlands', *Quart. J. geol. Soc., Lond.*, 122, 69—93.

Whitaker, A. (1973). 'Geology of Bredon Hill', *Bull. Geol. Surv. Gt. Brit.*, Soil Survey, Harpenden.

Wibberley, G. P. (1959). *Agriculture and Urban Growth*, Michael Joseph, London.

Williams, M. (1970). *The Draining of the Somerset Levels*, Cambridge University Press.

Williams, P. W. (1966). 'Limestone pavements, with special reference to Western Ireland', *Trans. Inst. Brit. Geogr.*, 40, 155—72.

Williams, R. B. G. (1964). 'Fossil patterned ground in eastern England', *Biul. peryglac.*, 14, 337—49.

Williams, R. B. G. (1968). 'Some estimates of periglacial erosion in Southern and Eastern England', *Biul. peryglac.*, 17, 311—35.

Willis, A. J. (1973). *Introduction to Plant Ecology*, George Allen and Unwin Ltd, London.

Wills, L. J. (1948). *The Palaeography of the Midlands*, Hodder and Stoughton, London.

Wills, L. J. (1951). *A Palaeogeographic Atlas*, Blackie, London.

Wirtz, D. (1953). 'Zur Stratigraphie des Pleistocans im Westen der Britischen Inseln', *Neues Jb. Geol. ä Palont*, 96, 267—303.

Wood, A. (1942). 'The development of hillside slopes', *Proc. Geol. Ass.*, 43, 128—40.

Wooldridge, S. W. and Linton, D. L. (1955). *Structure, Surface and Drainage in South-east England*, Philip, London.

Woytinsky, W. S. and Woytinsky, E. S. (1955). *World Population and Production, Trends and Outlook*, The Twentieth Century Fund, New York.

Young, A. (1972). 'Against land classification', in W. P. Adams and F. M. Helleiner (eds), *International Geography*, University of Toronto Press, 1184—6.

Young, A. (1973). 'Soil survey procedures in land development planning', *Geogrl. J.*, 139, 53—64.

Zimmerman, E. W. (1951). *World Resources and Industries*, Harper and Row, New York.

Zingg, A. W. (1940). 'Degree and length of land slope as it affects soil loss in runoff', *Agricultural Engineering*, 21, 59—64.

INDEX OF SUBJECTS

(*Note*: Soil series, associations and mapping units are in **bold type**; botanical names in *italic*. Page numbers in *italic* refer to figures.)

INDEX OF AUTHORS

Explanatory note to the Soil Map of England and Wales, 1 : 2 000 000

B. W. Avery, D. C. Findlay and D. Mackney, Soil Survey of England and Wales

This map is based on detailed and reconnaissance soil surveys, geological and relief maps. Soil maps at 1 : 25 000, and 1 : 63 360, cover some 20 per cent of the land including parts of most counties. Extrapolation of boundaries into unmapped terrain relies chiefly on inferred relationships between soil pattern and geology or relief.

The twenty-two map units shown are geographic soil associations identified by dominant (most frequently occurring) soil groups. Associated soils (sub-dominant soils) are indicated and the important agricultural properties of both dominant and associated soils listed. In most map units the dominant soil is associated with differing combinations of sub-dominant soils, reflecting variations in lithology and relief. This method of grouping has allowed the construction of a map which is simple in having a small number of units. More detailed information is presented on the soil map of England and Wales at the scale of 1 : 1 000 000 (Avery *et al.* in press), which displays seventy-one units.

SOIL CLASSIFICATION

The soil groups are differentiated primarily by observable or measurable characteristics of the soil profile, including distinctive surface and sub-surface horizons resulting from alteration of the original material by pedogenic processes, and proportions of organic matter, calcium carbonate and differently sized mineral particles within specified depths.

Soil materials containing more than 20–30 per cent organic matter, depending on clay content, are classed as *organic*. Mineral or organo-mineral soil materials, containing less organic matter, and less than 70 per cent by volume of stones (2–200 mm), are classed as sandy, clayey or loamy according to mass-percentages of sand (2 000–60 μm), silt (60–2 μm) and clay-sized particles (<2 μm) in the inorganic fraction (<2 mm), as follows:

Sandy: percentage silt + twice percentage clay less than 30,
Clayey: more than 35 per cent clay,
Loamy: other materials of intermediate composition.

Loamy soils with less than 18 per cent clay and more than 20 per cent sand are differentiated as coarse loamy and those with less than 20 per cent sand as silty.

Soil materials containing 5–35 per cent by volume of stones are described additionally as *stony*, those with 35–70 per cent stones as *very stony*, and water-sorted materials with more than 70 per cent stones as *gravel* or *shingle*.

Organo-mineral materials with more than 8–12 per cent organic matter, depending on clay content, are further distinguished as *humose*.

Soil profiles are also classed as organic, sandy, clayey, etc., with the general implication that at least the upper 40 cm, a similar thickness starting directly below the topsoil, or most of the profile if bedrock supervenes, has the composition named. Similarly, a predominantly mineral profile is considered to have a *humose topsoil* if at least the upper 15 cm is humose, and a *peaty topsoil* if it has a superficial organic layer 7.5—40 cm thick, formed under wet conditions.

The characteristics of the **Lithomorphic (A/C)** soils, **Podzolic** soils, **Pelosols**, **Gley** soils and **Peat** soils are summarised in Appendix 4, and are described in greater detail by Avery (1973).

Raw soils are those with no distinct pedogenic horizons other than a superficial organo-mineral or organic layer less than 7.5 cm thick, or a buried horizon below 30 cm depth. They are usually sparsely vegetated and of negligible value for agriculture or forestry.

Raw sands: Raw sandy soils, chiefly dune sands.
Raw skeletal soils: Extremely stony and/or very shallow raw soils, including screes, mountain-top detritus and shingle beaches.

ACKNOWLEDGEMENTS

The map was compiled from information supplied by regional Soil Survey staff and drawn by the Cartographic Section. The Clarendon Press Atlas was an invaluable aid in the preparation of the map.

REFERENCES

Avery, B. W. (1973). Soil classification in the Soil Survey of England and Wales, *J. Soil Sci.*, 24, 324—38.
Avery, B. W., Findlay, D. C. and Mackney, D. (in press). *Soil Map of England and Wales*, 1 : 1 000 000, Southampton, Ordnance Survey.

Note

This map can be purchased separately from Soil Survey of England and Wales, Rothamsted Experimental Station, Harpenden, Herts.